Fuzzy Logic: Theory and Applications

Fuzzy Logic: Theory and Applications

Edited by
Lukas Brooks

Larsen & Keller
www.larsen-keller.com

Fuzzy Logic: Theory and Applications
Edited by Lukas Brooks
ISBN: 978-1-63549-129-6 (Hardback)

© 2017 Larsen & Keller

⊟ Larsen & Keller

Published by Larsen and Keller Education,
5 Penn Plaza,
19th Floor,
New York, NY 10001, USA

Cataloging-in-Publication Data

Fuzzy logic : theory and applications / edited by Lukas Brooks.
    p. cm.
Includes bibliographical references and index.
ISBN 978-1-63549-129-6
1. Fuzzy logic. 2. Fuzzy mathematics. 3. Fuzzy sets.
I. Brooks, Lukas.

QA9.64 .F89 2017
511.313--dc23

The publisher's policy is to use permanent paper from mills that operate a sustainable forestry policy. Furthermore, the publisher ensures that the text paper and cover boards used have met acceptable environmental accreditation standards.

Printed and bound in the United States of America.

For more information regarding Larsen and Keller Education and its products, please visit the publisher's website www.larsen-keller.com

Table of Contents

Preface

Fuzzy logic refers to a mathematical many-valued logic which considers truth value to be more of one range of values than the other. The main use of fuzzy logic is in fields like artificial intelligence and control theory. This book provides comprehensive insights into this field. It is a compilation of chapters that discuss the most vital concepts in the field of fuzzy logic. The topics included in this text are of utmost significance and are bound to provide incredible insights to readers. This textbook, with its detailed analyses and data, will prove immensely beneficial to students involved in this area at various levels. It will be of great help to those studying artificial intelligence, computer science and probability.

A foreword of all chapters of the book is provided below:

Chapter 1 - Aristotle's logic has only two truth-values, true or false. Logic that has more than two truth-values, true, false and unknown is known as many-valued logic. Fuzzy logic and probability logic are examples of many valued logic. The following chapter provides the reader with an in-depth understanding of fuzzy logic; **Chapter 2** - A fuzzy number is when a real number does not refer to a single value but is connected to a number of possible values. The value is anywhere between 0 and 1. This text elucidates theories such as fuzzy numbers and fuzzy finite element. The chapter strategically encompasses and incorporates the major components and concepts of fuzzy logic; **Chapter 3** - The likelihood of an event occurring is known as possibility. Possibility is anywhere between 0 and 1, where 1 is certainty and 0 is impossibility. A situation where indeterminate information is involved is known as uncertainty. Uncertainty theory is a branch of mathematics that was founded very recently. The section will not only provide an overview, it will also delve deep into the topics related to it. The chapter on possibility offers an insightful focus, keeping in mind the complex subject matter; **Chapter 4** - In mathematics, Boolean algebra is a theory where truth-values are either true or false. 1 and 0 denotes truth and falsehood. Systems that are used in mathematical concepts are known as mathematical models. Fuzzy control systems, Fril, De Morgan algebra and fuzzy subalgebra are significant topics related to fuzzy logic. The following section unfolds its crucial aspects in a critical yet systematic manner; **Chapter 5** - The logical aspects of fuzzy sets include the principle of bivalence, three-valued logic, canonical form and probilistic logic. Three-valued logic is a logical system in which there are three truth-values, true, false and an unknown value. Three valued logic is in contrast to Boolean logic, which is binary and has only true and false. The aspects explicated are of vital importance, and help in the better comprehension of fuzzy logic; **Chapter 6** - Logic helps in analyzing and structuring arguments. A valid argument is considered to be an argument that has a specific relation to the arguments and to its conclusion. This section is an overview of the subject matter incorporating all the major aspects to explain logic; **Chapter 7** - Informal logic is a branch that helps in developing criteria for analyzing and interpreting the construction of an argument. Mathematical logic, term logic, BL and noise-based logic are some of the aspects elucidated in the section. The chapter provides a plethora of themes on the types of logic for a better comprehension; **Chapter 8** - The applications of fuzzy logic explained in the chapter are neuro-fuzzy, fuzzy electronics, fuzzy mathematics, fuzzy extractor, distribution management system etc. Neuro-fuzzy is referred to as the combination of artificial neural networks and fuzzy logic. The aspects elucidated in this chapter are of vital importance, and provides a better understanding of fuzzy logic.

At the end, I would like to thank all the people associated with this book devoting their precious time and providing their valuable contributions to this book. I would also like to express my gratitude to my fellow colleagues who encouraged me throughout the process.

Editor

Introduction to Fuzzy Logic

Aristotle's logic has only two truth-values, true or false. Logic that has more than two truth-values, true, false and unknown is known as many-valued logic. Fuzzy logic and probability logic are examples of many valued logic. The following chapter provides the reader with an in-depth understanding of fuzzy logic.

Fuzzy logic is a form of many-valued logic in which the truth values of variables may be any real number between 0 and 1, considered to be "fuzzy". By contrast, in Boolean logic, the truth values of variables may only be the "crisp" values 0 or 1. Fuzzy logic has been employed to handle the concept of partial truth, where the truth value may range between completely true and completely false. Furthermore, when linguistic variables are used, these degrees may be managed by specific (membership) functions.

The term *fuzzy logic* was introduced with the 1965 proposal of fuzzy set theory by Lotfi Zadeh. Fuzzy logic had however been studied since the 1920s, as infinite-valued logic—notably by Łukasiewicz and Tarski.

Fuzzy logic has been applied to many fields, from control theory to artificial intelligence.

Overview

Classical logic only permits conclusions which are either true or false. However, there are also propositions with variable answers, such as one might find when asking a group of people to identify a colour. In such instances, the truth appears as the result of reasoning from inexact or partial knowledge in which the sampled answers are mapped on a spectrum.

Humans and animals often operate using fuzzy evaluations in many everyday situations. In the case where someone is tossing an object into a container from a distance, the person does not compute exact values for the object weight, density, distance, direction, container height and width, and air resistance to determine the force and angle to toss the object. Instead the person instinctively applies quick "fuzzy" estimates, based upon previous experience, to determine what output values of force, direction and vertical angle to use to make the toss.

Both degrees of truth and probabilities range between 0 and 1 and hence may seem similar at first, but fuzzy logic uses degrees of truth as a mathematical model of *vagueness*, while probability is a mathematical model of *ignorance*.

Take, for example, the concepts of "empty" and "full". The meaning of each of them can be represented by a certain fuzzy set. The concept of emptiness would be subjective and thus would depend on the observer or designer. A 100 ml glass containing 30 ml of water may be defined as being 0.7 empty and 0.3 full, but another designer might, equally well, design a set membership function where the glass would be considered full for all values down to 50 ml.

Applying Truth Values

A basic application might characterize various sub-ranges of a continuous variable. For instance, a temperature measurement for anti-lock brakes might have several separate membership functions defining particular temperature ranges needed to control the brakes properly. Each function maps the same temperature value to a truth value in the 0 to 1 range. These truth values can then be used to determine how the brakes should be controlled.

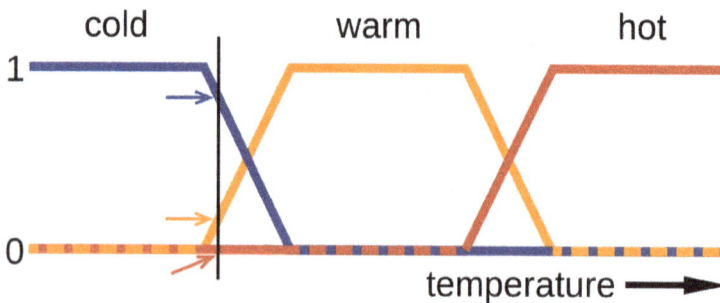

Fuzzy logic temperature

In this image, the meanings of the expressions *cold*, *warm*, and *hot* are represented by functions mapping a temperature scale. A point on that scale has three "truth values"—one for each of the three functions. The vertical line in the image represents a particular temperature that the three arrows (truth values) gauge. Since the red arrow points to zero, this temperature may be interpreted as "not hot". The orange arrow (pointing at 0.2) may describe it as "slightly warm" and the blue arrow (pointing at 0.8) "fairly cold".

Linguistic Variables

While variables in mathematics usually take numerical values, in fuzzy logic applications non-numeric values are often used to facilitate the expression of rules and facts.

A linguistic variable such as *age* may accept values such as *young* and its antonym *old*. Because natural languages do not always contain enough value terms to express a fuzzy value scale, it is common practice to modify linguistic values with adjectives or adverbs. For example, we can use the hedges *rather* and *somewhat* to construct the additional values *rather old* or *somewhat young*.

Fuzzification operations can map mathematical input values into fuzzy membership functions. And the opposite de-fuzzifying operations can be used to map a fuzzy output membership functions into a "crisp" output value that can be then used for decision or control purposes.

Forming a Consensus of Inputs and Fuzzy Rules

Since the fuzzy system output is a consensus of all of the inputs and all of the rules, fuzzy logic systems can be well behaved when input values are not available or are not trustworthy. Weightings can be optionally added to each rule in the rulebase and weightings can be used to regulate the degree to which a rule affects the output values. These rule weightings can be based upon the priority, reliability or consistency of each rule. These rule weightings may be static or can be changed dynamically, even based upon the output from other rules.

Process

1. Fuzzify all input values into fuzzy membership functions.

2. Execute all applicable rules in the rulebase to compute the fuzzy output functions.

3. De-fuzzify the fuzzy output functions to get "crisp" output values.

Synthesis of Fuzzy Logic Functions Given in Tabular Form

Fuzzy logic function of Zadeh takes value of one of its arguments or a negation of an argument. Thus, a fuzzy logic function can be given by a choice table where all the variants of ordering of arguments and their negations are listed. For example, a row of a choice table of a two arguments function can look as follows:

$$x_1 \leq x_2 \leq \overline{x}_2 \leq \overline{x}_1 : x_2$$

However, an arbitrary choice table not always defines a fuzzy logic function. In the paper, a criterion has been formulated to recognize whether a given choice table defines a fuzzy logic function and a simple algorithm of fuzzy logic function synthesis has been proposed based on introduced concepts of constituents of minimum and maximum. A fuzzy logic function represents a disdjunction of constituents of minimum, where a constituent of minimum is a conjunction of variables of the current area greater than or equal to the function value in this area (to the right of the function value in the inequality, including the function value). For the above example row, the constituent of minimum is $x_2\overline{x_2}\overline{x_1}$.

Early Applications

Many of the early successful applications of fuzzy logic were implemented in Japan. The first notable application was on the high-speed train in Sendai, in which fuzzy logic was able to improve the economy, comfort, and precision of the ride. It has also been used in recognition of hand written symbols in Sony pocket computers, flight aid for helicopters, controlling of subway systems in order to improve driving comfort, precision of halting, and power economy, improved fuel consumption for automobiles, single-button control for washing machines, automatic motor control for vacuum cleaners with recognition of surface condition and degree of soiling, and prediction systems for early recognition of earthquakes through the Institute of Seismology Bureau of Meteorology, Japan.

Example

Hard Science with IF-THEN Rules

Fuzzy set theory defines fuzzy operators on fuzzy sets. The problem in applying this is that the appropriate fuzzy operator may not be known. For example, the logic for a simple temperature regulator that uses a fan might look like this:

```
IF temperature IS very cold THEN stop fan
IF temperature IS cold THEN fan speed is slow
IF temperature IS warm THEN fan speed is moderate
IF temperature IS hot THEN fan speed is high
```

Using this rulebase and the previous image, we would expect the output fan speed to be a combination of zero and moderate, which would evaluated as some degree of slow when the input value is a combination of cold and warm and not hot. The fan speed will continue to get slower as the input temperature gets colder until the input temperature is 100% cold, 0% warm and 0% hot, at which point the output fan speed will be zero. As the temperature input gets warmer and hotter, the output fan speed will continue to get faster until the input temperature is 0% cold, 0% warm and 100% hot, at which point the fan speed output will be high.

If the fuzzy membership functions cover 100% of the input variable domain, then it can be proven that the behavior of the fuzzy system is fully deterministic over the entire input domain and nowhere ambiguous. This determinism is very important for use in control and decision systems.

There is no "ELSE"—all of the rules are evaluated, because the temperature might be "cold" and "normal" at the same time to different degrees.

The AND, OR, and NOT operators of Boolean logic exist in fuzzy logic, usually defined as the minimum, maximum, and complement; when they are defined this way, they are called the *Zadeh operators*. So for the fuzzy variables x and y:

```
NOT x = (1 - truth(x))

x AND y = minimum(truth(x), truth(y))

x OR y = maximum(truth(x), truth(y))
```

There are also other operators, more linguistic in nature, called *hedges* that can be applied. These are generally adverbs such as *very*, or *somewhat*, which modify the meaning of a set using a mathematical formula.

Define with Multiply

```
x AND y = x*y

x OR y = 1-(1-x)*(1-y) = x+y-x*y
```

1-(1-x)*(1-y) comes from this:

```
x OR y = NOT( AND( NOT(x), NOT(y) ) )

x OR y = NOT( AND(1-x, 1-y) )

x OR y = NOT( (1-x)*(1-y) )

x OR y = 1-(1-x)*(1-y)
```

Define with Sigmoid

Fuzzy sets can be defined using the sigmoid function.

```
sigmoid(x)=1/(1+e^-x)

sigmoid(x)+sigmoid(-x) = 1

(sigmoid(x)+sigmoid(-x))*(sigmoid(y)+sigmoid(-y))*(sig-
moid(z)+sigmoid(-z)) = 1
```

Logical Analysis

In mathematical logic, there are several formal systems of "fuzzy logic"; most of them belong among so-called t-norm fuzzy logic.

Propositional Fuzzy Logics

The most important propositional fuzzy logics are:-

- Monoidal t-norm-based propositional fuzzy logic MTL is an axiomatization of logic where conjunction is defined by a left continuous t-norm and implication is defined as the residuum of the t-norm. Its models correspond to MTL-algebras that are pre-linear commutative bounded integral residuated lattices.

- Basic propositional fuzzy logic BL is an extension of MTL logic where conjunction is defined by a continuous t-norm, and implication is also defined as the residuum of the t-norm. Its models correspond to BL-algebras.

- Łukasiewicz fuzzy logic is the extension of basic fuzzy logic BL where standard conjunction is the Łukasiewicz t-norm. It has the axioms of basic fuzzy logic plus an axiom of double negation, and its models correspond to MV-algebras.

- Gödel fuzzy logic is the extension of basic fuzzy logic BL where conjunction is Gödel t-norm. It has the axioms of BL plus an axiom of idempotence of conjunction, and its models are called G-algebras.

- Product fuzzy logic is the extension of basic fuzzy logic BL where conjunction is product t-norm. It has the axioms of BL plus another axiom for cancellativity of conjunction, and its models are called product algebras.

- Fuzzy logic with evaluated syntax (sometimes also called Pavelka's logic), denoted by EVŁ, is a further generalization of mathematical fuzzy logic. While the above kinds of fuzzy logic have traditional syntax and many-valued semantics, in EVŁ is evaluated also syntax. This means that each formula has an evaluation. Axiomatization of EVŁ stems from Łukasziewicz fuzzy logic. A generalization of classical Gödel completeness theorem is provable in EVŁ.

Predicate Fuzzy Logics

These extend the above-mentioned fuzzy logics by adding universal and existential quantifiers in a manner similar to the way that predicate logic is created from propositional logic. The semantics of the universal (resp. existential) quantifier in t-norm fuzzy logics is the infimum (resp. supremum) of the truth degrees of the instances of the quantified subformula.

Decidability Issues for Fuzzy Logic

The notions of a "decidable subset" and "recursively enumerable subset" are basic ones for classical mathematics and classical logic. Thus the question of a suitable extension of these concepts to fuzzy set theory arises. A first proposal in such a direction was made by E.S. Santos by the notions of *fuzzy Turing machine*, *Markov normal fuzzy algorithm* and *fuzzy program*. Successively, L. Biacino and G. Gerla argued that the proposed definitions are rather questionable and therefore they proposed the following ones. Denote by $Ü$ the set of rational numbers in [0,1]. Then a fuzzy subset $s : S \rightarrow [0,1]$ of a set S is recursively enumerable if a recursive map $h : S \times N \rightarrow Ü$ exists such that, for every x in S, the function $h(x,n)$ is increasing with respect to n and $s(x) = \lim h(x,n)$. We say that *s* is *decidable* if both *s* and its complement −*s* are recursively enumerable. An extension of such a theory to the general case of the L-subsets is possible. The proposed definitions are well related with fuzzy logic. Indeed, the following theorem holds true (provided that the deduction apparatus of the considered fuzzy logic satisfies some obvious effectiveness property).

Theorem. Any axiomatizable fuzzy theory is recursively enumerable. In particular, the fuzzy set of logically true formulas is recursively enumerable in spite of the fact that the crisp set of valid formulas is not recursively enumerable, in general. Moreover, any axiomatizable and complete theory is decidable.

It is an open question to give supports for a "Church thesis" for fuzzy mathematics, the proposed notion of recursive enumerability for fuzzy subsets is the adequate one. To this aim, an extension of the notions of fuzzy grammar and fuzzy Turing machine should be necessary. Another open question is to start from this notion to find an extension of Gödel's theorems to fuzzy logic.

Fuzzy Databases

Once fuzzy relations are defined, it is possible to develop fuzzy relational databases. The first fuzzy relational database, FRDB, appeared in Maria Zemankova's dissertation (1983). Later, some other models arose like the Buckles-Petry model, the Prade-Testemale Model, the Umano-Fukami model or the GEFRED model by J.M. Medina, M.A. Vila et al.

Fuzzy querying languages have been defined, such as the SQLf by P. Bosc et al. and the FSQL by J. Galindo et al. These languages define some structures in order to include fuzzy aspects in the SQL statements, like fuzzy conditions, fuzzy comparators, fuzzy constants, fuzzy constraints, fuzzy thresholds, linguistic labels etc.

Comparison to Probability

Fuzzy logic and probability address different forms of uncertainty. While both fuzzy logic and probability theory can represent degrees of certain kinds of subjective belief, fuzzy set theory uses the concept of fuzzy set membership, i.e., *how much* a variable is in a set (there is not necessarily any uncertainty about this degree), and probability theory uses the concept of subjective probability, i.e., *how probable* is it that a variable is in a set (it either entirely is or entirely is not in the set in reality, but there is uncertainty around whether it is or is not). The technical consequence of this distinction is that fuzzy set theory relaxes the axioms of classical probability, which are themselves derived from adding uncertainty, but not degree, to the crisp true/false distinctions of classical Aristotelian logic.

Bruno de Finetti controversially argues that only one kind of mathematical uncertainty, probability, is needed, and thus fuzzy logic is superfluous. However, Bart Kosko shows in Fuzziness vs. Probability that probability theory is a subtheory of fuzzy logic, as questions of degrees of belief in mutually-exclusive set membership in probability theory can be represented as certain cases of non-mutually-exclusive graded membership in fuzzy theory. In that context, he also derives Bayes' theorem from the concept of fuzzy subsethood. Lotfi A. Zadeh argues that fuzzy logic is different in character from probability, and is not a replacement for it. He fuzzified probability to fuzzy probability and also generalized it to possibility theory. (cf.)

More generally, fuzzy logic is one of many different extensions to classical logic intended to deal with issues of uncertainty outside of the scope of classical logic, the inapplicability of probability theory in many domains, and the paradoxes of Dempster-Shafer theory.

Relation to Ecorithms

Computational theorist Leslie Valiant uses the term *ecorithms* to describe how many less exact systems and techniques like fuzzy logic (and "less robust" logic) can be applied to learning algorithms. Valiant essentially redefines machine learning as evolutionary. In general use, ecorithms are algorithms that learn from their more complex environments (hence *eco-*) to generalize, approximate and simplify solution logic. Like fuzzy logic, they are methods used to overcome continuous variables or systems too complex to completely enumerate or understand discretely or exactly. Ecorithms and fuzzy logic also have the common property of dealing with possibilities more than probabilities, although feedback and feed forward, basically stochastic weights, are a feature of both when dealing with, for example, dynamical systems.

Compensatory Fuzzy Logic

Compensatory fuzzy logic (CFL) is a branch of fuzzy logic with modified rules for conjunction and disjunction. When the truth value of one component of a conjunction or disjunction is increased or decreased, the other component is decreased or increased to compensate. This increase or decrease in truth value may be offset by the increase or decrease in another component. An offset may be blocked when certain thresholds are met. Proponents claim that CFL allows for better computational semantic behaviors.

Compensatory Fuzzy Logic consists of four continuous operators: conjunction (c); disjunction (d); fuzzy strict order (or); and negation (n). The conjunction is the geometric mean and its dual as conjunctive and disjunctive operators.

References

- Novák, V., Perfilieva, I. and Močkoř, J. (1999) Mathematical principles of fuzzy logic Dodrecht: Kluwer Academic. ISBN 0-7923-8595-0

- Valiant, Leslie, (2013) Probably Approximately Correct: Nature's Algorithms for Learning and Prospering in a Complex World New York: Basic Books. ISBN 978-0465032716

- Kreiser, Lothar; Gottwald, Siegfried; Stelzner, Werner (1990). Nichtklassische Logik. Eine Einführung. Berlin: Akademie-Verlag. pp. 41ff -- 45ff. ISBN 978-3-05-000274-3.

- Wierman, Mark J. "An Introduction to the Mathematics of Uncertainty: including Set Theory, Logic, Probability, Fuzzy Sets, Rough Sets, and Evidence Theory"(PDF). Creighton University. Retrieved 16 July 2016.

- Cejas, Jesús, (2011) Compensatory Fuzzy Logic. La Habana: Revista de Ingeniería Industrial. ISSN 1815-5936

2

Concepts of Fuzzy Logic

A fuzzy number is when a real number does not refer to a single value but is connected to a number of possible values. The value is anywhere between 0 and 1. This text elucidates theories such as fuzzy numbers and fuzzy finite element. The chapter strategically encompasses and incorporates the major components and concepts of fuzzy logic.

Fuzzy Concept

A fuzzy concept is a concept of which the boundaries of application can vary considerably according to context or conditions, instead of being fixed once and for all. This means the concept is *vague* in some way, lacking a fixed, precise meaning, without however being unclear or meaningless altogether. It has a definite meaning, which can become more precise only through further elaboration and specification, including a closer definition of the context in which the concept is used.

A fuzzy concept is understood by scientists as a concept which is "to an extent applicable" in a situation, and it therefore implies gradations of meaning. The best known example of a fuzzy concept around the world is an amber traffic light, and indeed fuzzy concepts are nowadays widely used in traffic control systems.

The Nordic myth of Loki's wager suggests that concepts which lack a precise meaning or precise boundaries of application cannot be usefully discussed at all. However, the idea of "fuzzy concepts" proposes that "somewhat vague terms" can be operated with, since we can explicate and define the variability of their application, by assigning numbers to it.

Origin and Etymology

The intellectual origins of the idea of fuzzy concepts have been traced to a diversity of famous and less well known thinkers including Plato, Georg Wilhelm Friedrich Hegel, Karl Marx, Friedrich Engels, Friedrich Nietzsche, Jan Łukasiewicz, Alfred Tarski, Stanisław Jaśkowski and Donald Knuth. This suggests that the idea of fuzzy concepts has, in one form or another, a very long history in human thought.

However, usually the Iranian born, American computer scientist Lotfi A. Zadeh is credited with inventing the specific idea of a "fuzzy concept" in his seminal 1965 paper on fuzzy sets, because he gave a formal mathematical presentation of the phenomenon which was widely accepted by scholars. It was also Zadeh who played a decisive role in developing the field of fuzzy logic, fuzzy sets and fuzzy systems, with a large number of scholarly papers.

In fact, the German scholar Dieter Klaua also published a German-language paper on fuzzy sets in 1965, but he used a different terminology (he referred to "many-valued sets"). An earlier attempt to create a theory of sets where set membership is a matter of degree was made by Abraham Kaplan and Hermann Schott in 1951. They intended to apply the idea to empirical research. Kaplan and Schott measured the degree of membership of empirical classes using real numbers between 0 and 1, and they defined corresponding notions of intersection, union, complementation and subset. However, at the time, their idea "fell on stony ground".

Radim Belohlavek explains:

"There exists strong evidence, established in the 1970s in the psychology of concepts… that human concepts have a graded structure in that whether or not a concept applies to a given object is a matter of degree, rather than a yes-or-no question, and that people are capable of working with the degrees in a consistent way. This finding is intuitively quite appealing, because people say "this product is more or less good" or "to a certain degree, he is a good athlete", implying the graded structure of concepts. In his classic paper, Zadeh called the concepts with a graded structure *fuzzy concepts* and argued that these concepts are a rule rather than an exception when it comes to how people communicate knowledge. Moreover, he argued that to model such concepts mathematically is important for the tasks of control, decision making, pattern recognition, and the like. Zadeh proposed the notion of a *fuzzy set* that gave birth to the field of *fuzzy logic*…"

Hence, a concept is regarded as "fuzzy" in a logical sense if:

- defining characteristics of the concept apply to it "to a certain degree or extent" (or, more unusually, "with a certain magnitude of likelihood")

- or, the boundaries of applicability (the truth-value) of a concept can vary according to different conditions.

- or, the fuzzy concept itself straightforwardly consists of a fuzzy set, or a combination of such sets.

The fact that a concept is fuzzy does not prevent its use in logical reasoning, it merely affects the type of reasoning which can be applied. If the concept has gradations of meaningful significance, it is necessary to specify and formalize what those gradations are. Not all fuzzy concepts have the same logical structure, but they can often be formally described or reconstructed using fuzzy logic.

Zadeh's seminal 1965 paper is acknowledged to be one of the most-cited scholarly articles in the 20th century.

Applications

In philosophical logic, fuzzy concepts are often regarded as concepts which in their application, or formally speaking, are neither completely true nor completely false, or which are partly true and partly false; they are ideas which require further elaboration, specification or qualification to understand their applicability (the conditions under which they truly make sense).

In mathematics and statistics, a fuzzy variable (such as "the temperature", "hot" or "cold") is a value which could lie in a probable range defined by quantitative limits or parameters, and which can be usefully described with imprecise categories (such as "high", "medium" or "low") using some kind of qualitative scale.

In mathematics and computer science, the gradations of applicable meaning of a fuzzy concept are described in terms of *quantitative* relationships defined by logical operators. Such an approach is sometimes called "degree-theoretic semantics" by logicians and philosophers, but the more usual term is fuzzy logic or many-valued logic. The novelty of fuzzy logic is, that it "breaks with the traditional principle that formalisation should correct and avoid, but not compromise with, vagueness".

The basic idea of fuzzy logic is, that a real number is assigned to each statement written in a language, within a range from 0 to 1, where 1 means that the statement is completely true, and 0 means that the statement is completely false, while values less than 1 but greater than 0 represent that the statements are "partly true", to a given, quantifiable extent. Susan Haack comments:

"Whereas in classical set theory an object either is or is not a member of a given set, in fuzzy set theory membership is a matter of degree; the degree of membership of an object in a fuzzy set is represented by some real number between 0 and 1, with 0 denoting *no* membership and 1 *full* membership."

"Truth" in this mathematical context usually means simply that "something is the case", or that "something is applicable". This makes it possible to analyze a distribution of statements for their truth-content, identify data patterns, make inferences and predictions, and model how processes operate. Fuzzy logic in principle allows us to give a definite, precise answer to the question: "to what extent is something the case?", or "to what extent is something applicable?". Via a series of switches, this kind of reasoning can be built into electronic devices. That was already happening before fuzzy logic was invented, but using fuzzy logic in modelling has become an important aid in design, which creates many new technical possibilities.

Fuzzy reasoning (i.e., reasoning with graded concepts) turns out to have many practical uses. It is nowadays widely used in the programming of vehicle and transport electronics, household appliances, video games, language filters, robotics, all kinds of control systems, and various kinds of electronic equipment used for pattern recognition, surveying and monitoring (such as radars). Fuzzy reasoning is also used in artificial intelligence, virtual intelligence and soft computing research. "Fuzzy risk scores" are used by project managers and portfolio managers to express risk assessments. Fuzzy logic has even been applied to the problem of predicting cement strength. It looks like fuzzy logic will eventually be applied in almost every aspect of life, even if people are not aware of it, and in that sense fuzzy logic is an astonishingly successful invention.

A lot of research on fuzzy logic was done by Japanese researchers inventing new machinery, electronic equipment and appliances. The North American Fuzzy Information Processing Society (NAFIPS) was founded in 1981. There also exists an International Fuzzy Systems Association (IFSA). In Europe, there is a European Society for Fuzzy Logic and Technology (EUSFLAT).

Lotfi Zadeh estimates there are more than 50,000 fuzzy logic-related, patented inventions. He lists 28 journals dealing with fuzzy reasoning, and 21 journal titles on soft computing. There are now close to 100,000 publications with the word "fuzzy" in their titles, or maybe even 300,000.

Fuzzy Concept Lattices and Big Data

According to the computer scientist Andrei Popescu at Middlesex University London, a concept can be operationally defined to consist of (1) an *intent*, which is a description or specification stated in a language, and (2) an *extent*, which is the collection of all the objects to which the description refers. Additionally the concept can be defined only with reference to a *context*, which is stated by (i) the universe of all possible objects, (ii) the universe of all possible attributes of objects, and (iii) the formal definition of the relation whereby an object possesses an attribute. Once the context is defined, we can then specify relationships between sets of objects and sets of attributes which they do, or do not share.

However, whether an object belongs to a concept, and whether an object does, or does not have an attribute, can often be a matter of degree. Thus, for example, "many attributes are fuzzy rather than crisp". To overcome this issue, a numerical value is assigned to each object or attribute along a scale, and the results are placed in a table which links each assigned value within the given range to a numerical value denoting a given degree of applicability.

This is the basic idea of a "fuzzy concept lattice", which can also be graphed; different fuzzy concept lattices can be connected to each other as well. Fuzzy concept lattices are a useful programming tool for the exploratory analysis of big data, for example in cases where sets of linked behavioural responses are broadly similar, but can nevertheless vary in important ways, within certain limits. It can help to find out what the structure and dimensions are, of a behaviour that occurs with an important but limited amount of variation within a large population.

Fuzzy lattices can be applied, for instance, in the psephological analysis of big data about voter behaviour, where researchers want to explore the characteristics and associations involved in "somewhat vague" opinions; patterns of variation in voter attitudes; and variability in voter behaviour (or personal characteristics) within a set of parameters. The basic programming techniques for this kind of fuzzy concept mapping and deep learning are by now well-established and big data analytics had a strong influence on the US elections of 2016. A US study concluded in 2015 that for 20% of undecided voters, Google's secret search algorithm had the power to change the way they voted.

Very large quantities of data can now be explored using computers with fuzzy logic systems and open-source architectures such as Apache Hadoop, Apache Spark, and MongoDB. One author claimed in 2016 that it is now possible to obtain and analyze "400 data points" for each voter in a population, using Oracle systems. However, NBC News reported that the British firm Cambridge Analytica which profiled voters for Donald Trump has not 400, but 4,000 data points for each of 230 million US adults. Cambridge Analytica's own website claims that "up to 5,000 data points" are collected for each of 220 million Americans. Harvard University Professor Latanya Sweeney calculated, that if a company knows just your date of birth, your ZIP code and sex, the company can identify you by name 87% of the time - simply by using linked data from various sources. With 4,000-5,000 data points instead of three, a very comprehensive personal profile becomes possible for almost every voter, and many behavioural patterns can be inferred by linking together different data sets. It also becomes possible to identify and measure gradations in personal characteristics which, in aggregate, have very large effects.

Some researchers argue that this kind of big data analysis has severe limitations, and that the analytical results can only be regarded as indicative, and not as definitive. This was confirmed by Kellyanne Conway, Donald Trump's campaign manager, who emphasized the importance of human judgement and common sense in drawing conclusions from fuzzy data. Conway candidly admitted that much of her own research would "never see the light of day", because it was client confidential. Another Trump adviser criticized Conway, claiming that she "produces an analysis that buries every terrible number and highlights every positive number". A traditional objection to big data is, that it cannot cope with rapid change, but the technology now exists for corporations like Amazon, Google and Microsoft to pump cloud-based data streams from app-users straight into big data analytics programmes, in real time.

Definitional Controversies

Some scientists claimed that in reality fuzzy concepts do not exist. For example, Rudolf E. Kálmán stated in 1972 that "there is no such thing as a fuzzy concept... We do talk about fuzzy things but they are not scientific concepts". The suggestion is that a concept, to qualify as a concept, must be clear and precise. A vague notion would be at best a prologue to formulating a concept. However, there is no general agreement how the notion of a "concept", or a scientific concept in particular, should be defined, and of course scientists also quite often do use imprecise analogies in their models to help understanding an issue.

Susan Haack once claimed that a many-valued logic requires neither intermediate terms between true and false, nor a rejection of bivalence. Her suggestion was, that the intermediate terms (i.e. the gradations of truth) can always be restated as conditional if-then statements, and by implication, that fuzzy logic is fully reducible to binary true-or-false logic. This interpretation is disputed, but even if it was correct, the ability to assign a numerical value to the applicability of a statement is often enormously more efficient than a long sequence of if-then statements that would have the same meaning. That point is obviously of great importance to computer programmers seeking to code a process or operation according to logical rules.

- Not all scholars would agree that a concept is equal to, or reducible to, a mathematical set.

- Qualities may not be fully reducible to quantities - if there are no qualities, it becomes impossible to say what the numbers are numbers of, or what they refer to, except that they refer to other numbers.

- In creating a formalization or formal specification of a concept, for example for the purpose of measurements, administrative procedure or programming, part of the meaning of the concept may be changed or lost.

Programmers, statisticians or logicians are concerned in their work with the main operational or technical significance of a concept which is specifiable in objective, quantifiable terms. They are not primarily concerned with all kinds of interpretive frameworks associated with the concept, or with those aspects of the concept which seem to have no particular functional purpose - however entertaining they might be. However, some of the qualitative characteristics of the concept may not be quantifiable or measurable at all, at least not directly. The temptation exists to ignore them, or try to infer them from data results. A concept may also be created as an ideal type to understand something imaginatively, without any strong claim that it is a "true and complete description" or a "true and complete reflection" of whatever is being conceptualized.

Philosophers regard fuzziness as a particular kind of vagueness, and consider that "no specific assignment of semantic values to vague predicates, not even a fuzzy one, can fully satisfy our conception of what the extensions of vague predicates are like". However, Lotfi Zadeh claims that "vagueness connotes insufficient specificity, whereas fuzziness connotes unsharpness of class boundaries". Thus, he argues, a sentence like "I will be back in a few minutes" is fuzzy *but not* vague, whereas a sentence such as "I will be back sometime", is fuzzy *and* vague. His suggestion is that fuzziness and vagueness are logically quite different qualities, rather than fuzziness being a type or subcategory of vagueness. Zadeh claims that "inappropriate use of the term 'vague' is still a common practice in the literature of philosophy".

The definitional disputes remain unresolved, mainly because, as anthropologists and psychologists have documented, different languages (or symbol systems) that have been created by people to signal meanings suggest different ontologies. Put simply: it is not merely that describing "what is there" involves symbolic representations of some kind. How distinctions are drawn, influences perceptions of "what is there", and vice versa, perceptions of "what is there" influence how distinctions are drawn. For example, cosmologist Max Tegmark argues the universe consists of math: "If you accept the idea that both space itself, and all the stuff in space, have no properties at all except mathematical properties," then the idea that everything is mathematical "starts to sound a little bit less insane."

Sociology and Journalism

The idea of fuzzy concepts has also been applied in the philosophical, sociological and linguistic analysis of human behaviour. In a 1973 paper, George Lakoff for example analyzed hedges in the interpretation of the meaning of categories. Charles Ragin and others have applied the idea to sociological analysis.

In a more general sociological or journalistic sense, a "fuzzy concept" has come to mean a concept which is meaningful but inexact, implying that it does not exhaustively or completely define the meaning of the phenomenon to which it refers - often because it is too abstract. In this context, it is said that fuzzy concepts "lack clarity and are difficult to test or operationalize". To specify the relevant meaning more precisely, additional distinctions, conditions and/or qualifiers would be required. Thus, for example, in a handbook of sociology we find a statement such as "The theory of interaction rituals contains some gaps that need to be filled and some fuzzy concepts that need to be differentiated." The idea is that if finer distinctions are introduced, then the fuzziness or vagueness would be eliminated.

The main reason why the term is now often used in describing human behaviour, is that human interaction has many characteristics which are difficult to quantify and measure precisely, although we know that they have magnitudes, among other things because they are interactive and reflexive (the observers and the observed mutually influence the meaning of events). Those human characteristics can be usefully expressed only in an *approximate* way.

Newspaper stories frequently contain fuzzy concepts, which are readily understood and used, even although they are far from exact. Thus, many of the meanings which people ordinarily use to negotiate their way through life in reality turn out to be "fuzzy concepts". While people often do need to be exact about some things (e.g. money or time), many areas of their lives involve expressions which are far from exact.

Sometimes the term is also used in a pejorative sense. For example, a New York Times journalist wrote that Prince Sihanouk "seems unable to differentiate between friends and enemies, a disturbing trait since it suggests that he stands for nothing beyond the fuzzy concept of peace and prosperity in Cambodia".

The use of fuzzy logic in the social sciences and humanities has remained limited. Lotfi Zadeh said in a 1994 interview that:

"I expected people in the social sciences - economics, psychology, philosophy, linguistics, politics, sociology, religion and numerous other areas to pick up on it. It's been somewhat of a mystery to me why even to this day, so few social scientists have discovered how useful it could be."

Two decades later, fuzzy concepts and fuzzy logic are being widely applied in big data analysis of social, commercial and psychological phenomena. Jaakko Hintikka once claimed that "the logic of natural language we are in effect already using can serve as a "fuzzy logic" better than its trade name variant without any additional assumptions or constructions." That might help to explain why fuzzy logic is not much used to formalize concepts in the "soft" social sciences. However Lotfi Zadeh rejected such an interpretation, on the ground that in many human endeavours as well as technologies it is highly important to define more exactly "to what extent" something is applicable or true, when it is known that its applicability can vary to some important extent. Reasoning which accepts and uses fuzzy concepts can be shown to be perfectly valid with the aid of fuzzy logic, because the degrees of applicability of a concept can be more precisely and efficiently defined with the aid of numerical notation.

Another possible explanation for the lack of use of fuzzy logic by social scientists is simply that, beyond basic statistical analysis (using programs such as SPSS and Excel) the mathematical knowledge of social scientists is often rather limited; they may not know how to formalize a fuzzy concept using the conventions of fuzzy logic. But Hintikka may be correct, in the sense that it can be much more efficient to use natural language to denote a complex idea, than to formalize it. The quest for formalization might introduce much more complexity, which is not wanted, and which detracts from communicating the relevant issue.

Uncertainty

Fuzzy concepts can generate uncertainty because they are imprecise (especially if they refer to a process in motion, or a process of transformation where something is "in the process of turning into something else"). In that case, they do not provide a clear orientation for action or decision-making ("what does X really mean or imply?"); reducing fuzziness, perhaps by applying fuzzy logic, would generate more certainty.

However, this is not necessarily always so. A concept, even although it is not fuzzy at all, and even though it is very exact, could equally well fail to capture the meaning of something adequately. That is, a concept can be very precise and exact, but not - or insufficiently - *applicable* or *relevant* in the situation to which it refers. In this sense, a definition can be "very precise", but "miss the point" altogether.

A fuzzy concept may indeed provide *more* security, because it provides a meaning for something when an exact concept is unavailable - which is better than not being able to denote it at all. A concept such as God, although not easily definable, for instance can provide security to the believer.

Language

Ordinary language, which uses symbolic conventions and associations which are often not logical, inherently contains many fuzzy concepts - "knowing what you mean" in this case depends on knowing the context or being familiar with the way in which a term is normally used, or what it is associated with. This can be easily verified for instance by consulting a dictionary, a thesaurus or an encyclopedia which show the multiple meanings of words, or by observing the behaviours involved in ordinary relationships which rely on mutually understood meanings. Bertrand Russell regarded language as intrinsically vague.

To communicate, receive or convey a message, an individual somehow has to bridge his own intended meaning and the meanings which are understood by others, i.e., the message has to be conveyed in a way that it will be socially understood, preferably in the intended manner. Thus, people might state: "you have to say it in a way that I understand".

This may be done instinctively, habitually or unconsciously, but it usually involves a choice of terms, assumptions or symbols whose meanings may often not be completely fixed, but which depend among other things on how the receiver of the message responds to it, or the context. In this sense, meaning is often "negotiated" or "interactive" (or, more cynically, manipulated). This gives rise to many fuzzy concepts.

But even using ordinary set theory and binary logic to reason something out, logicians have discovered that it is possible to generate statements which are logically speaking not completely true or imply a paradox, even although in other respects they conform to logical rules.

Psychology

The formation of fuzzy concepts is partly due to the fact that the human brain does not operate like a computer .

- While ordinary computers use strict binary logic gates, the brain does not; i.e., it is capable of making all kinds of neural associations according to all kinds of ordering principles (or fairly chaotically) in associative patterns which are not logical but nevertheless meaningful. For example, a work of art can be meaningful without being logical. A pattern can be regular and non-arbitrary, hence meaningful, without it being possible to describe it completely or exhaustively in formal-logical terms.

- Something can be meaningful although we cannot name it, or we might only be able to name it and nothing else.

- The human brain can also interpret the same phenomenon in several different but interacting frames of reference, at the same time, or in quick succession, without there necessarily being an explicit logical connection between the frames.

In part, fuzzy concepts arise also because learning or the growth of understanding involves a transition from a vague awareness, which cannot orient behaviour greatly, to clearer insight, which can orient behaviour.

Some logicians argue that fuzzy concepts are a necessary consequence of the reality that any kind of distinction we might like to draw has *limits of application*. At a certain level of generality, a distinction works fine. But if we pursued its application in a very exact and rigorous manner, or overextend its application, it appears that the distinction simply does not apply in some areas or contexts, or that we cannot fully specify how it should be drawn. An analogy might be that zooming a telescope, camera, or microscope in and out reveals that a pattern which is sharply focused at a certain distance disappears at another distance (or becomes blurry).

Faced with any large, complex and continually changing phenomenon, any short statement made about that phenomenon is likely to be "fuzzy", i.e., it is meaningful, but - strictly speaking - incorrect and imprecise. It will not really do full justice to the reality of what is happening with the phenomenon. A correct, precise statement would require a lot of elaborations and qualifiers. Nevertheless, the "fuzzy" description turns out to be a useful shorthand that saves a lot of time in communicating what is going on ("you know what I mean").

In psychophysics it has been discovered that the perceptual distinctions we draw in the mind are often more sharply defined than they are in the real world. Thus, the brain actually tends to "sharpen up" our perceptions of differences in the external world. Between black and white, we are able to detect only a limited number of shades of gray, or colour gradations. If there are more gradations and transitions in reality, than our conceptual or perceptual distinctions can capture, then it could be argued that how those distinctions will actually apply, must *necessarily* become vaguer at some point. If, for example, one wants to count and quantify distinct objects using numbers, one needs to be able to distinguish between those separate objects, but if this is difficult or impossible, then, although this may not invalidate a quantitative procedure as such, quantification is not really possible in practice; at best, we may be able to assume or infer indirectly a certain distribution of quantities that must be there.

Finally, in interacting with the external world, the human mind may often encounter new, or partly *new phenomena or relationships* which cannot (yet) be sharply defined given the background knowledge available, and by known distinctions, associations or generalizations.

"Crisis management plans cannot be put 'on the fly' after the crisis occurs. At the outset, information is often vague, even contradictory. Events move so quickly that decision makers experience a sense of loss of control. Often denial sets in, and managers unintentionally cut off information flow about the situation" - L. Paul Bremer, "Corporate governance and crisis management", in: *Directors & Boards*, Winter 2002

It also can be argued that fuzzy concepts are generated by a certain sort of lifestyle or way of working which evades definite distinctions, makes them impossible or inoperable, or which is in some way chaotic. To obtain concepts which are not fuzzy, it must be possible to test out their application in some way. But in the absence of any relevant clear distinctions, or when everything is "in a state of flux" or in transition, it may not be possible to do so, so that the amount of fuzziness increases.

Everyday Occurrence

Fuzzy concepts often play a role in the creative process of forming new concepts to understand something. In the most primitive sense, this can be observed in infants who, through practical experience, learn to identify, distinguish and generalise the correct application of a concept, and relate it to other concepts.

However, fuzzy concepts may also occur in scientific, journalistic, programming and philosophical activity, when a thinker is in the process of clarifying and defining a newly emerging concept which is based on distinctions which, for one reason or another, cannot (yet) be more exactly specified or validated. Fuzzy concepts are often used to denote complex phenomena, or to describe something which is developing and changing, which might involve shedding some old meanings and acquiring new ones.

- In politics, it can be highly important and problematic how exactly a conceptual distinction is drawn, or indeed whether a distinction is drawn at all; distinctions used in administration may be deliberately sharpened, or kept fuzzy, due to some political motive or power relationship. Politicians may be deliberately vague about some things, and very clear and explicit about others; if there is information that proves their case, they become very precise, but if the information doesn't prove their case, they become vague or saying nothing. The "fuzzy area" can also refer simply to a *residual* number of cases which cannot be allocated to a known and identifiable group, class or set if strict criteria are used.

- In administration and accounting, fuzziness problems of interpretation and boundary problems can arise, because it is not clear to what category exactly a case, item, transaction or piece of information belongs. In principle, each case, event or item must be allocated to the correct category in a procedure, but it may be, that it is difficult to make the appropriate or relevant distinctions.

- In translation work, fuzzy concepts are analyzed for the purpose of good translation. A concept in one language may not have quite the same meaning or significance in another language, or it may not be feasible to translate it literally, or at all. Some languages have concepts which do not exist in another language, raising the problem of how one would most easily render their meaning. In computer-assisted translation, a technique called fuzzy matching is used to find the most likely translation of a piece of text, using previous translated texts as a basis.

- In information services fuzzy concepts are frequently encountered because a customer or client asks a question about something which could be interpreted in many different ways, or, a document is transmitted of a type or meaning which cannot be easily allocated to a known type or category, or to a known procedure. It might take considerable inquiry to "place" the information, or establish in what framework it should be understood.

- In the legal system, it is essential that rules are interpreted and applied in a standard way, so that the same cases and the same circumstances are treated equally. Otherwise one would be accused of arbitrariness, which would not serve the interests of justice. Consequently, lawmakers aim to devise definitions and categories which are sufficiently precise, so that they are not open to different interpretations. For this purpose, it is critically important to remove fuzziness, and differences of interpretation are typically resolved through a court ruling based on evidence. Alternatively, some other procedure is devised which permits the correct distinction to be discovered and made.

- In medical diagnosis, the assessment of what the symptoms of a patient are often cannot be very exactly specified, since there are many possible qualitative and quantitative gradations in severity, incidence or frequency that could occur. Different symptoms may also overlap to some extent. These gradations can be difficult to measure, and so the medical professionals use approximate "fuzzy" categories in their judgement of a medical condition or a patient's state of health. Although it may not be exact, the diagnosis is often useful enough for treatment purposes.

- In statistical research, it is an aim to measure the magnitudes of phenomena. For this purpose, phenomena have to be grouped and categorized so that distinct and discrete counting units can be defined. It must be possible to allocate all observations to mutually exclusive categories so that they are properly quantifiable. Survey observations do not spontaneously transform themselves into countable data; they have to be identified, categorized and classified in such a way, that identical observations can be grouped together, and that observations are not counted twice or more. Again, for this purpose it is a requirement that the concepts used are exactly defined, and not fuzzy. There could be a margin of error, but the amount of error must be kept within tolerable limits, and preferably its magnitude should be known.

- In hypnotherapy, fuzzy language is deliberately used for the purpose of trance induction. Hypnotic suggestions are often couched in a somewhat vague, general or ambiguous language requiring interpretation by the subject. The intention is to distract and shift the conscious awareness of the subject away from external reality to his own internal state. In response to the somewhat confusing signals he gets, the awareness of the subject spontaneously tends to withdraw inward, in search of understanding or escape.

- In biology, protein complexes with multiple structural forms are called fuzzy complexes. The different conformations can result in different, even opposite functions. The conformational ensemble is modulated by the environmental conditions. Post-translational modifications or alternative splicing can also impact the ensemble and thereby affinity or specificity of interactions.

- In theology an attempt is made to define more precisely the meaning of spiritual concepts, which refer to how human beings construct the meaning of human existence, and, often, the relationship people have with a supernatural world. Many spiritual concepts and beliefs are fuzzy, to the extent that, although abstract, they often have a highly personalized meaning, or involve personal interpretation of a type that is not easy to define in a cut-and-dried way. A similar situation occurs in psychotherapy. The Dutch theologian Kees de Groot has explored the imprecise notion that psychotherapy is like an "implicit religion", defined as a "fuzzy concept" (it all depends on what one means by "psychotherapy" and "religion").

- In meteorology, where changes and effects of complex interactions in the atmosphere are studied, the weather reports often use fuzzy expressions indicating a broad trend, likelihood or level. The main reason is that the forecast can rarely be totally exact for any given location.

- In phenomenology which studies the structure of subjective experience, an important insight is that how someone experiences something can be influenced *both* by the influence of the thing being experienced itself, but *also* by how the person responds to it. Thus, the actual experience the person has, is shaped by an "interactive object-subject relationship". To describe this experience, fuzzy categories are often necessary, since it is often impossible to predict or describe with great exactitude what the interaction will be, and how it is experienced.

It could be argued that many concepts used fairly universally in daily life (e.g. "love" or "God" or "health" or "social") are *inherently or intrinsically* fuzzy concepts, to the extent that their meaning can never be completely and exactly specified with logical operators or objective terms, and can have multiple interpretations, which are in part exclusively subjective. Yet despite this limitation, such concepts are not meaningless. People keep using the concepts, even if they are difficult to define precisely.

It may also be possible to specify one personal meaning for the concept, without however placing restrictions on a different use of the concept in other contexts (as when, for example, one says "this is what I mean by X" in contrast to other possible meanings). In ordinary speech, concepts may sometimes also be uttered purely randomly; for example a child may repeat the same idea in completely unrelated contexts, or an expletive term may be uttered arbitrarily. A feeling or sense is conveyed, without it being fully clear what it is about.

Fuzzy concepts can be used deliberately to create ambiguity and vagueness, as an evasive tactic, or to bridge what would otherwise be immediately recognized as a contradiction of terms. They might be used to indicate that there is definitely a connection between two things, without giving a complete specification of what the connection is, for some or other reason. This could be due to a failure or refusal to be more precise. But it could also could be a prologue to a more exact formulation of a concept, or a better understanding.

Economy of Distinctions

Fuzzy concepts can be used as a practical method to describe something of which a complete description would be an unmanageably large undertaking, or very time-consuming; thus, a simplified indication of what is at issue is regarded as sufficient, although it is not exact. There is also such a thing as an "economy of distinctions", meaning that it is not helpful or efficient to use more detailed definitions than are really necessary for a given purpose. The provision of "too many details" could be disorienting and confusing, instead of being enlightening, while a fuzzy term might be sufficient to provide an orientation. The reason for using fuzzy concepts can therefore be purely pragmatic, if it is not feasible or desirable (for practical purposes) to provide "all the details" about the meaning of a shared symbol or sign. Thus people might say "I realize this is not exact, but you know what I mean" - they assume practically that stating all the details is not required for the purpose of the communication.

Lotfi Zadeh has picked up this point, and draws attention to a "major misunderstanding" about applying fuzzy logic. It is true that the basic aim of fuzzy logic is to make what is imprecise more precise. Yet in many cases, fuzzy logic is used paradoxically to "imprecisiate what is precise", meaning that there is a deliberate tolerance for imprecision for the sake of simplicity of procedure and economy of expression. In such uses, there is a tolerance for imprecision, because making ideas more precise would be unnecessary and costly, while "imprecisiation reduces cost and enhances tractability" (tractability means "being easy to manage or operationalize"). Zadeh calls this approach the "Fuzzy Logic Gambit" (a gambit means giving up something now, to achieve a better position later). In the Fuzzy Logic Gambit, "what is sacrificed is precision in [quantitative] value, but not precision in meaning", and more concretely, "imprecisiation in value is followed by precisiation in meaning". He cites as example Takeshi Yamakawa's programming for an inverted pendulum, where differential equations are replaced by fuzzy if-then rules in which words are used in place of numbers.

Analysis

In mathematical logic, computer programming, philosophy and linguistics fuzzy concepts can be analyzed and defined more accurately or comprehensively, by describing or modelling the concepts using the terms of fuzzy logic. More generally, clarification techniques can be used such as:

- concretizing the concept by finding specific examples, illustrations, details or cases to which it applies.

- specifying a range of *conditions* to which the concept applies (for example, in computer programming of a procedure).

- *classifying* or *categorizing* all or most cases or uses to which the concept applies (taxonomy).

- *probing the assumptions* on which a concept is based, or which are associated with its use (Critical thought).

- identifying *operational rules* for the use of the concept, which cover all or most cases.

- allocating different applications of the concept to different but related *sets* (e.g. using Boolean logic).

- examining how *probable* it is that the concept applies, statistically or intuitively (Probability theory).

- examining the *distribution* or distributional frequency of (possibly different) uses of the concept (statistics).

- some other kind of *measure or scale* of the degree to which the concept applies.

- specifying a series of logical operators (an inferential system or algorithm) which captures all or most cases to which the concept applies.

- *mapping or graphing* the applications of the concept using some basic parameters.

- applying a *meta-language* which includes fuzzy concepts in a more inclusive categorical system which is not fuzzy.

- *reducing or restating* fuzzy concepts in terms which are simpler or similar, and which are not fuzzy or less fuzzy.

- *relating* the fuzzy concept to other concepts which are not fuzzy or less fuzzy, or simply by *replacing* the fuzzy concept altogether with another, alternative concept which is not fuzzy yet "works exactly the same way".

Operationalization of "Free and Fair Judiciary"

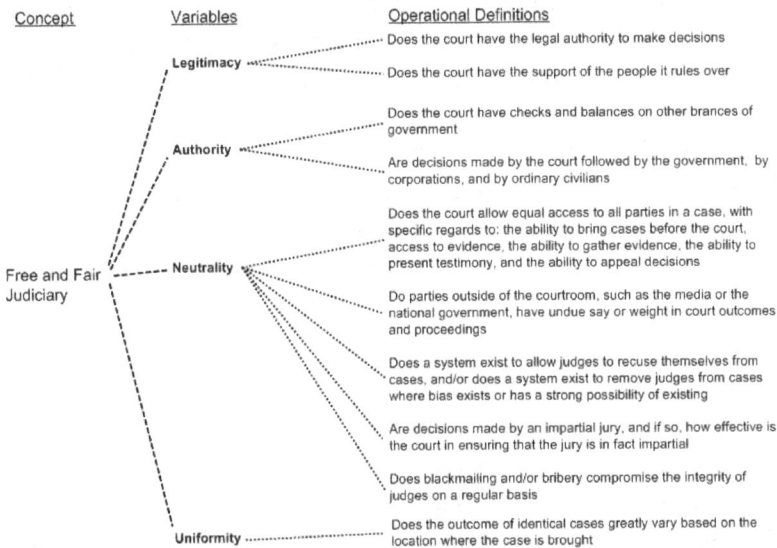

Concept	Variables	Operational Definitions
	Legitimacy	Does the court have the legal authority to make decisions
		Does the court have the support of the people it rules over
	Authority	Does the court have checks and balances on other brances of government
		Are decisions made by the court followed by the government, by corporations, and by ordinary civilians
Free and Fair Judiciary	Neutrality	Does the court allow equal access to all parties in a case, with specific regards to: the ability to bring cases before the court, access to evidence, the ability to gather evidence, the ability to present testimony, and the ability to appeal decisions
		Do parties outside of the courtroom, such as the media or the national government, have undue say or weight in court outcomes and proceedings
		Does a system exist to allow judges to recuse themselves from cases, and/or does a system exist to remove judges from cases where bias exists or has a strong possibility of existing
		Are decisions made by an impartial jury, and if so, how effective is the court in ensuring that the jury is in fact impartial
		Does blackmailing and/or bribery compromise the integrity of judges on a regular basis
	Uniformity	Does the outcome of identical cases greatly vary based on the location where the case is brought

An operationalization diagram, one method of clarifing fuzzy concepts.

In this way, we can obtain a more exact understanding of the use of a fuzzy concept, and possibly decrease the amount of fuzziness. It may not be possible to specify all the possible meanings or applications of a concept completely and exhaustively, but if it is possible to capture the majority of them, statistically or otherwise, this may be useful enough for practical purposes.

A process of defuzzification is said to occur, when fuzzy concepts can be logically described in terms of (the relationships between) fuzzy sets, which makes it possible to define variations in the meaning or applicability of concepts as quantities. Effectively, qualitative differences may then be described more precisely as quantitative variations or quantitative variability (assigning a numerical value then denotes the magnitude of variation).

The difficulty that can occur in judging the fuzziness of a concept can be illustrated with the question *"Is this one of those?"*. If it is not possible to clearly answer this question, that could be because "this" (the object) is itself fuzzy and evades definition, or because "one of those" (the concept of the object) is fuzzy and inadequately defined.

Thus, the source of fuzziness may be in the nature of the reality being dealt with, the concepts used to interpret it, or the way in which the two are being related by a person. It may be that the personal meanings which people attach to something are quite clear to the persons themselves, but that it is not possible to communicate those meanings to others except as fuzzy concepts.

Fuzzy Number

A fuzzy number is a generalization of a regular, real number in the sense that it does not refer to one single value but rather to a connected set of possible values, where each possible value has its own weight between 0 and 1. This weight is called the membership function. A fuzzy number is thus a special case of a convex, normalized fuzzy set of the real line. Just like Fuzzy logic is an extension of Boolean logic (which uses absolute truth and falsehood only, and nothing in between), fuzzy numbers are an extension of real numbers. Calculations with fuzzy numbers allow the incorporation of uncertainty on parameters, properties, geometry, initial conditions, etc.

Fuzzy Finite Element

The fuzzy finite element method combines the well-established finite element method with the concept of fuzzy numbers, the latter being a special case of a fuzzy set. The advantage of using fuzzy numbers instead of real numbers lies in the incorporation of uncertainty (on material properties, parameters, geometry, initial conditions, etc.) in the finite element analysis.

One way to establish a fuzzy finite element (FE) analysis is to use existing FE software (in-house or commercial) as an inner-level module to compute a deterministic result, and to add an outer-level loop to handle the fuzziness (uncertainty). This outer-level loop comes down to solving an optimization problem. If the inner-level deterministic module produces monotonic behavior with respect to the input variables, then the outer-level optimization problem is greatly simplified, since in this case the extrema will be located at the vertices of the domain.

Traditional finite element method is a well-established method to solve various problems of science and engineering. Different authors have used various methods to solve governing differential equation of heat conduction problem. In this study, heat conduction in a circular rod has been considered which is made up of two different materials viz. aluminum and copper. In earlier studies parameters in the differential equation have been taken as fixed numbers which actually may not. Those parameters are found in general by some measurements or experiments. So the material properties are actually uncertain and may be considered to vary in an interval or as fuzzy and in that case complex interval arithmetic or fuzzy arithmetic has to be considered in the analysis.Hence interval/fuzzy arithmetic is applied in the finite element method to solve a steady state heat conduction problem. Application of fuzzy finite element method in the said problem gives fuzzy system of linear equations in general. Here we have also proposed new methods to handle such type of fuzzy system of linear equations. Corresponding results are computed and has been reported here.

Fuzzy Associative Matrix

A fuzzy associative matrix expresses fuzzy logic rules in tabular form. These rules usually take two variables as input, mapping cleanly to a two-dimensional matrix, although theoretically a matrix of any number of dimensions is possible.

Suppose a professional is tasked with writing fuzzy logic rules for a video game monster. In the game being built, entities have two variables: hit points (HP) and firepower (FP):

HP/FP	Very low HP	Low HP	Medium HP	High HP	Very high HP
Very weak FP	Retreat!	Retreat!	Defend	Defend	Defend
Weak FP	Retreat!	Defend	Defend	Attack	Attack
Medium FP	Retreat!	Defend	Attack	Attack	Full attack!
High FP	Retreat!	Defend	Attack	Attack	Full attack!
Very high FP	Defend	Attack	Attack	Full attack!	Full attack!

This translates to:

```
IF MonsterHP IS VeryLowHP AND MonsterFP IS VeryWeakFP THEN Retreat

IF MonsterHP IS LowHP AND MonsterFP IS VeryWeakFP THEN Retreat

IF MonsterHP IS MediumHP AND MonsterFP is VeryWeakFP THEN Defend
```

Multiple rules can fire at once, and often will, because the distinction between "very low" and "low" is fuzzy. If it is more "very low" than it is low, then the "very low" rule will generate a stronger response. The program will evaluate all the rules that fire and use an appropriate defuzzification method to generate its actual response.

An implementation of this system might use either the matrix or the explicit IF/THEN form. The matrix makes it easy to visualize the system, but it also makes it impossible to add a third variable just for one rule, so it is less flexible.

There is no inherent pattern in the matrix. It appears as if the rules were just made up, and indeed they were. This is both a strength and a weakness of fuzzy logic in general. It is often impractical or impossible to find an exact set of rules or formulae for dealing with a specific situation. For a sufficiently complex game, a mathematician would not be able to study the system and figure out a mathematically accurate set of rules. However, this weakness is intrinsic to the realities of the situation, not of fuzzy logic itself. The strength of the system is that even if one of the rules is wrong, even greatly wrong, other rules that are correct are likely to fire as well and they may compensate for the error.

This does not mean a fuzzy system should be sloppy. Depending on the system, it might get away with being sloppy, but it will underperform. While the rules are fairly arbitrary, they should be chosen carefully. If possible, an expert should decide on the rules, and the sets and rules should be tested vigorously and refined as needed. In this way, a fuzzy system is like an expert system. (Fuzzy logic is used in many true expert systems, as well.)

Element (Mathematics)

In mathematics, an element, or member, of a set is any one of the distinct objects that make up that set.

Sets

Writing $A = \{1, 2, 3, 4\}$ means that the elements of the set A are the numbers 1, 2, 3 and 4. Sets of elements of A, for example $\{1, 2\}$, are subsets of A.

Sets can themselves be elements. For example, consider the set $B = \{1, 2, \{3, 4\}\}$. The elements of B are *not* 1, 2, 3, and 4. Rather, there are only three elements of B, namely the numbers 1 and 2, and the set $\{3, 4\}$.

The elements of a set can be anything. For example, $C = \{$ red, green, blue $\}$, is the set whose elements are the colors red, green and blue.

Notation and Terminology

IV. *De classibus.*

Signo K significatur *classis*, sive entium aggregatio.
Signum ε significat *est*. Ita $a \, \epsilon \, b$ legitur a est quoddam b ; $a \, \epsilon$ K significat a est quaedam *classis*; $a \, \epsilon$ P significat a est quaedam *propositio.*

First usage of the symbol ε in the work Arithmetices principia nova methodo exposita by Giuseppe Peano.

The relation "is an element of", also called set membership, is denoted by the symbol "$\in$". Writing

$$x \in A$$

means that "x is an element of A". Equivalent expressions are "x is a member of A", "x belongs to A", "x is in A" and "x lies in A". The expressions "A includes x" and "A contains x" are also used to mean set membership, however some authors use them to mean instead "x is a subset of A". Logician George Boolos strongly urged that "contains" be used for membership only and "includes" for the subset relation only.

Another possible notation for the same relation is

$$A \ni x,$$

meaning "A contains x", though it is used less often.

The negation of set membership is denoted by the symbol "$\notin$". Writing

$$x \notin A$$

means that "x is not an element of A".

The symbol $\in$ was first used by Giuseppe Peano 1889 in his work Arithmetices principia nova methodo exposita. Here he wrote on page X:

"Signum $\in$ significat est. Ita a $\in$ b legitur a est quoddam b; ..."

which means

"The symbol $\in$ means *is*. So a $\in$ b is read as a *is a* b; ..."

The symbol itself is a stylized lowercase Greek letter epsilon ("ϵ"), the first letter of the word, which means "is".

The Unicode characters for these symbols are U+2208 ('element of'), U+220B ('contains as member') and U+2209 ('not an element of'). The equivalent LaTeX commands are "\in", "\ni" and "\notin". Mathematica has commands "\[Element]" and "\[NotElement]".

Cardinality of Sets

The number of elements in a particular set is a property known as cardinality; informally, this is the size of a set. In the above examples the cardinality of the set A is 4, while the cardinality of either of the sets B and C is 3. An infinite set is a set with an infinite number of elements, while a finite set is a set with a finite number of elements. The above examples are examples of finite sets. An example of an infinite set is the set of positive integers = { 1, 2, 3, 4, ... }.

Examples

Using the sets defined above, namely $A = \{1, 2, 3, 4\}$, $B = \{1, 2, \{3, 4\}\}$ and $C = \{$ red, green, blue $\}$:

- $2 \in A$

- $\{3,4\} \in B$

- $3,4 \notin B$

- $\{3,4\}$ is a member of B

- Yellow $\notin C$

- The cardinality of $D = \{2, 4, 8, 10, 12\}$ is finite and equal to 5.

- The cardinality of $P = \{2, 3, 5, 7, 11, 13, ...\}$ (the prime numbers) is infinite (this was proven by Euclid).

Vagueness

In analytic philosophy and linguistics, a concept may be considered **vague** if its extension is deemed lacking in clarity, if there is uncertainty about which objects belong to the concept or which exhibit characteristics that have this predicate (so-called "border-line cases"), or if the Sorites paradox applies to the concept or predicate.

In everyday speech, vagueness is an inevitable, often even desired effect of language usage. However, in most specialized texts (e.g., legal documents), vagueness is often regarded as problematic and undesirable.

Importance

Vagueness is philosophically important. Suppose one wants to come up with a definition of "right" in the moral sense. One wants a definition to cover actions that are clearly right and exclude actions that are clearly wrong, but what does one do with the borderline cases? Surely, there are such cases. Some philosophers say that one should try to come up with a definition that is itself unclear on just those cases. Others say that one has an interest in making his or her definitions more precise than ordinary language, or his or her ordinary concepts, themselves allow; they recommend one advances precising definitions.

Vagueness is also a problem which arises in law, and in some cases judges have to arbitrate regarding whether a borderline case does, or does not, satisfy a given vague concept. Examples include disability (how much loss of vision is required before one is legally blind?), human life (at what point from conception to birth is one a legal human being, protected for instance by laws against murder?), adulthood (most familiarly reflected in legal ages for driving, drinking, voting, consensual sex, etc.), race (how to classify someone of mixed racial heritage), etc. Even such apparently unambiguous concepts such as gender can be subject to vagueness problems, not just from transsexuals' gender transitions but also from certain genetic conditions which can give an individual mixed male and female biological traits.

Many scientific concepts are of necessity vague, for instance species in biology cannot be precisely defined, owing to unclear cases such as ring species. Nonetheless, the concept of species can be clearly applied in the vast majority of cases. As this example illustrates, to say that a definition is "vague" is not necessarily a criticism. Consider those animals in Alaska that are the result of breeding huskies and wolves: are they dogs? It is not clear: they are borderline cases of dogs. This means one's ordinary concept of doghood is not clear enough to let us rule conclusively in this case.

Approaches

The philosophical question of what the best theoretical treatment of vagueness is - which is closely related to the problem of the paradox of the heap, a.k.a. sorites paradox - has been the subject of much philosophical debate.

Fuzzy Logic

One theoretical approach is that of fuzzy logic, developed by American mathematician Lotfi Zadeh. Fuzzy logic proposes a gradual transition between "perfect falsity", for example, the statement "Bill Clinton is bald", to "perfect truth", for, say, "Patrick Stewart is bald". In ordinary logics, there are only two truth-values: "true" and "false". The fuzzy perspective differs by introducing *an infinite number of truth-values* along a spectrum between perfect truth and perfect falsity. Perfect truth may be represented by "1", and perfect falsity by "0". Borderline cases are thought of as having a "truth-value" anywhere between 0 and 1 (for example, 0.6). Advocates of the fuzzy logic approach have included K. F. Machina (1976) and Dorothy Edgington (1993).

Supervaluationism

Another theoretical approach is known as "supervaluationism". This approach has been defended by Kit Fine and Rosanna Keefe. Fine argues that borderline applications of vague predicates are neither true nor false, but rather are instances of "truth value gaps". He defends an interesting and sophisticated system of vague semantics, based on the notion that a vague predicate might be "made precise" in many alternative ways. This system has the consequence that borderline cases of vague terms yield statements that are neither true, nor false.

Given a supervaluationist semantics, one can define the predicate 'supertrue' as meaning "true on all precisifications". This predicate will not change the semantics of atomic statements (e.g. 'Frank is bald', where Frank is a borderline case of baldness), but does have consequences for logically complex statements. In particular, the tautologies of sentential logic, such as 'Frank is bald or Frank is not bald', will turn out to be supertrue, since on any precisification of baldness, either 'Frank is bald' or 'Frank is not bald' will be true. Since the presence of borderline cases seems to threaten principles like this one (excluded middle), the fact that supervaluationism can "rescue" them is seen as a virtue.

The Epistemicist View

A third approach, known as the "epistemicist view", has been defended by Timothy Williamson (1994), R. A. Sorensen (1988) and (2001), and Nicholas Rescher (2009). They maintain that vague predicates do, in fact, draw sharp boundaries, but that one cannot know where these boundaries lie. One's confusion about whether some vague word does or does not apply in a borderline case is explained as being due to one's ignorance. For example, on the epistemicist view, there is a fact of the matter, for every person, about whether that person is old or not old. It is just that one may sometimes be ignorant of this fact.

Vagueness as a Property of Objects

One possibility is that one's words and concepts are perfectly precise, but that objects themselves are vague. Consider Peter Unger's example of a cloud (from his famous 1980 paper, "The Problem of the Many") : it's not clear where the boundary of a cloud lies; for any given bit of water vapor, one can ask whether it's part of the cloud or not, and for many such bits, one won't know how to answer. So perhaps one's term 'cloud' denotes a vague object precisely. This strategy has been poorly received, in part due to Gareth Evans's short paper "Can There Be Vague Objects?" (1978). Evans's argument appears to show that there can be no vague identities (e.g. "Princeton = Princeton Borough"), but as Lewis (1988) makes clear, Evans takes for granted that there are in fact vague identities, and that any proof to the contrary cannot be right. Since the proof Evans produces relies on the assumption that terms precisely denote vague objects, the implication is that the assumption is false, and so the vague-objects view is wrong.

Still by, for instance, proposing alternative deduction rules involving Leibniz's law or other rules for validity some philosophers are willing to defend vagueness as some kind of metaphysical phenomenon. One has, for example, Peter van Inwagen (1990), Trenton Merricks and Terence Parsons (2000).

Legal Principle

In the common law system, vagueness is a possible legal defence against by-laws and other regulations. The legal principle is that delegated power cannot be used more broadly than the delegator intended. Therefore, a regulation may not be so vague as to regulate areas beyond what the law allows. Any such regulation would be "void for vagueness" and unenforceable. This principle is sometimes used to strike down municipal by-laws that forbid "explicit" or "objectionable" contents from being sold in a certain city; courts often find such expressions to be too vague, giving municipal inspectors discretion beyond what the law allows. In the US this is known as the vagueness doctrine and in Europe as the principle of legal certainty.

Linear Partial Information

Linear partial information (LPI) is a method of making decisions based on insufficient or fuzzy information. LPI was introduced in 1970 by Polish - Swiss mathematician Edward Kofler (1911–2007) to simplify decision processes. Comparing to other methods the LPI-fuzziness is algorithmically simple and particularly in decision making, more practically oriented. Instead of an indicator function the decision maker linearizes any fuzziness by establishing of linear restrictions for fuzzy probability distributions or normalized weights. In the LPI-procedure the decision maker linearizes any fuzziness instead of applying a membership function. This can be done by establishing stochastic and non-stochastic LPI-relations. A mixed stochastic and non-stochastic fuzzification is often a basis for the LPI-procedure. By using the LPI-methods any fuzziness in any decision situation can be considered on the base of the linear fuzzy logic.

Definition

Any Stochastic Partial Information SPI(p), which can be considered as a solution of a linear inequality system, is called Linear Partial Information LPI(p) about probability p. It can be considered as an LPI-fuzzification of the probability p corresponding to the concepts of linear fuzzy logic.

Applications

The MaxEmin Principle

> To obtain the maximally warranted expected value, the decision maker has to choose the strategy which maximizes the minimal expected value. This procedure leads to the MaxEmin - Principle and is an extension of the Bernoulli's principle.

The MaxWmin Principle

> This principle leads to the maximal guaranteed weight function, regarding the extreme weights.

The Prognostic Decision Principle (PDP)

> This principle is based on the prognosis interpretation of strategies under fuzziness.

Fuzzy Equilibrium and Stability

Despite the fuzziness of information, it is often necessary to choose the optimal, most cautious strategy, for example in economic planning, in conflict situations or in daily decisions. This is impossible without the concept of fuzzy equilibrium. The concept of fuzzy stability is considered as an extension into a time interval, taking into account the corresponding stability area of the decision maker. The more complex is the model, the softer a choice has to be considered. The idea of fuzzy equilibrium is based on the optimization principles. Therefore the MaxEmin-, MaxGmin- and PDP-stability have to be analyzed. The violation of these principles leads often to wrong predictions and decisions.

LPI Equilibrium Point

Considering a given LPI-decision model, as a convolution of the corresponding fuzzy states or a disturbance set, the fuzzy equilibrium strategy remains the most cautious one, despite of the presence of the fuzziness. Any deviation from this strategy can cause a loss for the decision maker.

References

- Michael Hanss, 2005. Applied Fuzzy Arithmetic, An Introduction with Engineering Applications. Springer, ISBN 3-540-24201-5

- Daniel Kreiss, Prototype Politics: Technology-Intensive Campaigning and the Data of Democracy. Oxford University Press, 2016.

- Kate Brannely, "Trump Campaign Pays Millions to Overseas Big Data Firm." NBC News, 4 November 2016.

- Steve Lohr and Natasha Singernov, "How Data Failed Us in Calling an Election." New York Times, 10 November 2016.

- Robert M. Wachter, "How Measurement Fails Doctors and Teachers". New York Times, 16 January 2016.

- Adam Tanner, "Nine Things You Don't Know About The Gathering Of Your Personal Data." Forbes Magazine, 4 November 2014.

- Lotfi Zadeh, "What is fuzzy logic?". IFSA Newsletter (International Fuzzy Systems Association), Vol. 10, No. 1, March 2013.

- Stosberg, Mark (16 December 1996). "The Role of Fuzziness in Artifical [sic] Intelligence". Minds and Machines. Retrieved 19 April 2013.

3

Theories Related to Fuzzy Logic

The likelihood of an event occurring is known as possibility. Possibility is anywhere between 0 and 1, where 1 is certainty and 0 is impossibility. A situation where indeterminate information is involved is known as uncertainty. Uncertainty theory is a branch of mathematics that was founded very recently. The section will not only provide an overview, it will also delve deep into the topics related to it. The chapter on possibility offers an insightful focus, keeping in mind the complex subject matter.

Fuzzy Measure Theory

In mathematics, fuzzy measure theory considers generalized measures in which the additive property is replaced by the weaker property of monotonicity. The central concept of fuzzy measure theory is the fuzzy measure which was introduced by Choquet in 1953 and independently defined by Sugeno in 1974 in the context of fuzzy integrals. There exists a number of different classes of fuzzy measures including plausibility/belief measures; possibility/necessity measures; and probability measures which are a subset of classical measures.

Definitions

Let $\mathbf{X}$ be a universe of discourse, $\mathcal{C}$ be a class of subsets of $\mathbf{X}$, and $E, F \in \mathcal{C}$. A function $g : \mathcal{C} \to \mathbb{R}$ where

1. $\varnothing \in \mathcal{C} \Rightarrow g(\varnothing) = 0$
2. $E \subseteq F \Rightarrow g(E) \leq g(F)$

is called a *fuzzy measure*. A fuzzy measure is called *normalized* or *regular* if $g(\mathbf{X}) = 1$.

Properties of Fuzzy Measures

For any $E, F \in \mathcal{C}$, a fuzzy measure is:

- additive if $g(E \cup F) = g(E) + g(F)$. for all $E \cap F = \varnothing$;

- supermodular if $g(E \cup F) + g(E \cap F) \geq g(E) + g(F)$;

- submodular if $g(E \cup F) + g(E \cap F) \leq g(E) + g(F)$;

- superadditive if $g(E \cup F) \geq g(E) + g(F)$ for all $E \cap F = \varnothing$;
- subadditive if $g(E \cup F) \leq g(E) + g(F)$ for all $E \cap F = \varnothing$;;
- symmetric if $|E| = |F|$ implies $g(E) = g(F)$;
- Boolean if $g(E) = 0$ or $g(E) = 1$.

Understanding the properties of fuzzy measures is useful in application. When a fuzzy measure is used to define a function such as the Sugeno integral or Choquet integral, these properties will be crucial in understanding the function's behavior. For instance, the Choquet integral with respect to an additive fuzzy measure reduces to the Lebesgue integral. In discrete cases, a symmetric fuzzy measure will result in the ordered weighted averaging (OWA) operator. Submodular fuzzy measures result in convex functions, while supermodular fuzzy measures result in concave functions when used to define a Choquet integral.

Möbius Representation

Let g be a fuzzy measure, the Möbius representation of g is given by the set function M, where for every $E, F \subseteq X,,$

$$M(E) = \sum_{F \subseteq E} (-1)^{|E \setminus F|} g(F).$$

The equivalent axioms in Möbius representation are:

1. $M(\varnothing) = 0$.

2. $\sum_{F \subseteq E | i \in F} M(F) \geq 0$, for all $E \subseteq X$ and all $i \in E$

A fuzzy measure in Möbius representation M is called *normalized* if $\sum_{\subseteq} M(E) = 1$.

Möbius representation can be used to give an indication of which subsets of X interact with one another. For instance, an additive fuzzy measure has Möbius values all equal to zero except for singletons. The fuzzy measure g in standard representation can be recovered from the Möbius form using the Zeta transform:

$$g(E) = \sum_{F \subseteq E} M(F), \forall E \subseteq X.$$

Simplification Assumptions for Fuzzy Measures

Fuzzy measures are defined on a semiring of sets or monotone class which may be as granular as the power set of X, and even in discrete cases the number of variables can as large as $2^{|X|}$. For this reason, in the context of multi-criteria decision analysis and other disciplines, simplification assumptions on the fuzzy measure have been introduced so that it is less computationally expensive to determine and use. For instance, when it is assumed the fuzzy measure is *additive*, it will hold that $g(E) = \sum_{i \in E} g(\{i\})$

and the values of the fuzzy measure can be evaluated from the values on **X**. Similarly, a *symmetric* fuzzy measure is defined uniquely by |**X**| values. Two important fuzzy measures that can be used are the Sugeno- or $\lambda -$ fuzzy measure and k-additive measures, introduced by Sugeno and Grabisch respectively.

Sugeno λ-measure

The Sugeno λ -measure is a special case of fuzzy measures defined iteratively. It has the following definition:

Definition

Let $\mathbf{X} = \{x_1, \ldots, x_n\}$ be a finite set and let $\in (-1, +\infty)$. A Sugeno -measure is a function $g : 2^X \to [0,1]$ such that

1. $g(X) = 1$ ·
2. if $A, B \subseteq \mathbf{X}$ (alternatively $A, B \in 2^{\mathbf{X}}$) with $A \cap B = \varnothing$ then
   $g(A \cup B) = g(A) + g(B) + \lambda g(A)g(B)$.

As a convention, the value of g at a singleton set $\{x_i\}$ is called a density and is denoted by $g_i = g(\{x_i\})$. In addition, we have that λ satisfies the property

$$\lambda + 1 = \prod_{i=1}^{n} (1 + \lambda g_i). .$$

Tahani and Keller as well as Wang and Klir have showed that once the densities are known, it is possible to use the previous polynomial to obtain the values of λ uniquely.

k-additive Fuzzy Measure

The k-additive fuzzy measure limits the interaction between the subsets $E \subseteq X$ to size $|E| = k$. This drastically reduces the number of variables needed to define the fuzzy measure, and as k can be anything from 1 (in which case the fuzzy measure is additive) to **X**, it allows for a compromise between modelling ability and simplicity.

Definition

A discrete fuzzy measure g on a set **X** is called *k-additive* ($1 \le k \le |\mathbf{X}|$) if its Möbius representation verifies $M(E) = 0$, whenever $|E| > k$ for any $E \subseteq \mathbf{X}$, and there exists a subset F with k elements such that $M(F) \ne 0$.

Shapley and Interaction Indices

In game theory, the Shapley value or Shapley index is used to indicate the weight of a game. Shapley values can be calculated for fuzzy measures in order to give some indication of the importance of each singleton. In the case of additive fuzzy measures, the Shapley value will be the same as each singleton.

For a given fuzzy measure g, and $|\mathbf{X}| = n$, the Shapley index for every $i, \ldots, n \in X$ is:

$$\phi(i) = \sum_{E \subseteq \mathbf{X} \setminus \{i\}} \frac{(n - |E| - 1)! \, |E|!}{n!} [g(E \cup \{i\}) - g(E)].$$

The Shapley value is the vector $\phi(g) = (\psi(1), \ldots, \psi(n))$.

Uncertainty Theory

Uncertainty theory is a branch of mathematics based on normality, monotonicity, self-duality, countable subadditivity, and product measure axioms. It was founded by Baoding Liu in 2007 and refined in 2009.

Mathematical measures of the likelihood of an event being true include probability theory, capacity, fuzzy logic, possibility, and credibility, as well as uncertainty.

Five Axioms

Axiom 1. (Normality Axiom) $\mathcal{M}\{\Gamma\} = 1$ for the universal set Γ.

Axiom 2. (Monotonicity Axiom) $\mathcal{M}\{\Lambda_1\} \leq \mathcal{M}\{\Lambda_2\}$ whenever $\Lambda_1 \subset \Lambda_2$.

Axiom 3. (Self-Duality Axiom) $\mathcal{M}\{\Lambda\} + \mathcal{M}\{\Lambda^c\} = 1$ for any event Λ..

Axiom 4. (Countable Subadditivity Axiom) For every countable sequence of events Λ_1, Λ_2, ..., we have

$$\mathcal{M}\left\{ \bigcup_{i=1}^{\infty} \Lambda_i \right\} \leq \sum_{i=1}^{\infty} \mathcal{M}\{\Lambda_i\}.$$

Axiom 5. (Product Measure Axiom) Let $(\Gamma_k, \mathcal{L}_k, \mathcal{M}_k)$ be uncertainty spaces for $k = 1, 2, \cdots, n.$. Then the product uncertain measure $\mathcal{M}$ is an uncertain measure on the product σ-algebra satisfying

$$\mathcal{M}\left\{ \prod_{i=1}^{n} \Lambda_i \right\} = \min_{1 \leq i \leq n} \mathcal{M}_i\{\Lambda_i\}.$$

Principle. (Maximum Uncertainty Principle) For any event, if there are multiple reasonable values that an uncertain measure may take, then the value as close to 0.5 as possible is assigned to the event.

Uncertain Variables

An uncertain variable is a measurable function ξ from an uncertainty space (Γ, L, M) to the set of real numbers, i.e., for any Borel set $\mathbf{B}$ of real numbers, the set $\{\xi \in B\} = \{\gamma \in \Gamma \mid \xi(\gamma) \in B\}$ is an event.

Uncertainty Distribution

Uncertainty distribution is inducted to describe uncertain variables.

Definition:The uncertainty distribution $\Phi(x): R \to [0,1]$ of an uncertain variable ξ is defined by $\Phi(x) = M\{\xi \leq x\}$.

Theorem(Peng and Iwamura, Sufficient and Necessary Condition for Uncertainty Distribution) A function $\Phi(x): R \to [0,1]$ is an uncertain distribution if and only if it is an increasing function except $\Phi(x) \equiv 0$ and $\Phi(x) \equiv 1$.

Independence

Definition: The uncertain variables $\xi_1, \xi_2, \ldots, \xi_m$ are said to be independent if

$$M\{\cap_{i=1}^{m}(\xi \in B_i)\} = \min_{1 \leq i \leq m} M\{\xi_i \in B_i\}$$

for any Borel sets $B_1, B_2, \ldots, B_m$ of real numbers.

Theorem 1: The uncertain variables $\xi_1, \xi_2, \ldots, \xi_m$ are independent if

$$M\{\cup_{i=1}^{m}(\xi \in B_i)\} = \max_{1 \leq i \leq m} M\{\xi_i \in B_i\}$$

for any Borel sets $B_1, B_2, \ldots, B_m$ of real numbers.

Theorem 2: Let $\xi_1, \xi_2, \ldots, \xi_m$ be independent uncertain variables, and $f_1, f_2, \ldots, f_m$ measurable functions. Then $f_1(\xi_1), f_2(\xi_2), \ldots, f_m(\xi_m)$ are independent uncertain variables.

Theorem 3: Let Φ_i be uncertainty distributions of independent uncertain variables ξ_i, $i = 1, 2, \ldots, m$ respectively, and Φ the joint uncertainty distribution of uncertain vector $(\xi_1, \xi_2, \ldots, \xi_m)$. If $\xi_1, \xi_2, \ldots, \xi_m$ are independent, then we have

$$\Phi(x_1, x_2, \ldots, x_m) = \min_{1 \leq i \leq m} \Phi_i(x_i)$$

for any real numbers $x_1, x_2, \ldots, x_m$.

Operational Law

Theorem: Let $\xi_1, \xi_2, \ldots, \xi_m$ be independent uncertain variables, and $f: R^n \to R$ a measurable function. Then $\xi = f(\xi_1, \xi_2, \ldots, \xi_m)$ is an uncertain variable such that

$$M\{\xi \in B\} = \begin{cases} \sup_{f(B_1, B_2, \cdots, B_n) \subset B} \min_{1 \leq k \leq n} \mathcal{M}_k\{\xi_k \in B_k\}, & \text{if } \sup_{f(B_1, B_2, \cdots, B_n) \subset B} \min_{1 \leq k \leq n} \mathcal{M}_k\{\xi_k \in B_k\} > 0.5 \\ 1 - \sup_{f(B_1, B_2, \cdots, B_n) \subset B^c} \min_{1 \leq k \leq n} \mathcal{M}_k\{\xi_k \in B_k\}, & \text{if } \sup_{f(B_1, B_2, \cdots, B_n) \subset B^c} \min_{1 \leq k \leq n} \mathcal{M}_k\{\xi_k \in B_k\} > 0.5 \\ 0.5, & \text{otherwise} \end{cases}$$

where $B, B_1, B_2, \ldots, B_m$ are Borel sets, and $f(B_1, B_2, \ldots, B_m) \subset B$ means $f(x_1, x_2, \ldots, x_m) \in B$ for any $x_1 \in B_1, x_2 \in B_2, \ldots, x_m \in B_m$..

Expected Value

Definition: Let ξ be an uncertain variable. Then the expected value of ξ is defined by

$$E[\xi] = \int_0^{+\infty} M\{\xi \geq r\} dr - \int_{-\infty}^{0} M\{\xi \leq r\} dr$$

provided that at least one of the two integrals is finite.

Theorem 1: Let ξ be an uncertain variable with uncertainty distribution Φ. If the expected value exists, then

$$E[\xi] = \int_0^{+\infty} (1 - \Phi(x)) dx - \int_{-\infty}^{0} \Phi(x) dx \ .$$

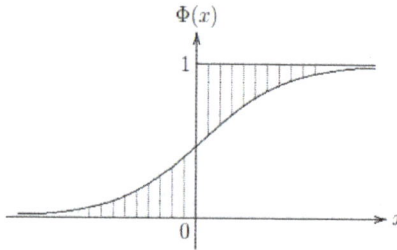

Expected Value $E[\xi] = \int_0^{+\infty} (1 - \Phi(x)) \mathrm{d}x - \int_{-\infty}^{0} \Phi(x) \mathrm{d}x$

Theorem 2: Let ξ be an uncertain variable with regular uncertainty distribution Φ. If the expected value exists, then

$$E[\xi] = \int_0^1 \Phi^{-1}(\alpha) d\alpha..$$

Theorem 3: Let ξ and η be independent uncertain variables with finite expected values. Then for any real numbers a and b, we have

$$E[a\xi + b\eta] = aE[\xi] + b[\eta]..$$

Variance

Definition: Let be an uncertain variable with finite expected value . Then the variance of is defined by

$$V[\xi] = E[(\xi - e)^2]..$$

Theorem: If be an uncertain variable with finite expected value, and are real numbers, then

$$V[a\xi + b] = a^2 V[\xi].$$

Critical Value

Definition: Let ξ be an uncertain variable, and $\alpha \in (0,1]$. Then

$$\xi_{sup}(\alpha) = \sup\{r \mid M\{\xi \geq r\} \geq \alpha\}$$

is called the α-optimistic value to ξ, and

$$\xi_{inf}(\alpha) = \inf\{r \mid M\{\xi \leq r\} \geq \alpha\}$$

is called the α-pessimistic value to ξ.

Theorem 1: Let be an uncertain variable with regular uncertainty distribution . Then its α-optimistic value and α-pessimistic value are

$$\xi_{sup}(\alpha) = \Phi^{-1}(1-\alpha),,$$

$$\xi_{inf}(\alpha) = \Phi^{-1}(\alpha).$$

Theorem 2: Let be an uncertain variable, and $\alpha \in (0,1]$. Then we have

- if $\alpha > 0.5$,, then $\xi_{inf}(\alpha) \geq \xi_{sup}(\alpha)$;

- if $\alpha \leq 0.5$, then $\xi_{inf}(\alpha) \leq \xi_{sup}(\alpha)$.

Theorem 3: Suppose that ξ and η are independent uncertain variables, and $\alpha \in (0,1]$. Then we have

$$(\xi+\eta)_{sup}(\alpha) = \xi_{sup}(\alpha) + \eta_{sup}\alpha,$$

$$(\xi+\eta)_{inf}(\alpha) = \xi_{inf}(\alpha) + \eta_{inf}\alpha,$$

$$(\xi\vee\eta)_{sup}(\alpha) = \xi_{sup}(\alpha) \vee \eta_{sup}\alpha,$$

$$(\xi\vee\eta)_{inf}(\alpha) = \xi_{inf}(\alpha) \vee \eta_{inf}\alpha,$$

$$(\xi\wedge\eta)_{sup}(\alpha) = \xi_{sup}(\alpha) \wedge \eta_{sup}\alpha,$$

$$(\xi\wedge\eta)_{inf}(\alpha) = \xi_{inf}(\alpha) \wedge \eta_{inf}\alpha.$$

Entropy

Definition: Let ξ be an uncertain variable with uncertainty distribution Φ. Then its entropy is defined by

$$H[\xi] = \int_{-\infty}^{+\infty} S(\Phi(x))dx$$

where $S(x) = -t\ln(t) - (1-t)\ln(1-t)$.

Theorem 1(Dai an*d Chen*): Let ξ be an uncertain variable with regular uncertainty distribution Φ. Then

$$H[\xi] = \int_0^1 \Phi^{-1}(\alpha)\ln\frac{\alpha}{1-\alpha}d\alpha..$$

Theorem 2: Let and be independent uncertain variables. Then for any real numbers and , we have

$$H[a\xi + b\eta] = |a| E[\xi] + |b| E[\eta]..$$

Theorem 3: Let be an uncertain variable whose uncertainty distribution is arbitrary but the expected value and variance . Then

$$H[\xi] \le \frac{\pi\sigma}{\sqrt{3}}..$$

Inequalities

Theorem 1(Liu, Markov Inequality): Let be an uncertain variable. Then for any given numbers and , we have

$$M\{|\xi| \ge t\} \le \frac{E[|\xi|^p]}{t^p}..$$

Theorem 2 (Liu, Chebyshev Inequality) Let ξ be an uncertain variable whose variance $V[\xi]$ exists. Then for any given number $t > 0$, we have

$$M\{|\xi - E[\xi]| \ge t\} \le \frac{V[\xi]}{t^2}..$$

Theorem 3 (Liu, Holder's Inequality) Let p and q be positive numbers with , and let and be independent uncertain variables with $E[|\xi|^p] < \infty$ and $E[|\eta|^q] < \infty$. Then we have

$$E[|\xi\eta|] \le \sqrt[p]{E[|\xi|^p]}\sqrt[p]{E[\eta|^p]}.$$

Theorem 4:(Liu , Minkowski Inequality) Let p be a real number with $p \le 1$, and let ξ and η be independent uncertain variables with and . Then we have

$$\sqrt[p]{E[|\xi+\eta|^p]} \le \sqrt[p]{E[|\xi|^p]} + \sqrt[p]{E[\eta|^p]}..$$

Convergence Concept

Definition 1: Suppose that are uncertain variables defined on the uncertainty space . The sequence is said to be convergent a.s. to if there exists an event with such that

for every . In that case we write ,a.s. $\lim_{i\to\infty} |\xi_i(\gamma) - \xi(\gamma)| = 0$

Definition 2: Suppose that $\xi, \xi_1, \xi_2, \ldots$ are uncertain variables. We say that the sequence $\{\xi_i\}$ converges in measure to ξ if

$$\lim_{i\to\infty} M\{|\xi_i - \xi| \le \varepsilon\} = 0$$

for every $\varepsilon > 0$.

Definition 3: Suppose that $\xi, \xi_1, \xi_2, \ldots$ are uncertain variables with finite expected values. We say that the sequence $\{\xi_i\}$ converges in mean to ξ if

$$\lim_{i \to \infty} E[|\xi_i - \xi|] = 0..$$

Definition 4: Suppose that $\Phi, \phi_1, \Phi_2, \ldots$ are uncertainty distributions of uncertain variables $\xi, \xi_1, \xi_2, \ldots,$, respectively. We say that the sequence $\{\xi_i\}$ converges in distribution to ξ if $\Phi_i \to \Phi$ at any continuity point of $\ddot{O}$.

Theorem 1: Convergence in Mean $\Rightarrow$ Convergence in Measure $\Rightarrow$ Convergence in Distribution. However, Convergence in Mean $\Rightarrow$ Convergence Almost Surely $\Rightarrow$ Convergence in Distribution.

Conditional Uncertainty

Definition 1: Let (Γ, L, M) be an uncertainty space, and $A, B \in L.$. Then the conditional uncertain measure of A given B is defined by

$$M\{A \mid B\} = \begin{cases} \dfrac{M\{A \cap B\}}{M\{B\}}, & \text{if } \dfrac{M\{A \cap B\}}{M\{B\}} < 0.5 \\[3ex] 1 - \dfrac{M\{A^c \cap B\}}{M\{B\}}, & \text{if } \dfrac{M\{A^c \cap B\}}{M\{B\}} < 0.5 \\[3ex] 0.5, & \text{otherwise} \end{cases}$$

provided that $M\{B\} > 0$

Theorem 1: Let (Γ, L, M) be an uncertainty space, and B an event with $M\{B\} > 0$. Then $M\{\cdot|B\}$ defined by Definition 1 is an uncertain measure, and $(\Gamma, L, M\{\cdot \mid B\})$ is an uncertainty space.

Definition 2: Let ξ be an uncertain variable on (Γ, L, M). A conditional uncertain variable of ξ given B is a measurable function $\xi|_B$ from the conditional uncertainty space $(\Gamma, L, M\{\cdot|_B\})$ to the set of real numbers such that

$$\xi|_B(\gamma) = \xi(\gamma), \forall \gamma \in \Gamma..$$

Definition 3: The conditional uncertainty distribution $\Phi \to [0,1]$ of an uncertain variable given B is defined by

$$\Phi(x \mid B) = M\{\xi \leq x \mid B\}$$

provided that $M\{B\} > 0$.

Theorem 2: Let ξ be an uncertain variable with regular uncertainty distribution $\Phi(x)$, and t a real number with $\Phi(t) < 1$. Then the conditional uncertainty distribution of ξ given $\xi > t$ is

$$\Phi(x \mid (t, +\infty)) = \begin{cases} 0, & \text{if } \Phi(x) \leq \Phi(t) \\ \dfrac{\Phi(x)}{1 - \Phi(t)} \wedge 0.5, & \text{if } \Phi(t) < \Phi(x) \leq (1 + \Phi(t))/2 \\ \dfrac{\Phi(x) - \Phi(t)}{1 - \Phi(t)}, & \text{if } (1 + \Phi(t))/2 \leq \Phi(x) \end{cases}$$

Theorem 3: Let ξ be an uncertain variable with regular uncertainty distribution $\Phi(x)$, and t a real number with $\Phi(t) > 0$. Then the conditional uncertainty distribution of ξ given $\xi \leq t$ is

$$\Phi(x \mid (-\infty, t]) = \begin{cases} \dfrac{\Phi(x)}{\Phi(t)}, & \text{if } \Phi(x) \leq \Phi(t)/2 \\ \dfrac{\Phi(x) + \Phi(t) - 1}{\Phi(t)} \vee 0.5, & \text{if } \Phi(t)/2 \leq \Phi(x) < \Phi(t) \\ 1, & \text{if } \Phi(t) \leq \Phi(x) \end{cases}$$

Definition 4: Let ξ be an uncertain variable. Then the conditional expected value of ξ given B is defined by

$$E[\xi \mid B] = \int_0^{+\infty} M\{\xi \geq r \mid B\}dr - \int_{-\infty}^0 M\{\xi \leq r \mid B\}dr$$

provided that at least one of the two integrals is finite.

Possibility Theory

Possibility theory is a mathematical theory for dealing with certain types of uncertainty and is an alternative to probability theory. Professor Lotfi Zadeh first introduced possibility theory in 1978 as an extension of his theory of fuzzy sets and fuzzy logic. Didier Dubois and Henri Prade further contributed to its development. Earlier in the 1950s, economist G. L. S. Shackle proposed the min/max algebra to describe degrees of potential surprise.

Formalization of Possibility

For simplicity, assume that the universe of discourse Ω is a finite set, and assume that all subsets are measurable. A distribution of possibility is a function pos from 2^Ω to [0, 1] such that:

Axiom 1: $pos(\varnothing) = 0$

Axiom 2: $\mathrm{pos}(\Omega) = 1$

Axiom 3: $pos(U \cup V) = \max\big(pos(U), pos(V)\big)$ for any disjoint subsets U and .

It follows that, like probability, the possibility measure is determined by its behavior on singletons:

$$\mathrm{pos}(U) = \max_{\omega \in U} \mathrm{pos}(\{\omega\})$$

provided that U is finite or countably infinite.

Axiom 1 can be interpreted as the assumption that Ω is an exhaustive description of future states of the world, because it means that no belief weight is given to elements outside Ω.

Axiom 2 could be interpreted as the assumption that the evidence from which pos was constructed is free of any contradiction. Technically, it implies that there is at least one element in Ω with possibility 1.

Axiom 3 corresponds to the additivity axiom in probabilities. However there is an important practical difference. Possibility theory is computationally more convenient because Axioms 1–3 imply that:

$$\mathrm{pos}(U \cup V) = \max\big(\mathrm{pos}(U), \mathrm{pos}(V)\big) \text{ for } any \text{ subsets } U \text{ and } V.$$

Because one can know the possibility of the union from the possibility of each component, it can be said that possibility is *compositional* with respect to the union operator. Note however that it is not compositional with respect to the intersection operator. Generally:

$$\mathrm{pos}(U \cap V) \le \min\big(\mathrm{pos}(U), \mathrm{pos}(V)\big)$$

When Ω is not finite, Axiom 3 can be replaced by:

For all index sets I, if the subsets $U_{i, i \in I}$ are pairwise disjoint, $\mathrm{pos}\big(\cup_{i \in I} U_i\big) = \sup_{i \in I} \mathrm{pos}(U_i).$

Necessity

Whereas probability theory uses a single number, the probability, to describe how likely an event is to occur, possibility theory uses two concepts, the *possibility* and the *necessity* of the event. For any set U, the necessity measure is defined by

$$\mathrm{nec}(U) = 1 - \mathrm{pos}(\overline{U})$$

In the above formula, $\overline{U}$ denotes the complement of U, that is the elements of Ω that do not belong to U. It is straightforward to show that:

$$nec(U) \leq pos(U) \text{ for any } U$$

and that:

$$nec(U \cap V) = \min(nec(U), nec(V))$$

Note that contrary to probability theory, possibility is not self-dual. That is, for any event U, we only have the inequality:

$$pos(U) + pos(\bar{U}) \geq 1$$

However, the following duality rule holds:

For any event U,, either $pos(U) = 1$, or $nec(U) = 0$

Accordingly, beliefs about an event can be represented by a number and a bit.

Interpretation

There are four cases that can be interpreted as follows:

The intersection of the last two cases is $nec(U) = 0$ and $pos(U) = 1$ meaning that I believe nothing at all about U.. Because it allows for indeterminacy like this, possibility theory relates to the graduation of a many-valued logic, such as intuitionistic logic, rather than the classical two-valued logic.

Note that unlike possibility, fuzzy logic is compositional with respect to both the union and the intersection operator. The relationship with fuzzy theory can be explained with the following classical example.

- Fuzzy logic: When a bottle is half full, it can be said that the level of truth of the proposition "The bottle is full" is 0.5. The word "full" is seen as a fuzzy predicate describing the amount of liquid in the bottle.

- Possibility theory: There is one bottle, either completely full or totally empty. The proposition "the possibility level that the bottle is full is 0.5" describes a degree of belief. One way to interpret 0.5 in that proposition is to define its meaning as: I am ready to bet that it's empty as long as the odds are even (1:1) or better, and I would not bet at any rate that it's full.

Possibility Theory as an Imprecise Probability Theory

There is an extensive formal correspondence between probability and possibility theories, where the addition operator corresponds to the maximum operator.

A possibility measure can be seen as a consonant plausibility measure in Dempster–

Shafer theory of evidence. The operators of possibility theory can be seen as a hyper-cautious version of the operators of the transferable belief model, a modern development of the theory of evidence.

Possibility can be seen as an upper probability: any possibility distribution defines a unique set of admissible probability distributions by

$$\{p : \forall S \ p(S) \le \text{pos}(S)\}.$$

This allows one to study possibility theory using the tools of imprecise probabilities.

Necessity Logic

We call *generalized possibility* every function satisfying Axiom 1 and Axiom 3. We call *generalized necessity* the dual of a generalized possibility. The generalized necessities are related with a very simple and interesting fuzzy logic we call *necessity logic*. In the deduction apparatus of necessity logic the logical axioms are the usual classical tautologies. Also, there is only a fuzzy inference rule extending the usual Modus Ponens. Such a rule says that if α and $\alpha \to \beta$ are proved at degree λ and μ, respectively, then we can assert β at degree $\min\{\lambda,\mu\}$. It is easy to see that the theories of such a logic are the generalized necessities and that the completely consistent theories coincide with the necessities.

4

Fuzzy Sets and Algebra

In mathematics, Boolean algebra is a theory where truth-values are either true or false. 1 and 0 denotes truth and falsehood. Systems that are used in mathematical concepts are known as mathematical models. Fuzzy control systems, Fril, De Morgan algebra and fuzzy subalgebra are significant topics related to fuzzy logic. The following section unfolds its crucial aspects in a critical yet systematic manner.

Fuzzy Set

In mathematics, fuzzy sets are sets whose elements have degrees of membership. Fuzzy sets were introduced by Lotfi A. Zadeh and Dieter Klaua in 1965 as an extension of the classical notion of set. At the same time, Salii (1965) defined a more general kind of structure called an L-relation, which he studied in an abstract algebraic context. Fuzzy relations, which are used now in different areas, such as linguistics (De Cock, Bodenhofer & Kerre 2000) decision-making (Kuzmin 1982) and clustering (Bezdek 1978), are special cases of L-relations when L is the unit interval [0, 1].

In classical set theory, the membership of elements in a set is assessed in binary terms according to a bivalent condition — an element either belongs or does not belong to the set. By contrast, fuzzy set theory permits the gradual assessment of the membership of elements in a set; this is described with the aid of a membership function valued in the real unit interval [0, 1]. Fuzzy sets generalize classical sets, since the indicator functions of classical sets are special cases of the membership functions of fuzzy sets, if the latter only take values 0 or 1. In fuzzy set theory, classical bivalent sets are usually called *crisp* sets. The fuzzy set theory can be used in a wide range of domains in which information is incomplete or imprecise, such as bioinformatics.

Definition

Sometimes, more general variants of the notion of fuzzy set are used, with membership functions taking values in a (fixed or variable) algebra or structure L of a given kind; usually it is required that L be at least a poset or lattice. These are usually called L-fuzzy sets, to distinguish them from those valued over the unit interval. The usual membership functions with values in [0, 1] are then called [0, 1]-valued membership functions. These kinds of generalizations were first considered in 1967 by Joseph Goguen, who was a student of Zadeh.

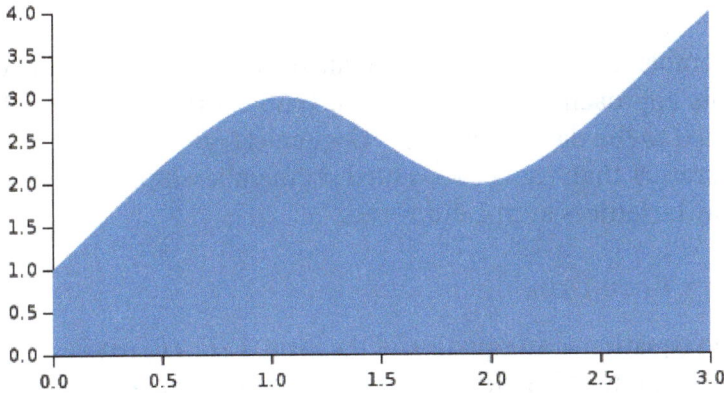

Fuzzy Logic

As an extension of the case of multi-valued logic, valuations ($\mu : V_o \rightarrow W$) of propositional variables (V_o) into a set of membership degrees (W) can be thought of as membership functions mapping predicates into fuzzy sets (or more formally, into an ordered set of fuzzy pairs, called a fuzzy relation). With these valuations, many-valued logic can be extended to allow for fuzzy premises from which graded conclusions may be drawn.

This extension is sometimes called "fuzzy logic in the narrow sense" as opposed to "fuzzy logic in the wider sense," which originated in the engineering fields of automated control and knowledge engineering, and which encompasses many topics involving fuzzy sets and "approximated reasoning."

Industrial applications of fuzzy sets in the context of "fuzzy logic in the wider sense" can be found at fuzzy logic.

Fuzzy Number

A fuzzy number is a convex, normalized fuzzy set $\tilde{A} \subseteq \mathbb{R}$ whose membership function is at least segmentally continuous and has the functional value $\mu_A(x) = 1$ at at least one element.

This can be likened to the funfair game "guess your weight," where someone guesses the contestant's weight, with closer guesses being more correct, and where the guesser "wins" if he or she guesses near enough to the contestant's weight, with the actual weight being completely correct (mapping to 1 by the membership function).

Fuzzy Interval

A fuzzy interval is an uncertain set with a mean interval whose elements possess the $\tilde{A} \subseteq \mathbb{R}$ membership. As in fuzzy numbers, the membership function must be convex, normalized, at least segmentally continuous.

Fuzzy Categories

The use of set membership as a key components of category theory can be generalized to fuzzy sets. This approach which initiated in 1968 shortly after the introduction of fuzzy set theory led to the development of "Goguen categories" in the 21st century. In these categories, rather than using two valued set membership, more general intervals are used, and may be lattices as in L-fuzzy sets.

Fuzzy Relation Equation

The fuzzy relation equation is an equation of the form $A \cdot R = B$, where A and B are fuzzy sets, R is a fuzzy relation, and $A \cdot R$ stands for the composition of A with R.

Entropy

Let A be a fuzzy variable with a continuous membership function. Then its entropy is

$$H[A] = \int_{-\infty}^{\infty} S(\mathrm{Cr}\{A \geq t\})\,dt$$

where

$$S(y) = -y \ln y - (1-y)\ln(1-y).$$

Extensions

There are many mathematical constructions similar to or more general than fuzzy sets. Since fuzzy sets were introduced in 1965, a lot of new mathematical constructions and theories treating imprecision, inexactness, ambiguity, and uncertainty have been developed. Some of these constructions and theories are extensions of fuzzy set theory, while others try to mathematically model imprecision and uncertainty in a different way (Burgin & Chunihin 1997; Kerre 2001; Deschrijver and Kerre, 2003).

The diversity of such constructions and corresponding theories includes:

- interval sets (Moore, 1966),
- L-fuzzy sets (Goguen, 1967),
- flou sets (Gentilhomme, 1968),
- Boolean-valued fuzzy sets (Brown, 1971),
- type-2 fuzzy sets and type-n fuzzy sets (Zadeh, 1975),
- set-valued sets (Chapin, 1974; 1975),
- interval-valued fuzzy sets (Grattan-Guinness, 1975; Jahn, 1975; Sambuc, 1975; Zadeh, 1975),

- functions as generalizations of fuzzy sets and multisets (Lake, 1976),
- level fuzzy sets (Radecki, 1977)
- underdetermined sets (Narinyani, 1980),
- rough sets (Pawlak, 1982),
- intuitionistic fuzzy sets (Atanassov, 1983),
- fuzzy multisets (Yager, 1986),
- intuitionistic L-fuzzy sets (Atanassov, 1986),
- rough multisets (Grzymala-Busse, 1987),
- fuzzy rough sets (Nakamura, 1988),
- real-valued fuzzy sets (Blizard, 1989),
- named sets (Burgin, 1990),
- vague sets (Wen-Lung Gau and Buehrer, 1993),
- Q-sets (Gylys, 1994)
- shadowed sets (Pedrycz, 1998),
- α-level sets (Yao, 1997),
- genuine sets (Demirci, 1999),
- soft sets (Molodtsov, 1999),
- intuitionistic fuzzy rough sets (Cornelis, De Cock and Kerre, 2003)
- blurry sets (Smith, 2004)
- L-fuzzy rough sets (Radzikowska and Kerre, 2004),
- generalized rough fuzzy sets (Feng, 2010)
- rough intuitionistic fuzzy sets (Thomas and Nair, 2011),
- soft rough fuzzy sets (Meng, Zhang and Qin, 2011)
- soft fuzzy rough sets (Meng, Zhang and Qin, 2011)
- soft multisets (Alkhazaleh, Salleh and Hassan, 2011)
- fuzzy soft multisets (Alkhazaleh and Salleh, 2012)
- bipolar fuzzy sets (Wen-Ran Zhang, 1998)

While most of the above can be generally categorized as truth-based extensions to fuzzy sets, bipolar fuzzy set theory presents a philosophically and logically different, equilibrium-based generalization of fuzzy sets.

Rough Set

In computer science, a rough set, first described by Polish computer scientist Zdzisław I. Pawlak, is a formal approximation of a crisp set (i.e., conventional set) in terms of a pair of sets which give the *lower* and the *upper* approximation of the original set. In the standard version of rough set theory (Pawlak 1991), the lower- and upper-approximation sets are crisp sets, but in other variations, the approximating sets may be fuzzy sets.

Definitions

The following section contains an overview of the basic framework of rough set theory, as originally proposed by Zdzisław I. Pawlak, along with some of the key definitions. More formal properties and boundaries of rough sets can be found in Pawlak (1991) and cited references. The initial and basic theory of rough sets is sometimes referred to as *"Pawlak Rough Sets"* or *"classical rough sets"*, as a means to distinguish from more recent extensions and generalizations.

Information System Framework

Let $I = (\mathbb{U}, \mathbb{A})$ be an information system (attribute-value system), where $\mathbb{U}$ is a non-empty set of finite objects (the universe) and $\mathbb{A}$ is a non-empty, finite set of attributes such that $a : \mathbb{U} \to V_a$ for every $a \in \mathbb{A}$.. V_a is the set of values that attribute $a(x)$ may take. The information table assigns a value from V_a to each attribute a and object x in the universe $\mathbb{U}$.

With any $P \subseteq \mathbb{A}$ there is an associated equivalence relation $\mathrm{IND}(P)$:

$$\mathrm{IND}(P) = \left\{ (x, y) \in \mathbb{U}^2 \mid \forall a \in P, a(x) = a(y) \right\}$$

The relation $\mathrm{IND}(P)$ is called a $P-indiscernibility$ $relation$. The partition of $\mathbb{U}$ is a family of all equivalence classes of $\mathrm{IND}(P)$ and is denoted by $\mathbb{U} / \mathrm{IND}(P)$ (or $\mathbb{U} / P$).

If $(x, y) \in \mathrm{IND}(P)$, then x and y are *indiscernible* (or indistinguishable) by attributes from P.

Example: Equivalence-class Structure

For example, consider the following information table:

When the full set of attributes $P = \{P_1, P_2, P_3, P_4, P_5\}$ is considered, we see that we have the following seven equivalence classes:

$$\begin{cases} \{O_1, O_2\} \\ \{O_3, O_7, O_{10}\} \\ \{O_4\} \\ \{O_5\} \\ \{O_6\} \\ \{O_8\} \\ \{O_9\} \end{cases}$$

Thus, the two objects within the first equivalence class, $\{O_1, O_2\}$, cannot be distinguished from each other based on the available attributes, and the three objects within the second equivalence class, $\{O_3, O_7, O_{10}\}$,, cannot be distinguished from one another based on the available attributes. The remaining five objects are each discernible from all other objects. The equivalence classes of the P-indiscernibility relation are denoted $[x]_P$.

It is apparent that different attribute subset selections will in general lead to different indiscernibility classes. For example, if attribute $P = \{P_1\}$ alone is selected, we obtain the following, much coarser, equivalence-class structure:

$$\begin{cases} \{O_1, O_2\} \\ \{O_3, O_5, O_7, O_9, O_{10}\} \\ \{O_4, O_6, O_8\} \end{cases}$$

Definition of a Rough Set

Let $X \subseteq \mathbb{U}$ be a target set that we wish to represent using attribute subset P; that is, we are told that an arbitrary set of objects X comprises a single class, and we wish to express this class (i.e., this subset) using the equivalence classes induced by attribute subset P. In general, X cannot be expressed exactly, because the set may include and exclude objects which are indistinguishable on the basis of attributes P.

For example, consider the target set $X = \{O_1, O_2, O_3, O_4\}$, and let attribute subset $P = \{P_1, P_2, P_3, P_4, P_5\}$,, the full available set of features. It will be noted that the set X cannot be expressed exactly, because in $[x]_P$, objects $\{O_3, O_7, O_{10}\}$ are indiscernible. Thus, there is no way to represent any set X which *includes* O_3 but *excludes* objects O_7 and O_{10}.

However, the target set X can be *approximated* using only the information contained within P by constructing the P-lower and P-upper approximations of X:

$$\underline{P}X = \{x \,|\, [x]_P \subseteq X\}$$

$$\overline{P}X = \{x \,|\, [x]_P \cap X \neq \varnothing\}$$

Lower Approximation and Positive Region

The P-*lower approximation*, or *positive region*, is the union of all equivalence classes in $[x]_P$ which are contained by (i.e., are subsets of) the target set – in the example, $\underline{P}X = \{O_1, O_2\} \cup \{O_4\}$, the union of the two equivalence classes in $[x]_P$ which are contained in the target set. The lower approximation is the complete set of objects in $\mathbb{U}/P$ that can be *positively* (i.e., unambiguously) classified as belonging to target set X.

Upper Approximation and Negative Region

The P-*upper approximation* is the union of all equivalence classes in $[x]_P$ which have non-empty intersection with the target set – in the example, $\overline{P}X = \{O_1, O_2\} \cup \{O_4\} \cup \{O_3, O_7, O_{10}\}$, the union of the three equivalence classes in $[x]_P$ that have non-empty intersection with the target set. The upper approximation is the complete set of objects that in $\mathbb{U}/P$ that *cannot* be positively (i.e., unambiguously) classified as belonging to the *complement* ($\overline{X}$) of the target set X. In other words, the upper approximation is the complete set of objects that are *possibly* members of the target set X.

The set $\mathbb{U} - \overline{P}X$ therefore represents the *negative region*, containing the set of objects that can be definitely ruled out as members of the target set.

Boundary Region

The *boundary region*, given by set difference $\overline{P}X - \underline{P}X$, consists of those objects that can neither be ruled in nor ruled out as members of the target set X.

In summary, the lower approximation of a target set is a *conservative* approximation consisting of only those objects which can positively be identified as members of the set. (These objects have no indiscernible "clones" which are excluded by the target set.) The upper approximation is a *liberal* approximation which includes all objects that might be members of target set. (Some objects in the upper approximation may not be members of the target set.) From the perspective of $\mathbb{U}/P$, the lower approximation contains objects that are members of the target set with certainty (probability = 1), while the upper approximation contains objects that are members of the target set with non-zero probability (probability > 0).

The Rough Set

The tuple $\langle \underline{P}X, \overline{P}X \rangle$ composed of the lower and upper approximation is called a *rough set*; thus, a rough set is composed of two crisp sets, one representing a *lower boundary* of the target set X, and the other representing an *upper boundary* of the target set X.

The *accuracy* of the rough-set representation of the set X can be given (Pawlak 1991) by the following:

$$\alpha_P(X) = \frac{\left|\underline{P}X\right|}{\left|\overline{P}X\right|}$$

That is, the accuracy of the rough set representation of X, $\alpha_P(X)$, $0 \leq \alpha_P(X) \leq 1$, is the ratio of the number of objects which can *positively* be placed in X to the number of objects that can *possibly* be placed in X – this provides a measure of how closely the rough set is approximating the target set. Clearly, when the upper and lower approximations are equal (i.e., boundary region empty), then $\alpha_P(X) = 1$, and the approximation is perfect; at the other extreme, whenever the lower approximation is empty, the accuracy is zero (regardless of the size of the upper approximation).

Objective Analysis

Rough set theory is one of many methods that can be employed to analyse uncertain (including vague) systems, although less common than more traditional methods of probability, statistics, entropy and Dempster–Shafer theory. However a key difference, and a unique strength, of using classical rough set theory is that it provides an objective form of analysis (Pawlak et al. 1995). Unlike other methods, as those given above, classical rough set analysis requires no additional information, external parameters, models, functions, grades or subjective interpretations to determine set membership – instead it only uses the information presented within the given data (Düntsch and Gediga 1995). More recent adaptations of rough set theory, such as dominance-based, decision-theoretic and fuzzy rough sets, have introduced more subjectivity to the analysis.

Definability

In general, the upper and lower approximations are not equal; in such cases, we say that target set X is *undefinable* or *roughly definable* on attribute set P. When the upper and lower approximations are equal (i.e., the boundary is empty), $\overline{P}X = \underline{P}X$, then the target set X is *definable* on attribute set P. We can distinguish the following special cases of undefinability:

- Set X is *internally undefinable* if $\underline{P}X \neq \varnothing$ and $\overline{P}X = \mathbb{U}$. This means that on attribute set P, there *are* objects which we can be certain belong to target set X, but there are *no* objects which we can definitively exclude from set X.

- Set X is *externally undefinable* if $\underline{P}X = \varnothing$ and $\overline{P}X \neq \mathbb{U}$. This means that on attribute set P, there are *no* objects which we can be certain belong to target set X, but there *are* objects which we can definitively exclude from set X.

- Set X is *totally undefinable* if $\underline{P}X = \varnothing$ and $\overline{P}X \neq \mathbb{U}$.. This means that on attribute set P, there are *no* objects which we can be certain belong to target set X, and there are *no* objects which we can definitively exclude from set X.. Thus, on attribute set X. we cannot decide whether any object is, or is not, a member of X.

Reduct and Core

An interesting question is whether there are attributes in the information system (attribute-value table) which are more important to the knowledge represented in the equivalence class structure than other attributes. Often, we wonder whether there is a subset of attributes which can, by itself, fully characterize the knowledge in the database; such an attribute set is called a *reduct*.

Formally, a reduct is a subset of attributes $\text{RED} \subseteq P$ such that

- $[x]_{\text{RED}} = [x]_P$, that is, the equivalence classes induced by the reduced attribute set RED are the same as the equivalence class structure induced by the full attribute set P.

- the attribute set RED is *minimal*, in the sense that $[x]_{(\text{RED}-\{a\})} \neq [x]_P$ for any attribute $a \in \text{RED}$; in other words, no attribute can be removed from set RED without changing the equivalence classes $[x]_P$.

A reduct can be thought of as a *sufficient* set of features – sufficient, that is, to represent the category structure. In the example table above, attribute set $\{P_3, P_4, P_5\}$ is a reduct – the information system projected on just these attributes possesses the same equivalence class structure as that expressed by the full attribute set:

$$\begin{cases} \{O_1, O_2\} \\ \{O_3, O_7, O_{10}\} \\ \{O_4\} \\ \{O_5\} \\ \{O_6\} \\ \{O_8\} \\ \{O_9\} \end{cases}$$

Attribute set $\{P_3, P_4, P_5\}$ is a legitimate reduct because eliminating any of these attributes causes a collapse of the equivalence-class structure, with the result that $[x]_{\text{RED}} \neq [x]_P$..

The reduct of an information system is *not unique*: there may be many subsets of attributes which preserve the equivalence-class structure (i.e., the knowledge) expressed in the information system. In the example information system above, another reduct is $\{P_1, P_2, P_5\}$,, producing the same equivalence-class structure as $[x]_P$..

The set of attributes which is common to all reducts is called the *core*: the core is the set of attributes which is possessed by *every* legitimate reduct, and therefore consists of attributes which cannot be removed from the information system without causing collapse of the equivalence-class structure. The core may be thought of as the set of *necessary* attributes – necessary, that is, for the category structure to be represented. In the example, the only such attribute is $\{P_5\}$; any one of the other attributes can be removed singly without damaging the equivalence-class structure, and hence these are all *dispensable*. However, removing $\{P_5\}$ by itself *does* change the equivalence-class structure, and thus $\{P_5\}$ is the *indispensable* attribute of this information system, and hence the core.

It is possible for the core to be empty, which means that there is no indispensable attribute: any single attribute in such an information system can be deleted without altering the equivalence-class structure. In such cases, there is no *essential* or necessary attribute which is required for the class structure to be represented.

Attribute Dependency

One of the most important aspects of database analysis or data acquisition is the discovery of attribute dependencies; that is, we wish to discover which variables are strongly related to which other variables. Generally, it is these strong relationships that will warrant further investigation, and that will ultimately be of use in predictive modeling.

In rough set theory, the notion of dependency is defined very simply. Let us take two (disjoint) sets of attributes, set P and set Q, and inquire what degree of dependency obtains between them. Each attribute set induces an (indiscernibility) equivalence class structure, the equivalence classes induced by P given by $[x]_P$, and the equivalence classes induced by Q given by $[x]_Q$.

Let $[x]_Q = \{Q_1, Q_2, Q_3, \ldots, Q_N\}$, where Q_i is a given equivalence class from the equivalence-class structure induced by attribute set Q. Then, the *dependency* of attribute set Q on attribute set Q, P, is given by

$$\gamma_P(Q) = \frac{\sum_{i=1}^{N}|\underline{P}Q_i|}{|\mathbb{U}|} \leq 1$$

That is, for each equivalence class Q_i in $[x]_Q$, we add up the size of its lower approximation by the attributes in P, i.e., $\underline{P}Q_i.$. This approximation (as above, for arbitrary set X) is the number of objects which on attribute set P can be positively identified as belonging to target set Q_i. Added across all equivalence classes in $[x]_Q$, the numerator above represents the total number of objects which – based on attribute set P – can be positively categorized according to the classification induced by attributes Q. The dependency ratio therefore expresses the proportion (within the entire universe) of such classifiable objects. The dependency $\gamma_P(Q)$ "can be interpreted as a proportion of such objects in the information system for which it suffices to know the values of attributes in P to determine the values of attributes in Q".

Another, intuitive, way to consider dependency is to take the partition induced by Q as the target class C, and consider P as the attribute set we wish to use in order to "re-construct" the target class C. If P can completely reconstruct C, then Q depends totally upon P; if P results in a poor and perhaps a random reconstruction of C, then Q does not depend upon P at all.

Thus, this measure of dependency expresses the degree of *functional* (i.e., deterministic) dependency of attribute set Q on attribute set P; it is *not* symmetric. The relationship of this notion of attribute dependency to more traditional information-theoretic (i.e., entropic) notions of attribute dependence has been discussed in a number of sources (e.g., Pawlak, Wong, & Ziarko 1988; Yao & Yao 2002; Wong, Ziarko, & Ye 1986, Quafafou & Boussouf 2000).

Rule Extraction

The category representations discussed above are all *extensional* in nature; that is, a category or complex class is simply the sum of all its members. To represent a category is, then, just to be able to list or identify all the objects belonging to that category. However, extensional category representations have very limited practical use, because they provide no insight for deciding whether novel (never-before-seen) objects are members of the category.

What is generally desired is an *intentional* description of the category, a representation of the category based on a set of *rules* that describe the scope of the category. The choice of such rules is not unique, and therein lies the issue of inductive bias.

There are a few rule-extraction methods. We will start from a rule-extraction procedure based on Ziarko & Shan (1995).

Decision Matrices

Let us say that we wish to find the minimal set of consistent rules (logical implications) that characterize our sample system. For a set of *condition* attributes $\mathcal{P} = \{P_1, P_2, P_3, \ldots, P_n\}$ and a decision attribute $Q, Q \notin \mathcal{P}$, these rules should have the form $P_i^a P_j^b \ldots P_k^c \rightarrow Q^d$, or, spelled out,

$$(P_i = a) \wedge (P_j = b) \wedge \ldots \wedge (P_k = c) \rightarrow (Q = d)$$

where $\{a, b, c, \ldots\}$ are legitimate values from the domains of their respective attributes. This is a form typical of association rules, and the number of items in $\mathbb{U}$ which match the condition/antecedent is called the *support* for the rule. The method for extracting such rules given in Ziarko & Shan (1995) is to form a *decision matrix* corresponding to each individual value d of decision attribute Q. Informally, the decision matrix for value d of decision attribute Q lists all attribute–value pairs that *differ* between objects having $Q = d$ and $Q \neq d$..

This is best explained by example (which also avoids a lot of notation). Consider the table above, and let P_4 be the decision variable (i.e., the variable on the right side of the implications) and let $\{P_1, P_2, P_3\}$ be the condition variables (on the left side of the implication). We note that the decision variable P_4 takes on two different values, namely $\{1, 2\}$. We treat each case separately.

First, we look at the case $P_4 = 1,$, and we divide up $\mathbb{U}$ into objects that have $P_4 = 1$ and those that have $P_4 \neq 1.$. (Note that objects with $P_4 \neq 1$ in this case are simply the objects that have $P_4 = 2$, but in general, $P_4 \neq 1$ would include all objects having any value for P_4 other than $P_4 = 1,$, and there may be several such classes of objects (for example, those having $P_4 = 2, 3, 4, etc.$).) In this case, the objects having $P_4 = 1$ are $\{O_1, O_2, O_3, O_7, O_{10}\}$ while the objects which have $P_4 \neq 1$ are $\{O_4, O_5, O_6, O_8, O_9\}$. The decision matrix for $P_4 = 1$ lists all the differences between the objects having $P_4 = 1$ and those having $P_4 \neq 1$; that is, the decision matrix lists all the differences between $\{O_1, O_2, O_3, O_7, O_{10}\}$ and $\{O_4, O_5, O_6, O_8, O_9\}$.. We put the "positive" objects ($P_4 = 1$) as the rows, and the "negative" objects $P_4 \neq 1$ as the columns.

To read this decision matrix, look, for example, at the intersection of row O_3 and column , showing in the cell. This means that *with regard to* decision value object differs from object on attributes and , and the particular values on these attributes for the positive object are and . Thi tells us that the correct classification of as belonging to decision class rests on attributes and although one or the other might be dispensable, we know that *at least one* of these attributes is *in*dispensable.

Next, from each decision matrix we form a set of Boolean expressions, one expression for each row of the matrix. The items within each cell are aggregated disjunctively, and the individuals cells are then aggregated conjunctively. Thus, for the above table we have the following five Boolean expressions:

$$
\begin{cases}
(P_1^1 \vee P_2^2 \vee P_3^0) \wedge (P_1^1 \vee P_2^2) \wedge (P_1^1 \vee P_2^2 \vee P_3^0) \wedge (P_1^1 \vee P_2^2 \vee P_3^0) \wedge (P_1^1 \vee P_2^2) \\
(P_1^1 \vee P_2^2 \vee P_3^0) \wedge (P_1^1 \vee P_2^2) \wedge (P_1^1 \vee P_2^2 \vee P_3^0) \wedge (P_1^1 \vee P_2^2 \vee P_3^0) \wedge (P_1^1 \vee P_2^2) \\
(P_1^2 \vee P_3^0) \wedge (P_2^0) \wedge (P_1^2 \vee P_3^0) \wedge (P_1^2 \vee P_2^0 \vee P_3^0) \wedge (P_2^0) \\
(P_1^2 \vee P_3^0) \wedge (P_2^0) \wedge (P_1^2 \vee P_3^0) \wedge (P_1^2 \vee P_2^0 \vee P_3^0) \wedge (P_2^0) \\
(P_1^2 \vee P_3^0) \wedge (P_2^0) \wedge (P_1^2 \vee P_3^0) \wedge (P_1^2 \vee P_2^0 \vee P_3^0) \wedge (P_2^0)
\end{cases}
$$

Each statement here is essentially a highly specific (probably *too* specific) rule governing the membership in class of the corresponding object. For example, the last statement, corresponding to object , states that all the following must be satisfied:

1. Either must have value 2, or must have value 0, or both.

2. must have value 0.

3. Either must have value 2, or must have value 0, or both.

4. Either must have value 2, or must have value 0, or must have value 0, or any combination thereof.

5. must have value 0.

It is clear that there is a large amount of redundancy here, and the next step is to simplify using traditional Boolean algebra. The statement $(P_1^1 \vee P_2^2 \vee P_3^0) \wedge (P_1^1 \vee P_2^2) \wedge (P_1^1 \vee P_2^2 \vee P_3^0) \wedge (P_1^1 \vee P_2^2 \vee P_3^0) \wedge (P_1^1 \vee P_2^2)$ corresponding to objects $\{O_1, O_2\}$ simplifies to $P_1^1 \vee P_2^2$, which yields the implication

$$(P_1 = 1) \vee (P_2 = 2) \to (P_4 = 1)$$

Likewise, the statement $(P_1^2 \vee P_3^0) \wedge (P_2^0) \wedge (P_1^2 \vee P_3^0) \wedge (P_1^2 \vee P_2^0 \vee P_3^0) \wedge (P_2^0)$ corresponding to $\{O_3, O_7, O_{10}\}$ objects simplifies to $P_1^2 P_2^0 \vee P_3^0 P_2^0$... This gives us the implication

$$(P_1 = 2 \wedge P_2 = 0) \vee (P_3 = 0 \wedge P_2 = 0) \to (P_4 = 1)$$

The above implications can also be written as the following rule set:

$$\begin{cases} (P_1 = 1) \to (P_4 = 1) \\ (P_2 = 2) \to (P_4 = 1) \\ (P_1 = 2) \wedge (P_2 = 0) \to (P_4 = 1) \\ (P_3 = 0) \wedge (P_2 = 0) \to (P_4 = 1) \end{cases}$$

It can be noted that each of the first two rules has a *support* of 1 (i.e., the antecedent matches two objects), while each of the last two rules has a support of 2. To finish writing the rule set for this knowledge system, the same procedure as above (starting with writing a new decision matrix) should be followed for the case of $P_4 = 2$, thus yielding a new set of implications for that decision value (i.e., a set of implications with $P_4 = 2$ as the consequent). In general, the procedure will be repeated for each possible value of the decision variable.

LERS Rule Induction System

The data system LERS (Learning from Examples based on Rough Sets) Grzymala-Busse (1997) may induce rules from inconsistent data, i.e., data with conflicting objects. Two objects are conflicting when they are characterized by the same values of all attributes, but they belong to different concepts (classes). LERS uses rough set theory to compute lower and upper approximations for concepts involved in conflicts with other concepts.

Rules induced from the lower approximation of the concept *certainly* describe the concept, hence such rules are called *certain*. On the other hand, rules induced from the upper approximation of the concept describe the concept *possibly*, so these rules are called *possible*. For rule induction LERS uses three algorithms: LEM1, LEM2, and IRIM.

The LEM2 algorithm of LERS is frequently used for rule induction and is used not only in LERS but also in other systems, e.g., in RSES (Bazan et al. (2004). LEM2 explores the search space of attribute-value pairs. Its input data set is a lower or upper approximation of a concept, so its input data set is always consistent. In general, LEM2 computes a local covering and then converts it into a rule set. We will quote a few definitions to describe the LEM2 algorithm.

The LEM2 algorithm is based on an idea of an attribute-value pair block. Let X be a nonempty lower or upper approximation of a concept represented by a decision-value pair (d, w). Set X *depends* on a set T of attribute-value pairs $t = (a, v)$ if and only if

Set is a *minimal complex* of if and only if depends on and no proper subset of exists such that depends on . Let be a nonempty collection of nonempty sets of attribute-value pairs. Then is a *local covering* of if and only if the following three conditions are satisfied:

each member of is a minimal complex of ,

is minimal, i.e., has the smallest possible number of members.

For our sample information system, LEM2 will induce the following rules:

$$\begin{cases} (P_1, 1) \rightarrow (P_4, 1) \\ (P_5, 0) \rightarrow (P_4, 1) \\ (P_1, 0) \rightarrow (P_4, 2) \\ (P_2, 1) \rightarrow (P_4, 2) \end{cases}$$

Other rule-learning methods can be found, e.g., in Pawlak (1991), Stefanowski (1998), Bazan et al. (2004), etc.

Incomplete Data

Rough set theory is useful for rule induction from incomplete data sets. Using this approach we can distinguish between three types of missing attribute values: *lost values* (the values that were recorded but currently are unavailable), *attribute-concept values* (these missing attribute values may be replaced by any attribute value limited to the same concept), and *"do not care" conditions* (the original values were irrelevant). A *concept (class)* is a set of all objects classified (or diagnosed) the same way.

Two special data sets with missing attribute values were extensively studied: in the first case, all missing attribute values were lost (Stefanowski and Tsoukias, 2001), in the second case, all missing attribute values were "do not care" conditions (Kryszkiewicz, 1999).

In attribute-concept values interpretation of a missing attribute value, the missing attribute value may be replaced by any value of the attribute domain restricted to the concept to which the object with a missing attribute value belongs (Grzymala-Busse and Grzymala-Busse, 2007). For example, if for a patient the value of an attribute Temperature is missing, this patient is sick with flu, and all remaining patients sick with flu have values high or very-high for Temperature when using the interpretation of the missing attribute value as the attribute-concept value, we will replace the missing attribute value with high and very-high. Additionally, the *characteristic relation*, enables to process data sets with all three kind of missing attribute values at the same time: lost, "do not care" conditions, and attribute-concept values.

Applications

Rough set methods can be applied as a component of hybrid solutions in machine learning and data mining. They have been found to be particularly useful for rule induction and feature selection (semantics-preserving dimensionality reduction). Rough set-based data analysis methods have been successfully applied in bioinformatics, economics and finance, medicine, multimedia, web and text mining, signal and image processing, software engineering, robotics, and engineering (e.g. power systems and control engineering). Recently the three regions of rough sets are interpreted as regions of acceptance, rejection and deferment. This leads to three-way decision making approach with the model which can potentially lead to interesting future applications.

History

The idea of rough set was proposed by Pawlak (1981) as a new mathematical tool to deal with vague concepts. Comer, Grzymala-Busse, Iwinski, Nieminen, Novotny, Pawlak, Obtulowicz, and Pomykala have studied algebraic properties of rough sets. Different algebraic semantics have been developed by P. Pagliani, I. Duntsch, M. K. Chakraborty, M. Banerjee and A. Mani; these have been extended to more generalized rough sets by D. Cattaneo and A. Mani, in particular. Rough sets can be used to represent ambiguity, vagueness and general uncertainty.

Extensions and Generalizations

Since the development of rough sets, extensions and generalizations have continued to evolve. Initial developments focused on the relationship - both similarities and difference - with fuzzy sets. While some literature contends these concepts are different, other literature considers that rough sets are a generalization of fuzzy sets - as represented through either fuzzy rough sets or rough fuzzy sets. Pawlak (1995) considered that fuzzy and rough sets should be treated as being complementary to each other, addressing different aspects of uncertainty and vagueness.

Three notable extensions of classical rough sets are:

- Dominance-based rough set approach (DRSA) is an extension of rough set theory for multi-criteria decision analysis (MCDA), introduced by Greco, Matarazzo and Słowiński (2001). The main change in this extension of classical rough sets is the substitution of the indiscernibility relation by a *dominance* relation, which permits the formalism to deal with inconsistencies typical in consideration of criteria and preference-ordered decision classes.

- Decision-theoretic rough sets (DTRS) is a probabilistic extension of rough set theory introduced by Yao, Wong, and Lingras (1990). It utilizes a Bayesian decision procedure for minimum risk decision making. Elements are included into the lower and upper approximations based on whether their conditional probability is above thresholds and . These upper and lower thresholds determine region inclusion for elements. This model is unique and powerful since the thresholds themselves are calculated from a set of six loss functions representing classification risks.

- Game-theoretic rough sets (GTRS) is a game theory-based extension of rough set that was introduced by Herbert and Yao (2011). It utilizes a game-theoretic environment to optimize certain criteria of rough sets based classification or decision making in order to obtain effective region sizes.

Rough Membership

Rough sets can be also defined, as a generalisation, by employing a rough membership function instead of objective approximation. The rough membership function expresses a conditional probability that belongs to given . This can be interpreted as a degree that belongs to in terms of information about expressed by .

Rough membership primarily differs from the fuzzy membership in that the membership of union and intersection of sets cannot, in general, be computed from their constituent membership as is the case of fuzzy sets. In this, rough membership is a generalization of fuzzy membership. Furthermore, the rough membership function is grounded more in probability than the conventionally held concepts of the fuzzy membership function.

Other Generalizations

Several generalizations of rough sets have been introduced, studied and applied to solving problems. Here are some of these generalizations:

- rough multisets (Grzymala-Busse, 1987)

- fuzzy rough sets extend the rough set concept through the use of fuzzy equivalence classes(Nakamura, 1988)

- Alpha rough set theory (α-RST) - a generalization of rough set theory that allows approximation using of fuzzy concepts (Quafafou, 2000)

- intuitionistic fuzzy rough sets (Cornelis, De Cock and Kerre, 2003)

- generalized rough fuzzy sets (Feng, 2010)

- rough intuitionistic fuzzy sets (Thomas and Nair, 2011)

- soft rough fuzzy sets and soft fuzzy rough sets (Meng, Zhang and Qin, 2011)

- composite rough sets (Zhang, Li and Chen, 2014)

Vague Set

In mathematics, vague sets are an extension of fuzzy sets.

In a fuzzy set, each object is assigned a single value in the interval [0,1] reflecting its *grade of membership*. This single value does not allow a separation of evidence for membership and evidence against membership.

Gau et al. proposed the notion of vague sets, where each object is characterized by two different membership functions: a true membership function and a false membership function. This kind of reasoning is also called interval membership, as opposed to point membership in the context of fuzzy sets.

Mathematical Definition

A vague set V is characterized by

- its true membership function $t_v(x)$

- its false membership function $f_v(x)$

- with $0 \leq t_v(x) + f_v(x) \leq 1$

The *grade of membership* for x is not a crisp value anymore, but can be located in $[t_v(x), 1 - f_v(x)]$. This interval can be interpreted as an extension to the fuzzy membership function. The vague set degrades to a fuzzy set, if $1 - f_v(x) = t_v(x)$ for all x. The *uncertainty* of x is the difference between the upper and lower bounds of the membership interval; it can be computed as $(1 - f_v(x)) - t_v(x)$..

Fuzzy Classification

Fuzzy classification is the process of grouping elements into a fuzzy set whose membership function is defined by the truth value of a fuzzy propositional function.

A fuzzy class ~C = { i | ~Π(i) } is defined as a fuzzy set ~C of individuals i satisfying a fuzzy classification predicate ~Π which is a fuzzy propositional function. The domain of the fuzzy class operator ~{ .| .} is the set of variables V and the set of fuzzy propositional functions ~PF, and the range is the fuzzy powerset (the set of fuzzy subsets) of this universe, ~P(U):

$$\sim\{ .| .\}{:}V \times \sim PF \longrightarrow \sim P(U)$$

A fuzzy propositional function is, analogous to, an expression containing one or more variables, such that, when values are assigned to these variables, the expression becomes a fuzzy proposition in the sense of.

Accordingly, fuzzy classification is the process of grouping individuals having the same characteristics into a *fuzzy set*. A fuzzy classification corresponds to a membership function μ that indicates whether an individual is a member of a class, given its fuzzy classification predicate ~Π.

$$\mu{:}\sim PF \times U \longrightarrow \sim T$$

Here, ~T is the set of fuzzy truth values (the interval between zero and one). The fuzzy classification predicate ~Π corresponds to a fuzzy restriction "i is R" of U, where R is a fuzzy set defined by a truth function. The degree of membership of an individual i in the fuzzy class ~C is defined by the truth value of the corresponding fuzzy predicate.

$$\mu\sim C(i){:=} \tau(\sim\Pi(i))$$

Classification

Intuitively, a class is a set that is defined by a certain property, and all objects having that property are elements of that class. The process of classification evaluates for a given set of objects whether they fulfill the classification property, and consequentially are a member of the corresponding class. However, this intuitive concept has some logical subtleties that need clarification.

A class logic is a logical system which supports set construction using logical predicates with the class operator { .| .}. A *class*

$$C = \{ i | \Pi(i) \}$$

is defined as a set C of individuals i satisfying a classification predicate Π which is a propositional function. The domain of the class operator { .| .} is the set of variables V and the set of propositional functions PF, and the range is the powerset of this universe P(U) that is, the set of possible subsets:

$$\{ .| .\}{:}V \times PF \longrightarrow P(U)$$

Here is an explanation of the logical elements that constitute this definition:

- An individual is a real object of reference.

- A universe of discourse is the set of all possible individuals considered.

- A variable V:→R is a function which maps into a predefined range R without any given function arguments: a zero-place function.

- A propositional function is "an expression containing one or more undetermined constituents, such that, when values are assigned to these constituents, the expression becomes a proposition".

In contrast, *classification* is the process of grouping individuals having the same characteristics into a set. A classification corresponds to a membership function μ that indicates whether an individual is a member of a class, given its classification predicate Π.

$$\mu{:}PF \times U \longrightarrow T$$

The membership function maps from the set of propositional functions PF and the universe of discourse U into the set of truth values T. The membership μ of individual i in Class C is defined by the truth value τ of the classification predicate Π.

$$\mu C(i){:=}\tau(\Pi(i))$$

In classical logic the truth values are certain. Therefore a classification is crisp, since the truth values are either exactly true or exactly false.

Fuzzy Set Operations

A fuzzy set operation is an operation on fuzzy sets. These operations are generalization of crisp set operations. There is more than one possible generalization. The most widely used operations are called *standard fuzzy set operations*. There are three operations: fuzzy complements, fuzzy intersections, and fuzzy unions.

Standard Fuzzy Set Operations

Let A and B be fuzzy sets that A,B ∈ U, u is an element in the U universe (e.g. value)

Standard complement

$$\neg\mu_A(u) = 1 - \mu_A(u)$$

Standard intersection

$$\mu_{A \cap B}(u) = \min\{\mu_A(u), \mu_B(u)\}$$

Standard union

$$\mu_{A \cup B}(u) = \max\{\mu_A(u), \mu_B(u)\}$$

Fuzzy Complements

$A(x)$ is defined as the degree to which x belongs to A. Let cA denote a fuzzy complement of A of type c. Then $cA(x)$ is the degree to which x belongs to cA, and the degree to which x does not belong to A. ($A(x)$ is therefore the degree to which x does not belong to cA.) Let a complement cA be defined by a function

$$c : [0,1] \to [0,1]$$

$$c(A(x)) = cA(x)$$

Axioms for Fuzzy Complements

Axiom c1. *Boundary condition*

$$c(0) = 1 \text{ and } c(1) = 0$$

Axiom c2. *Monotonicity*

For all $a, b \in [0, 1]$, if $a < b$, then $c(a) > c(b)$

Axiom c3. *Continuity*

c is continuous function.

Axiom c4. *Involutions*

c is an involution, which means that $c(c(a)) = a$ for each $a \in [0,1]$

Fuzzy Intersections

The intersection of two fuzzy sets A and B is specified in general by a binary operation on the unit interval, a function of the form

$$i:[0,1]\times[0,1] \to [0,1].$$

$$(A \cap B)(x) = i[A(x), B(x)] \text{ for all } x.$$

Axioms for Fuzzy Intersection

Axiom i1. *Boundary condition*

$$i(a, 1) = a$$

Axiom i2. *Monotonicity*

$$b \le d \text{ implies } i(a, b) \le i(a, d)$$

Axiom i3. *Commutativity*

$$i(a, b) = i(b, a)$$

Axiom i4. *Associativity*

$$i(a, i(b, d)) = i(i(a, b), d)$$

Axiom i5. *Continuity*

i is a continuous function

Axiom i6. *Subidempotency*

$$i(a, a) \leq a$$

Fuzzy Unions

The union of two fuzzy sets A and B is specified in general by a binary operation on the unit interval function of the form

$$u:[0,1]\times[0,1] \rightarrow [0,1].$$

$$(A \cup B)(x) = u[A(x), B(x)] \text{ for all } x$$

Axioms for Fuzzy Union

Axiom u1. *Boundary condition*

$$u(a, 0) = u(0, a) = a$$

Axiom u2. *Monotonicity*

$b \leq d$ implies $u(a, b) \leq u(a, d)$

Axiom u3. *Commutativity*

$$u(a, b) = u(b, a)$$

Axiom u4. *Associativity*

$$u(a, u(b, d)) = u(u(a, b), d)$$

Axiom u5. *Continuity*

u is a continuous function

Axiom u6. *Superidempotency*

$$u(a, a) \geq a$$

Axiom u7. *Strict monotonicity*

$a_1 < a_2$ and $b_1 < b_2$ implies $u(a_1, b_1) < u(a_2, b_2)$

Aggregation Operations

Aggregation operations on fuzzy sets are operations by which several fuzzy sets are combined in a desirable way to produce a single fuzzy set.

Aggregation operation on n fuzzy set $(2 \le n)$ is defined by a function

$$h:[0,1]^n \rightarrow [0,1]$$

Axioms for Aggregation Operations Fuzzy Sets

Axiom h1. *Boundary condition*

$$h(0, 0, ..., 0) = 0 \text{ and } h(1, 1, ..., 1) = 1$$

Axiom h2. *Monotonicity*

For any pair $<a_1, a_2, ..., a_n>$ and $<b_1, b_2, ..., b_n>$ of n-tuples such that $a_i, b_i \in [0,1]$ for all $i \in N_n$, if $a_i \le b_i$ for all $i \in N_n$, then $h(a_1, a_2, ...,a_n) \le h(b_1, b_2, ..., b_n)$; that is, h is monotonic increasing in all its arguments.

Axiom h3. *Continuity*

h is a continuous function.

Fril

Fril is a programming language for first-order predicate calculus. It includes the semantics of Prolog as a subset, but takes its syntax from the micro-Prolog of Logic Programming Associates and adds support for fuzzy sets, support logic, and metaprogramming.

Fril was originally developed by Trevor Martin and Jim Baldwin at the University of Bristol around 1980. In 1986, it was picked up and further developed by Equipu A.I. Research, which later became Fril Systems Ltd. The name *Fril* was originally an acronym for *Fuzzy Relational Inference Language*.

Prolog and Fril Comparison

Aside from the uncertainty-management features of Fril, there are some minor differences in Fril's implementation of standard Prolog features.

Types

The basic types in Fril are similar to those in Prolog, with one important exception: Prolog's compound data type is the term, with lists defined as nested terms using the

. functor; in Fril, the compound type is the list itself, which forms the basis for most constructs. Variables are distinguished by identifiers containing only uppercase letters and underscores (whereas Prolog only requires the first character to be uppercase). As in Prolog, the name _ is reserved to mean "any value", with multiple occurrences of _ replaced by distinct variables.

Syntax

Prolog has a syntax with a typical amount of punctuation, whereas Fril has an extremely simple syntax similar to that of Lisp. A (propositional) clause is a list consisting of a predicate followed by its arguments (if any). Among the types of top-level constructs are rules and direct-mode commands.

Rule

A rule is a list consisting of a conclusion followed by the hypotheses (**goals**). The general forms look like this:

(fact)

(conclusion goal_1 ... goal_n)

These are equivalent to the respective Prolog constructions:

fact.

conclusion :- goal_1, ..., goal_n.

For example, consider the member predicate in Prolog:

member(E, [E|_]).

member(E, [_|T]) :- member(E, T).

In Fril, this becomes:

((member E (E|_)))

((member E (_|T)) (member E T))

Relation

Some data can be represented in the form of relations. A relation is equivalent to a set of facts with the same predicate name and of constant arity, except that none of the facts can be removed (other than by killing the relation); such a representation consumes less memory internally. A relation is written literally as a list consisting of the predicate name followed by one or more tuples of the relation (all of the arguments of the equivalent fact without the predicate name). A predicate can also be declared a relation by calling the def_rel predicate; this only works if the proposed name does not already exist in the knowledge base. Once a predicate is a relation, anything that would ordinarily add a rule (and does not violate the restrictions of relations) automatically adds a tuple to the relation instead.

Here is an example. The following set of facts:

> ((less-than 2 3))

> ((less-than 8 23))

> ((less-than 42 69))

can be rewritten as the relation:

(less-than

>> (2 3)

>> (8 23)

>> (42 69)

Direct Mode

A predicate may be called with exactly one argument using the syntax:

predicate argument

Queries are submitted using this syntax, with *predicate* being ? (or one of the other query-related predicates).

Fuzzy Sets

Fril supports both continuous and discrete fuzzy sets, each with their own special syntaxes. A discrete set (dtype) lists discrete values and their degrees of membership, with this syntax:

{*value*:*dom value*:*dom* ... *value*:*dom*}

value is an atom or number, and *dom* is a value in the interval [0, 1].

A continuous set (itype) lists real numbers and their degrees of membership; the degree-of-membership function is the linear interpolation over these mappings. The syntax is thus:

[*value*:*dom value*:*dom* ... *value*:*dom*]

where the values must be given in non-decreasing order.

Each dtype and itype may be constrained to a universe (a set of allowable values). Fril has predicates for fuzzy set operations (but does not directly support control through fuzzy logic). It is even possible to combine dtypes and itypes through some operations, as long as the dtypes contain only real numbers.

Support Pairs

Any rule may have a probability interval (called a support pair) associated with it by appending :(*min max*) to it, where *min* and *max* are the minimum and maximum probabilities. Fril includes predicates that calculate the support for a given query.

Disjunction

While Prolog uses punctuation — namely ; — for disjunction within clauses, Fril instead has a built-in predicate orr.

Merits

There are advantages and disadvantages to this simpler syntax. On the positive side, it renders predicates such as Prolog's =.. (which maps between lists and clauses) unnecessary, as a clause *is* a list. On the other hand, it is more difficult to read.

Behavior

As a logic programming environment, Fril is very similar to Prolog. Here are some of the differences:

- Both Prolog and Fril have shell applications, which serve as the standard way of interacting with them. Prolog reads commands in two modes: in source file-reading mode, it accepts directives and clauses; in user interaction mode, it accepts only queries (although it is possible to tell Prolog to read directives and clauses from standard input by using consult(user).). Fril makes no distinction: all types of commands can be given both from source files and on the command line.

- The Prolog shell automatically prints the values of all instantiated variables that appear in a query, along with a *yes* or *no* answer. The Fril shell only gives the answer; it is the user's responsibility to print variables if desired.

Fuzzy Control System

A fuzzy control system is a control system based on fuzzy logic—a mathematical system that analyzes analog input values in terms of logical variables that take on continuous values between 0 and 1, in contrast to classical or digital logic, which operates on discrete values of either 1 or 0 (true or false, respectively).

Overview

Fuzzy logic is widely used in machine control. The term "fuzzy" refers to the fact that the logic involved can deal with concepts that cannot be expressed as the "true" or "false" but rather as "partially true". Although alternative approaches such as genetic algorithms and neural networks can perform just as well as fuzzy logic in many cases, fuzzy logic has the advantage that the solution to the problem can be cast in terms that human operators can understand, so that their experience can be used in the design of the controller. This makes it easier to mechanize tasks that are already successfully performed by humans.

History and Applications

Fuzzy logic was first proposed by Lotfi A. Zadeh of the University of California at Berkeley in a 1965 paper. He elaborated on his ideas in a 1973 paper that introduced the concept of "linguistic variables", which in this article equates to a variable defined as a fuzzy set. Other research followed, with the first industrial application, a cement kiln built in Denmark, coming on line in 1975.

Fuzzy systems were initially implemented in Japan.

- Interest in fuzzy systems was sparked by Seiji Yasunobu and Soji Miyamoto of Hitachi, who in 1985 provided simulations that demonstrated the feasibility of fuzzy control systems for the Sendai railway. Their ideas were adopted, and fuzzy systems were used to control accelerating, braking, and stopping when the line opened in 1987.

- In 1987, Takeshi Yamakawa demonstrated the use of fuzzy control, through a set of simple dedicated fuzzy logic chips, in an "inverted pendulum" experiment. This is a classic control problem, in which a vehicle tries to keep a pole mounted on its top by a hinge upright by moving back and forth. Yamakawa subsequently made the demonstration more sophisticated by mounting a wine glass containing water and even a live mouse to the top of the pendulum: the system maintained stability in both cases. Yamakawa eventually went on to organize his own fuzzy-systems research lab to help exploit his patents in the field.

- Japanese engineers subsequently developed a wide range of fuzzy systems for both industrial and consumer applications. In 1988 Japan established the Laboratory for International Fuzzy Engineering (LIFE), a cooperative arrangement between 48 companies to pursue fuzzy research. The automotive company Volkswagen was the only foreign corporate member of LIFE, dispatching a researcher for a duration of three years.

- Japanese consumer goods often incorporate fuzzy systems. Matsushita vacuum cleaners use microcontrollers running fuzzy algorithms to interrogate dust sensors and adjust suction power accordingly. Hitachi washing machines use fuzzy controllers to load-weight, fabric-mix, and dirt sensors and automatically set the wash cycle for the best use of power, water, and detergent.

- Canon developed an autofocusing camera that uses a charge-coupled device (CCD) to measure the clarity of the image in six regions of its field of view and use the information provided to determine if the image is in focus. It also tracks the rate of change of lens movement during focusing, and controls its speed to prevent overshoot. The camera's fuzzy control system uses 12 inputs: 6 to obtain the current clarity data provided by the CCD and 6 to measure the rate of change of lens movement. The output is the position of the lens. The fuzzy control system uses 13 rules and requires 1.1 kilobytes of memory.

- An industrial air conditioner designed by Mitsubishi uses 25 heating rules and 25 cooling rules. A temperature sensor provides input, with control outputs fed to an inverter, a compressor valve, and a fan motor. Compared to the previous design, the fuzzy controller heats and cools five times faster, reduces power consumption by 24%, increases temperature stability by a factor of two, and uses fewer sensors.

- Other applications investigated or implemented include: character and handwriting recognition; optical fuzzy systems; robots, including one for making Japanese flower arrangements; voice-controlled robot helicopters (hovering is a "balancing act" rather similar to the inverted pendulum problem); rehabilitation robotics to provide patient-specific solutions (e.g. to control heart rate and blood pressure); control of flow of powders in film manufacture; elevator systems; and so on.

Work on fuzzy systems is also proceeding in the United State and Europe, although on a less extensive scale than in Japan.

- The US Environmental Protection Agency has investigated fuzzy control for energy-efficient motors, and NASA has studied fuzzy control for automated space docking: simulations show that a fuzzy control system can greatly reduce fuel consumption.

- Firms such as Boeing, General Motors, Allen-Bradley, Chrysler, Eaton, and Whirlpool have worked on fuzzy logic for use in low-power refrigerators, improved automotive transmissions, and energy-efficient electric motors.

- In 1995 Maytag introduced an "intelligent" dishwasher based on a fuzzy controller and a "one-stop sensing module" that combines a thermistor, for temperature measurement; a conductivity sensor, to measure detergent level from the ions present in the wash; a turbidity sensor that measures scattered and transmitted light to measure the soiling of the wash; and a magnetostrictive sensor to read spin rate. The system determines the optimum wash cycle for any load to obtain the best results with the least amount of energy, detergent, and water. It even adjusts for dried-on foods by tracking the last time the door was opened, and estimates the number of dishes by the number of times the door was opened.

Research and development is also continuing on fuzzy applications in software, as opposed to firmware, design, including fuzzy expert systems and integration of fuzzy logic with neural-network and so-called adaptive "genetic" software systems, with the ultimate goal of building "self-learning" fuzzy-control systems. These systems can be employed to control complex, nonlinear dynamic plants, for example, human body.

Fuzzy Sets

The input variables in a fuzzy control system are in general mapped by sets of membership functions similar to this, known as "fuzzy sets". The process of converting a crisp input value to a fuzzy value is called "fuzzification".

A control system may also have various types of switch, or "ON-OFF", inputs along with its analog inputs, and such switch inputs of course will always have a truth value equal to either 1 or 0, but the scheme can deal with them as simplified fuzzy functions that happen to be either one value or another.

Given "mappings" of input variables into membership functions and truth values, the microcontroller then makes decisions for what action to take, based on a set of "rules", each of the form:

IF brake temperature IS warm AND speed IS not very fast

THEN brake pressure IS slightly decreased.

In this example, the two input variables are "brake temperature" and "speed" that have values defined as fuzzy sets. The output variable, "brake pressure" is also defined by a fuzzy set that can have values like "static" or "slightly increased" or "slightly decreased" etc.

This rule by itself is very puzzling since it looks like it could be used without bothering with fuzzy logic, but remember that the decision is based on a set of rules:

- All the rules that apply are invoked, using the membership functions and truth values obtained from the inputs, to determine the result of the rule.

- This result in turn will be mapped into a membership function and truth value controlling the output variable.

- These results are combined to give a specific ("crisp") answer, the actual brake pressure, a procedure known as "defuzzification".

This combination of fuzzy operations and rule-based "inference" describes a "fuzzy expert system".

Traditional control systems are based on mathematical models in which the control system is described using one or more differential equations that define the system response to its inputs. Such systems are often implemented as "PID controllers" (proportional-integral-derivative controllers). They are the products of decades of development and theoretical analysis, and are highly effective.

If PID and other traditional control systems are so well-developed, why bother with fuzzy control? It has some advantages. In many cases, the mathematical model of the control process may not exist, or may be too "expensive" in terms of computer processing power and memory, and a system based on empirical rules may be more effective.

Furthermore, fuzzy logic is well suited to low-cost implementations based on cheap sensors, low-resolution analog-to-digital converters, and 4-bit or 8-bit one-chip microcontroller chips. Such systems can be easily upgraded by adding new rules to improve performance or add new features. In many cases, fuzzy control can be used to improve existing traditional controller systems by adding an extra layer of intelligence to the current control method.

Fuzzy Control in Detail

Fuzzy controllers are very simple conceptually. They consist of an input stage, a processing stage, and an output stage. The input stage maps sensor or other inputs, such as switches, thumbwheels, and so on, to the appropriate membership functions and truth values. The processing stage invokes each appropriate rule and generates a result for each, then combines the results of the rules. Finally, the output stage converts the combined result back into a specific control output value.

The most common shape of membership functions is triangular, although trapezoidal and bell curves are also used, but the shape is generally less important than the number of curves and their placement. From three to seven curves are generally appropriate to cover the required range of an input value, or the "universe of discourse" in fuzzy jargon.

As discussed earlier, the processing stage is based on a collection of logic rules in the form of IF-THEN statements, where the IF part is called the "antecedent" and the THEN part is called the "consequent". Typical fuzzy control systems have dozens of rules.

Consider a rule for a thermostat:

IF (temperature is "cold") THEN (heater is "high")

This rule uses the truth value of the "temperature" input, which is some truth value of "cold", to generate a result in the fuzzy set for the "heater" output, which is some value of "high". This result is used with the results of other rules to finally generate the crisp composite output. Obviously, the greater the truth value of "cold", the higher the truth value of "high", though this does not necessarily mean that the output itself will be set to "high" since this is only one rule among many. In some cases, the membership functions can be modified by "hedges" that are equivalent to adverbs. Common hedges include "about", "near", "close to", "approximately", "very", "slightly", "too", "extremely", and "somewhat". These operations may have precise definitions, though the definitions can vary considerably between different implementations. "Very", for one example, squares membership functions; since the membership values are always less than 1, this narrows the membership function. "Extremely" cubes the values to give greater narrowing, while "somewhat" broadens the function by taking the square root.

In practice, the fuzzy rule sets usually have several antecedents that are combined using fuzzy operators, such as AND, OR, and NOT, though again the definitions tend to vary: AND, in one popular definition, simply uses the minimum weight of all the antecedents, while OR uses the maximum value. There is also a NOT operator that subtracts a membership function from 1 to give the "complementary" function.

There are several ways to define the result of a rule, but one of the most common and simplest is the "max-min" inference method, in which the output membership function is given the truth value generated by the premise.

Rules can be solved in parallel in hardware, or sequentially in software. The results of all the rules that have fired are "defuzzified" to a crisp value by one of several methods. There are dozens, in theory, each with various advantages or drawbacks.

The "centroid" method is very popular, in which the "center of mass" of the result provides the crisp value. Another approach is the "height" method, which takes the value of the biggest contributor. The centroid method favors the rule with the output of greatest area, while the height method obviously favors the rule with the greatest output value.

The diagram below demonstrates max-min inferencing and centroid defuzzification for a system with input variables "x", "y", and "z" and an output variable "n". Note that "mu" is standard fuzzy-logic nomenclature for "truth value":

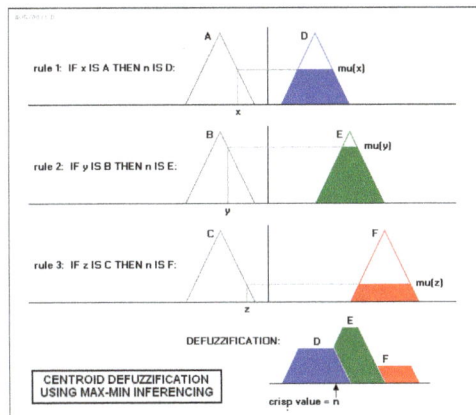

Notice how each rule provides a result as a truth value of a particular membership function for the output variable. In centroid defuzzification the values are OR'd, that is, the maximum value is used and values are not added, and the results are then combined using a centroid calculation.

Fuzzy control system design is based on empirical methods, basically a methodical approach to trial-and-error. The general process is as follows:

- Document the system's operational specifications and inputs and outputs.

- Document the fuzzy sets for the inputs.

- Document the rule set.

- Determine the defuzzification method.

- Run through test suite to validate system, adjust details as required.

- Complete document and release to production.

As a general example, consider the design of a fuzzy controller for a steam turbine. The block diagram of this control system appears as follows:

The input and output variables map into the following fuzzy set:

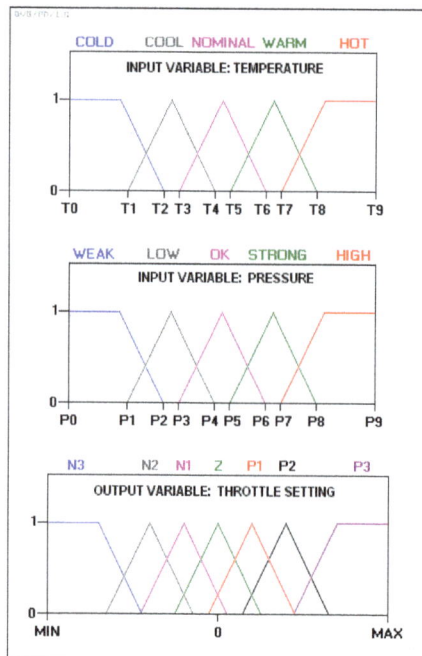

where:

N3: Large negative.

N2: Medium negative.

N1: Small negative.

Z: Zero.

P1: Small positive.

P2: Medium positive.

P3: Large positive.

The rule set includes such rules as:

rule 1: IF temperature IS cool AND pressure IS weak,

THEN throttle is P3.

rule 2: IF temperature IS cool AND pressure IS low,

THEN throttle is P2.

rule 3: IF temperature IS cool AND pressure IS ok,

THEN throttle is Z.

rule 4: IF temperature IS cool AND pressure IS strong,

THEN throttle is N2.

In practice, the controller accepts the inputs and maps them into their membership functions and truth values. These mappings are then fed into the rules. If the rule specifies an AND relationship between the mappings of the two input variables, as the examples above do, the minimum of the two is used as the combined truth value; if an OR is specified, the maximum is used. The appropriate output state is selected and assigned a membership value at the truth level of the premise. The truth values are then defuzzified. For an example, assume the temperature is in the "cool" state, and the pressure is in the "low" and "ok" states. The pressure values ensure that only rules 2 and 3 fire:

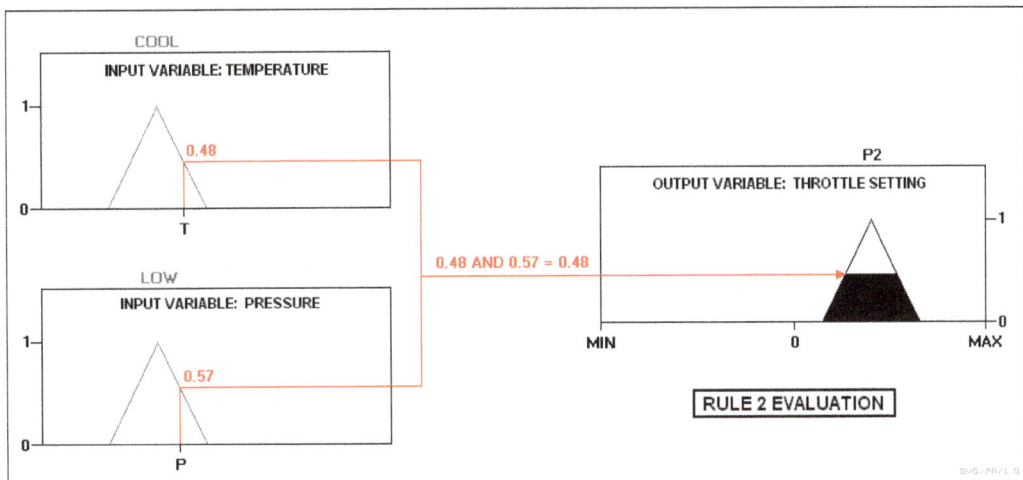

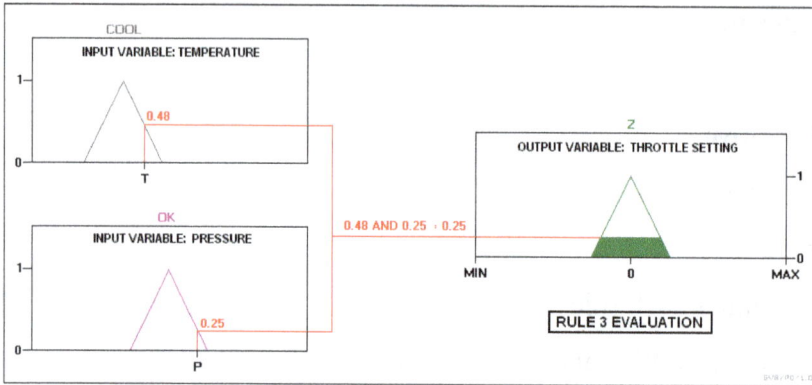

The two outputs are then defuzzified through centroid defuzzification:

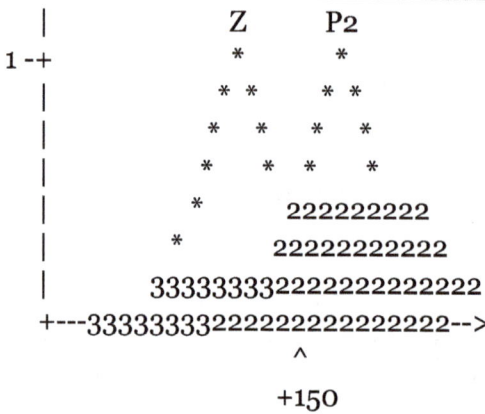

```
        |              Z         P2
   1 -+                *          *
        |             * *        * *
        |            *   *      *   *
        |           *     *    *     *
        |          *            222222222
        |         *            2222222222
        |        333333333222222222222
   +---333333333222222222222222-->
                      ^
                    +150
```

The output value will adjust the throttle and then the control cycle will begin again to generate the next value .

Building a fuzzy controller

Consider implementing with a microcontroller chip a simple feedback controller:

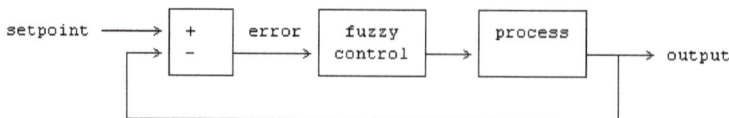

A fuzzy set is defined for the input error variable "e", and the derived change in error, "delta", as well as the "output", as follows:

LP: large positive

SP: small positive

ZE: zero

SN: small negative

LN: large negative

If the error ranges from -1 to +1, with the analog-to-digital converter used having a resolution of 0.25, then the input variable's fuzzy set (which, in this case, also applies to the output variable) can be described very simply as a table, with the error / delta / output values in the top row and the truth values for each membership function arranged in rows beneath:

	-1	-0.75	-0.5	-0.25	0	0.25	0.5	0.75	1
mu(LP)	0	0	0	0	0	0	0.3	0.7	1
mu(SP)	0	0	0	0	0.3	0.7	1	0.7	0.3
mu(ZE)	0	0	0.3	0.7	1	0.7	0.3	0	0
mu(SN)	0.3	0.7	1	0.7	0.3	0	0	0	0
mu(LN)	1	0.7	0.3	0	0	0	0	0	0

—or, in graphical form (where each "X" has a value of 0.1):

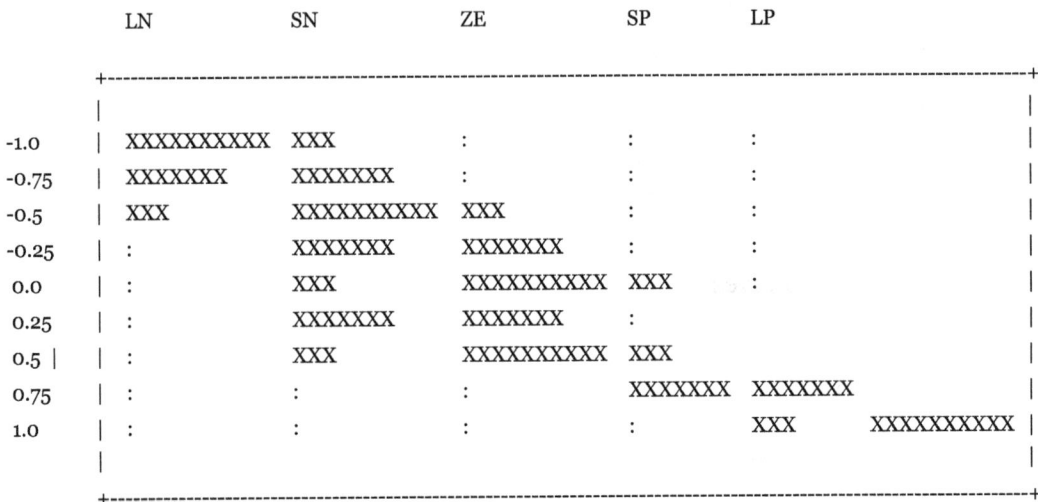

```
          LN              SN              ZE              SP              LP

        +----------------------------------------------------------------------------+
        |                                                                            |
 -1.0   |  XXXXXXXXXX  XXX             :               :               :             |
 -0.75  |  XXXXXXX     XXXXXXX         :               :               :             |
 -0.5   |  XXX         XXXXXXXXXX  XXX :               :               :             |
 -0.25  |  :           XXXXXXX     XXXXXXX             :               :             |
  0.0   |  :           XXX         XXXXXXXXXX  XXX     :                             |
  0.25  |  :           XXXXXXX     XXXXXXX         :                                 |
  0.5 | |  :           XXX         XXXXXXXXXX  XXX                                   |
  0.75  |  :             :               :       XXXXXXX  XXXXXXX                    |
  1.0   |  :             :               :               XXX         XXXXXXXXXX |
        |                                                                            |
        +----------------------------------------------------------------------------+
```

Suppose this fuzzy system has the following rule base:

```
rule 1:  IF e = ZE AND delta = ZE THEN output = ZE
rule 2:  IF e = ZE AND delta = SP THEN output = SN
rule 3:  IF e = SN AND delta = SN THEN output = LP
rule 4:  IF e = LP OR  delta = LP THEN output = LN
```

These rules are typical for control applications in that the antecedents consist of the logical combination of the error and error-delta signals, while the consequent is a control command output. The rule outputs can be defuzzified using a discrete centroid computation:

SUM(I = 1 TO 4 OF (mu(I) * output(I))) / SUM(I = 1 TO 4 OF mu(I))

Now, suppose that at a given time we have:

e   = 0.25

delta = 0.5

Then this gives:

	e	delta
mu(LP)	0	0.3
mu(SP)	0.7	1
mu(ZE)	0.7	0.3
mu(SN)	0	0
mu(LN)	0	0

Plugging this into rule 1 gives:

rule 1: IF e = ZE AND delta = ZE THEN output = ZE

mu(1)   = MIN(0.7, 0.3) = 0.3

output(1) = 0

-- where:

- mu(1): Truth value of the result membership function for rule 1. In terms of a centroid calculation, this is the "mass" of this result for this discrete case.

- output(1): Value (for rule 1) where the result membership function (ZE) is maximum over the output variable fuzzy set range. That is, in terms of a centroid calculation, the location of the "center of mass" for this individual result. This value is independent of the value of "mu". It simply identifies the location of ZE along the output range.

The other rules give:

rule 2: IF e = ZE AND delta = SP THEN output = SN

```
    mu(2)        = MIN( 0.7, 1 ) = 0.7
    output(2)  = -0.5
```

rule 3: IF e = SN AND delta = SN THEN output = LP

```
    mu(3)        = MIN( 0.0, 0.0 ) = 0
    output(3)  = 1
```

rule 4: IF e = LP OR delta = LP THEN output = LN

```
    mu(4)        = MAX( 0.0, 0.3 ) = 0.3
    output(4)  = -1
```

The centroid computation yields:

$$\frac{mu(1).output(1) + mu(2).output(2) + mu(3).output(3) + mu(4).output(4)}{mu(1) + mu(2) + mu(3) + mu(4)}$$

$$= \frac{(0.3*0) + (0.7*-0.5) + (0*1) + (0.3*-1)}{0.3 + 0.7 + 0 + 0.3}$$

$= -0.5$ — for the final control output. Simple. Of course the hard part is figuring out what rules actually work correctly in practice.

If you have problems figuring out the centroid equation, remember that a centroid is defined by summing all the moments (location times mass) around the center of gravity and equating the sum to zero. So if X_0 is the center of gravity, X_i is the location of each mass, and M_i is each mass, this gives:

$$0 = (X_1 - X_0)*M_1 + (X_2 - X_0)*M_2 + ... + (X_n - X_0)*M_n$$

$$0 = (X_1*M_1 + X_2*M_2 + ... + X_n*M_n) - X_0*(M_1 + M_2 + ... + M_n)$$

$$X_0*(M_1 + M_2 + ... + M_n) = X_1*M_1 + X_2*M_2 + ... + X_n*M_n$$

$$X_0 = \frac{X_1*M_1 + X_2*M_2 + ... + X_n*M_n}{M_1 + M_2 + ... + M_n}$$

In our example, the values of mu correspond to the masses, and the values of X to location of the masses (mu, however, only 'corresponds to the masses' if the initial 'mass' of the output functions are all the same/equivalent. If they are not the same, i.e. some are narrow triangles, while others maybe wide trapizoids or shouldered triangles, then the mass or area of the output function must be known or calculated. It is this mass that is then scaled by mu and multiplied by its location X_i).

This system can be implemented on a standard microprocessor, but dedicated fuzzy chips are now available. For example, Adaptive Logic INC of San Jose, California, sells a "fuzzy chip", the AL220, that can accept four analog inputs and generate four analog outputs. A block diagram of the chip is shown below:

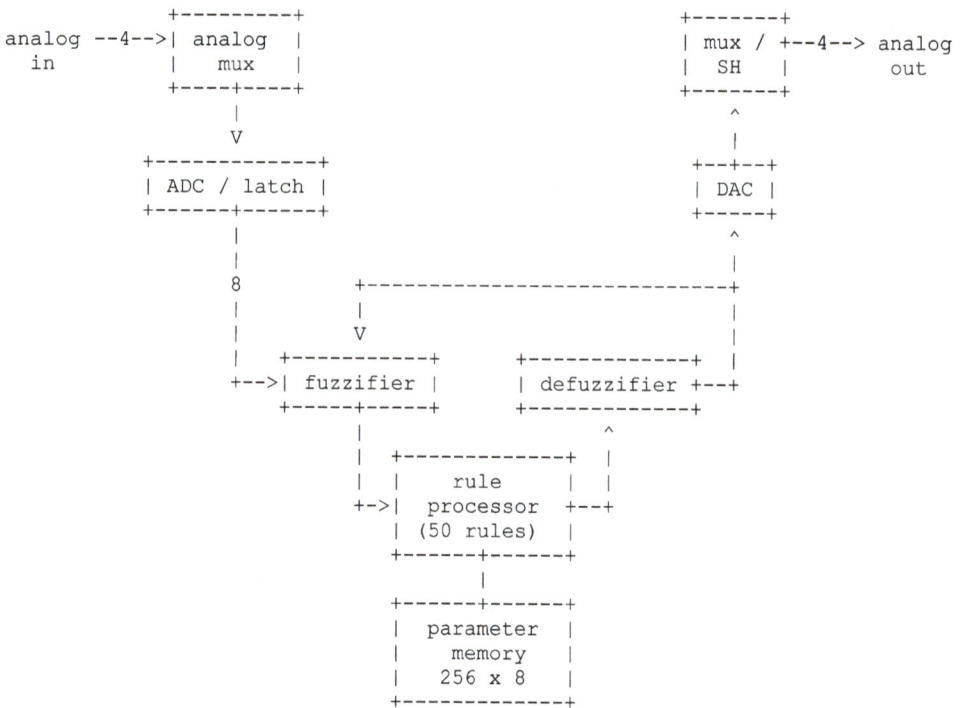

```
                +---------+                          +-------+
 analog --4-->| analog  |                          | mux / +--4--> analog
    in        |   mux   |                          |  SH   |          out
              +----+----+                          +-------+
                   |                                    ^
                   V                                    |
          +------------+                            +--+--+
          | ADC / latch |                           | DAC |
          +------+------+                            +-----+
                 |                                       ^
                 |                                       |
                 8                 +-----------------------------+
                 |                 |                             |
                 |                 V                             |
                 |   +-----------+        +-------------+        |
                 +-->| fuzzifier |        | defuzzifier +--+
                     +-----+-----+        +-------------+
                           |                      ^
                           |   +-------------+    |
                           |   |    rule     |    |
                           +-->|  processor  +--+
                               |  (50 rules) |
                               +------+------+
                                      |
                               +------+------+
                               | parameter   |
                               |   memory    |
                               |  256 x 8    |
                               +-------------+
```

```
ADC:   analog-to-digital converter
DAC:   digital-to-analog converter
SH:    sample/hold
```

Antilock Brakes

As a first example, consider an anti-lock braking system, directed by a microcontroller chip. The microcontroller has to make decisions based on brake temperature, speed, and other variables in the system.

The variable "temperature" in this system can be subdivided into a range of "states": "cold", "cool", "moderate", "warm", "hot", "very hot". The transition from one state to the next is hard to define.

An arbitrary static threshold might be set to divide "warm" from "hot". For example, at exactly 90 degrees, warm ends and hot begins. But this would result in a discontinuous change when the input value passed over that threshold. The transition wouldn't be smooth, as would be required in braking situations.

The way around this is to make the states *fuzzy*. That is, allow them to change gradually from one state to the next. In order to do this there must be a dynamic relationship established between different factors.

We start by defining the input temperature states using "membership functions":

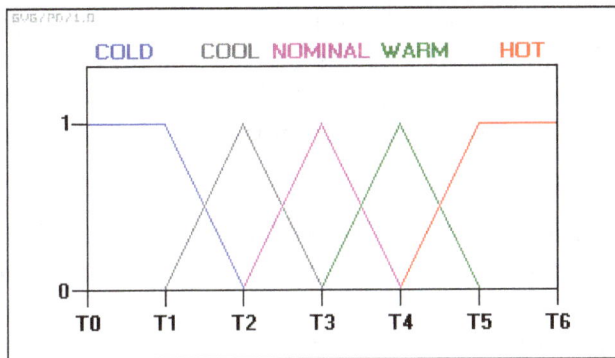

With this scheme, the input variable's state no longer jumps abruptly from one state to the next. Instead, as the temperature changes, it loses value in one membership function while gaining value in the next. In other words, its ranking in the category of cold decreases as it becomes more highly ranked in the warmer category.

At any sampled timeframe, the "truth value" of the brake temperature will almost always be in some degree part of two membership functions: i.e.: '0.6 nominal and 0.4 warm', or '0.7 nominal and 0.3 cool', and so on.

The above example demonstrates a simple application, using the abstraction of values from multiple values. This only represents one kind of data, however, in this case, temperature.

Adding additional sophistication to this braking system, could be done by additional factors such as traction, speed, inertia, set up in dynamic functions, according to the designed fuzzy system.

Logical Interpretation of Fuzzy Control

In spite of the appearance there are several difficulties to give a rigorous logical interpretation of the *IF-THEN* rules. As an example, interpret a rule as *IF (temperature is "cold") THEN (heater is "high")* by the first order formula $Cold(x) \rightarrow High(y)$ and assume that r is an input such that $Cold(r)$ is false. Then the formula $Cold(r) \rightarrow High(t)$ is true for any t and therefore any t gives a correct control given r. A rigorous logical justification of fuzzy control is given in Hájek's book where fuzzy control is represented as a theory of Hájek's basic logic. Also in Gerla 2005 another logical approach to fuzzy control is proposed based on fuzzy logic programming.Indeed, denote by f the fuzzy function arising of an IF-THEN systems of rules. Then we can translate this system into a fuzzy program P containing a series of rules whose head is "Good(x,y)". The interpretation of this predicate in the least fuzzy Herbrand model of P coincides with f. This gives further useful tools to fuzzy control.

Type-1 OWA Operators

The Yager's OWA (ordered weighted averaging) operators are used to aggregate the crisp values in decision making schemes (such as multi-criteria decision making, multi-expert decision making and multi-criteria/multi-expert decision making). It is widely accepted that Fuzzy sets are more suitable for representing preferences of criteria in decision making.

The type-1 OWA operators have been proposed for this purpose. The type-1 OWA operators provides a technique for directly aggregating uncertain information with uncertain weights via OWA mechanism in soft decision making and data mining, where these uncertain objects are modelled by fuzzy sets.

The two definitions for type-1 OWA operators are based on Zadeh's Extension Principle and $\alpha-$cuts of fuzzy sets. The two definitions lead to equivalent results.

Definitions

Definition 1

Let $F(X)$ be the set of fuzzy sets with domain of discourse X, a type-1 OWA operator is defined as follows:

Given n linguistic weights $\left\{W^i\right\}_{i=1}^n$ in the form of fuzzy sets defined on the domain of discourse $U=[0,1]$, a type-1 OWA operator is a mapping, Φ,

$$\Phi: F(X)\times\cdots\times F(X)\to F(X)$$

$$(A^1,\cdots,A^n)\mapsto Y$$

such that

$$\mu_Y(y)=\sup_{\sum_{k=1}^{n}\overline{w}_i a_{\sigma(i)}=y}\left(\mu_{W^1}(w_1)\wedge\cdots\wedge\mu_{W^n}(w_n)\wedge\mu_{A^1}(a_1)\wedge\cdots\wedge\mu_{A^n}(a_n)\right)$$

where $\overline{w}_i=\dfrac{w_i}{\displaystyle\sum_{i=1}^{n}w_i}$, ,and $\sigma:\{1,\cdots,n\}\to\{1,\cdots,n\}$ is a permutation function such that

$a_{\sigma(i)}\geq a_{\sigma(i+1)}$, $\forall i=1,\cdots,n-1$, , i.e., $a_{\sigma(i)}$ is the th highest element in the set $\left\{a_1,\cdots,a_n\right\}$..

Definition 2

Using the alpha-cuts of fuzzy sets:

Given the n linguistic weights $\left\{W^i\right\}_{i=1}^n$ in the form of fuzzy sets defined on the domain of discourse $U = [0, 1]$, , then for each $\alpha \in [0,1]$, an $\alpha-$level type-1 OWA operator with $\alpha-$level sets $\left\{W_\alpha^i\right\}_{i=1}^n$ to aggregate the α-cuts of fuzzy sets $\left\{A^i\right\}_{i=1}^n$ is:

$$\Phi_\alpha\left(A_\alpha^1,\ldots,A_\alpha^n\right) = \left\{ \frac{\displaystyle\sum_{i=1}^n w_i a_{\sigma(i)}}{\displaystyle\sum_{i=1}^n w_i} \middle| w_i \in W_\alpha^i, a_i \in A_\alpha^i, i = 1,\ldots,n \right\}$$

where $W_\alpha^i = \{w \mid \mu_{W_i}(w) \geq \alpha\}, A_\alpha^i = \{x \mid \mu_{A_i}(x) \geq \alpha\}$, and $\sigma : \{1,\cdots,n\} \to \{1,\cdots,n\}$ is a permutation function such that $a_{\sigma(i)} \geq a_{\sigma(i+1)}, \forall i = 1,\cdots,n-1$, , i.e., $a_{\sigma(i)}$ is the th largest element in the set $\{a_1,\cdots,a_n\}$..

Representation Theorem of Type-1 OWA Operators

Given the n linguistic weights $\left\{W^i\right\}_{i=1}^n$ in the form of fuzzy sets defined on the domain of discourse $U = [0, 1]$, , and the fuzzy sets $A^1,\cdots,A^n$, , then we have that

$$Y = G$$

where Y is the aggregation result obtained by Definition 1, and G is the result obtained by in Definition 2.

Programming Problems for Type-1 OWA Operators

According to the Representation Theorem of Type-1 OWA Operators, a general type-1 OWA operator can be decomposed into a series of $\alpha-$level type-1 OWA operators. In practice, this series of $\alpha-$level type-1 OWA operators is used to construct the resulting aggregation fuzzy set. So we only need to compute the left end-points and right end-points of the intervals $\Phi_\alpha\left(A_\alpha^1,\cdots,A_\alpha^n\right)$. Then, the resulting aggregation fuzzy set is constructed with the membership function as follows:

$$\mu_G(x) = \underset{\alpha : x \in \Phi_\alpha\left(A_\alpha^1,\cdots,A_\alpha^n\right)_\alpha}{\alpha}$$

For the left end-points, we need to solve the following programming problem:

$$\Phi_\alpha\left(A_\alpha^1,\cdots,A_\alpha^n\right)_- = \min_{\substack{W_{\alpha^-}^i \leq w_i \leq W_{\alpha+}^i, A_{\alpha-}^i \leq a_i \leq A_{\alpha+}^i}} \sum_{i=1}^n w_i a_{\sigma(i)} / \sum_{i=1}^n w_i$$

while for the right end-points, we need to solve the following programming problem:

$$\Phi_\alpha\left(A_\alpha^1,\cdots,A_\alpha^n\right)_+ = \max_{\substack{W_{\alpha^-}^i \leq w_i \leq W_{\alpha+}^i, A_{\alpha-}^i \leq a_i \leq A_{\alpha+}^i}} \sum_{i=1}^n w_i a_{\sigma(i)} / \sum_{i=1}^n w_i$$

A fast method has been presented to solve two programming problem so that the type-1 OWA aggregation operation can be performed efficiently.

Alpha-level Approach to Type-1 OWA Operation

Three-step process:

- Step 1—To set up the $\alpha-$ level resolution in [0, 1].

- Step 2—For each $\alpha \in [0,1]$,

  - Step 2.1—To calculate $\rho_{\alpha+}^{i_0^*}$

1. Let $i_0 = 1$;

2. If $\rho_{\alpha+}^{i_0} \geq A_{\alpha+}^{\sigma(i_0)}$, stop, $\rho_{\alpha+}^{i_0}$ is the solution; otherwise go to Step 2.1-3.

3. $i_0 \leftarrow i_0 + 1$, , go to Step 2.1-2.

  - Step 2.2 To calculate $\rho_{\alpha-}^{i_0^*}$

1. Let $i_0 = 1$;

2. If $\rho_{\alpha-}^{i_0} \geq A_{\alpha-}^{\sigma(i_0)}$, stop, $\rho_{\alpha-}^{i_0}$ is the solution; otherwise go to Step 2.2-3.

3. $i_0 \leftarrow i_0 + 1$, go to step Step 2.2-2.

- Step 3—To construct the aggregation resulting fuzzy set G based on all the available intervals $\left[\rho_{\alpha-}^{i_0^*}, \rho_{\alpha+}^{i_0^*} \right]$:

$$\mu_G(x) = \underset{\alpha: x \in \left[\rho_{\alpha-}^{i_0^*}, \rho_{\alpha+}^{i_0^*} \right]}{} \alpha$$

Special Cases

- Any OWA operators, like maximum, minimum, mean operators;

- Join operators of (type-1) fuzzy sets, i.e., fuzzy maximum operators;

- Meet operators of (type-1) fuzzy sets, i.e., fuzzy minimum operators;

- Join-like operators of (type-1) fuzzy sets;

- Meet-like operators of (type-1) fuzzy sets.

Generalizations

Type-2 OWA operators have been suggested to aggregate the type-2 fuzzy sets for soft decision making.

Type-2 Fuzzy Sets and Systems

Type-2 fuzzy sets and systems generalize Type-1 fuzzy sets and systems so that more uncertainty can be handled. From the very beginning of fuzzy sets, criticism was made about the fact that the membership function of a type-1 fuzzy set has no uncertainty associated with it, something that seems to contradict the word *fuzzy*, since that word has the connotation of lots of uncertainty. So, what does one do when there is uncertainty about the value of the membership function? The answer to this question was provided in 1975 by the inventor of fuzzy sets, Prof. Lotfi A. Zadeh , when he proposed more sophisticated kinds of fuzzy sets, the first of which he called a *type-2 fuzzy set*. A type-2 fuzzy set lets us incorporate uncertainty about the membership function into fuzzy set theory, and is a way to address the above criticism of type-1 fuzzy sets head-on. And, if there is no uncertainty, then a type-2 fuzzy set reduces to a type-1 fuzzy set, which is analogous to probability reducing to determinism when unpredictability vanishes,.

In order to symbolically distinguish between a type-1 fuzzy set and a type-2 fuzzy set, a tilde symbol is put over the symbol for the fuzzy set; so, A denotes a type-1 fuzzy set, whereas Ã denotes the comparable type-2 fuzzy set. When the latter is done, the resulting type-2 fuzzy set is called a *general type-2 fuzzy set* (to distinguish it from the special interval type-2 fuzzy set).

Prof. Zadeh didn't stop with type-2 fuzzy sets, because in that 1976 paper he also generalized all of this to type-n fuzzy sets. The present article focuses only on type-2 fuzzy sets because they are the *next step* in the logical progression from type-1 to type-n fuzzy sets, where $n = 1, 2, \dots$. Although some researchers are beginning to explore higher than type-2 fuzzy sets, as of early 2009, this work is in its infancy.

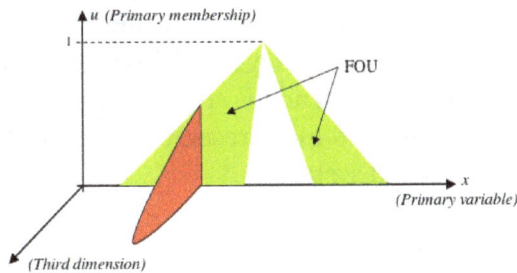

Figure 1. The membership function of a general type-2 fuzzy set is three-dimensional. A cross-section of one slice of the third dimension is shown. This cross-section, as well as all others, sits on the FOU. Only the boundary of the cross-section is used to describe the membership function of a general type-2 fuzzy set. It is shown filled-in for artistic purposes.

The membership function of a general type-2 fuzzy set, Ã, is three-dimensional (Fig. 1), where the third dimension is the value of the membership function at each point on its two-dimensional domain that is called its *footprint of uncertainty* (FOU).

For an interval type-2 fuzzy set that third-dimension value is the same (e.g., 1) everywhere, which means that no new information is contained in the third dimension of an interval type-2 fuzzy set. So, for such a set, the third dimension is ignored, and only the FOU is used to describe it. It is for this reason that an interval type-2 fuzzy set is sometimes called a *first-order uncertainty* fuzzy set model, whereas a general type-2 fuzzy set (with its useful third-dimension) is sometimes referred to as a *second-order uncertainty* fuzzy set model.

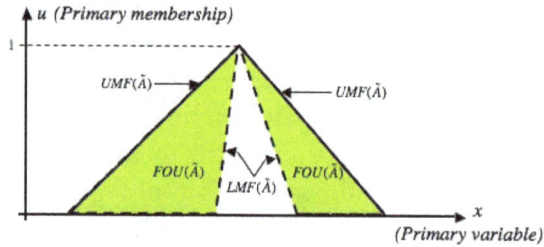

Figure 2. FOU for an interval type-2 fuzzy set. Many other shapes are possible for the FOU.

The FOU represents the blurring of a type-1 membership function, and is completely described by its two bounding functions (Fig. 2), a lower membership function (LMF) and an upper membership function (UMF), both of which are type-1 fuzzy sets! Consequently, it is possible to use type-1 fuzzy set mathematics to characterize and work with interval type-2 fuzzy sets. This means that engineers and scientists who already know type-1 fuzzy sets will not have to invest a lot of time learning about general type-2 fuzzy set mathematics in order to understand and use interval type-2 fuzzy sets.

Work on type-2 fuzzy sets languished during the 1980s and early-to-mid 1990's, although a small number of articles were published about them. People were still trying to figure out what to do with type-1 fuzzy sets, so even though Zadeh proposed type-2 fuzzy sets in 1976, the time was not right for researchers to drop what they were doing with type-1 fuzzy sets to focus on type-2 fuzzy sets. This changed in the latter part of the 1990s as a result of Prof. Jerry Mendel and his student's works on type-2 fuzzy sets and systems (e.g.,). Since then, more and more researchers around the world are writing articles about type-2 fuzzy sets and systems.

Interval Type-2 Fuzzy Sets

Interval type-2 fuzzy sets have received the most attention because the mathematics that is needed for such sets—primarily Interval arithmetic—is much simpler than the mathematics that is needed for general type-2 fuzzy sets. So, the literature about interval type-2 fuzzy sets is large, whereas the literature about general type-2 fuzzy sets is much smaller. Both kinds of fuzzy sets are being actively researched by an ever-growing number of researchers around the world.

Formilleri for the following have already been worked out for interval type-2 fuzzy sets:

- Fuzzy set operations: union, intersection and complement (,)

- Centroid (a very widely used operation by practitioners of such sets, and also an important uncertainty measure for them) (, ,)

- Other uncertainty measures [fuzziness, cardinality, variance and skewness and uncertainty bounds

- Similarity (, ,)

- Subsethood

- Embedded fuzzy sets (, ,)

- Fuzzy set ranking

- Fuzzy rule ranking and selection

- Type-reduction methods (,)

- Firing intervals for an interval type-2 fuzzy logic system (, ,)

- Fuzzy weighted average

- Linguistic weighted average

- Synthesizing an FOU from data that are collected from a group of subject

Interval Type-2 Fuzzy Logic Systems

Type-2 fuzzy sets are finding very wide applicability in *rule-based fuzzy logic systems* (FLSs) because they let uncertainties be modeled by them whereas such uncertainties cannot be modeled by type-1 fuzzy sets. A block diagram of a type-2 FLS is depicted in Fig. 3. This kind of FLS is used in fuzzy logic control, fuzzy logic signal processing, rule-based classification, etc., and is sometimes referred to as a *function approximation* application of fuzzy sets, because the FLS is designed to minimize an error function.

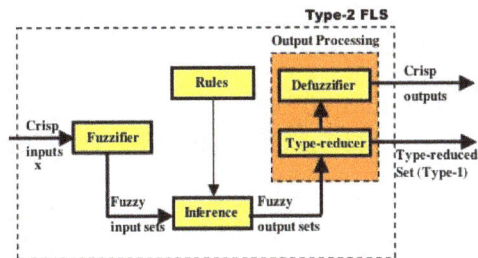

Figure 3. Type-2 FLS

The following discussions, about the four components in the Fig. 3 rule-based FLS, are given for an interval type-2 FLS, because to-date they are the most popular kind of type-2 FLS; however, most of the discussions are also applicable for a general type-2 FLS.

Rules, that are either provided by subject experts or are extracted from numerical data, are expressed as a collection of IF-THEN statements, e.g.,

> IF temperature is *moderate* and pressure is *high,* then rotate the valve *a bit to the right.*

Fuzzy sets are associated with the terms that appear in the antecedents (IF-part) or consequents (THEN-part) of rules, and with the inputs to and the outputs of the FLS. Membership functions are used to describe these fuzzy sets, and in a type-1 FLS they are all type-1 fuzzy sets, whereas in an interval type-2 FLS at least one membership function is an interval type-2 fuzzy set.

An interval type-2 FLS lets any one or all of the following kinds of uncertainties be quantified:

1. Words that are used in antecedents and consequents of rules—because words can mean different things to different people.

2. Uncertain consequents—because when rules are obtained from a group of experts, consequents will often be different for the same rule, i.e. the experts will not necessarily be in agreement.

3. Membership function parameters—because when those parameters are optimized using uncertain (noisy) training data, the parameters become uncertain.

4. Noisy measurements—because very often it is such measurements that activate the FLS.

In Fig. 3, measured (crisp) inputs are first transformed into fuzzy sets in the *Fuzzifier* block because it is fuzzy sets and not numbers that activate the rules which are described in terms of fuzzy sets and not numbers. Three kinds of fuzzifiers are possible in an interval type-2 FLS. When measurements are:

*   Perfect, they are modeled as a crisp set;

*   Noisy, but the noise is stationary, they are modeled as a type-1 fuzzy set; and,

*   Noisy, but the noise is non-stationary, they are modeled as an interval type-2 fuzzy set (this latter kind of fuzzification cannot be done in a type-1 FLS).

In Fig. 3, after measurements are fuzzified, the resulting input fuzzy sets are mapped into fuzzy output sets by the *Inference* block. This is accomplished by first quantifying each rule using fuzzy set theory, and by then using the mathematics of fuzzy sets to establish the output of each rule, with the help of an inference mechanism. If there are M rules then the fuzzy input sets to the Inference block will activate only a subset of those rules, where the subset contains at least one rule and usually way fewer than M rules. Inference is done one rule at a time. So, at the output of the Inference block, there will be one or more *fired-rule fuzzy output sets.*

In most engineering applications of a FLS, a number (and not a fuzzy set) is needed as its final output, e.g., the consequent of the rule given above is "Rotate the valve a bit to the right." No automatic valve will know what this means because "a bit to the right" is a linguistic expression, and a valve must be turned by numerical values, i.e. by a certain number of degrees. Consequently, the fired-rule output fuzzy sets have to be converted into a number, and this is done in the Fig. 3 *Output Processing* block.

In a type-1 FLS, output processing, called *Defuzzification*, maps a type-1 fuzzy set into a number. There are many ways for doing this, e.g., compute the union of the fired-rule output fuzzy sets (the result is another type-1 fuzzy set) and then compute the center of gravity of the membership function for that set; compute a weighted average of the center of gravities of each of the fired rule consequent membership functions; etc.

Things are somewhat more complicated for an interval type-2 FLS, because to go from an interval type-2 fuzzy set to a number (usually) requires two steps (Fig. 3). The first step, called *type-reduction*, is where an interval type-2 fuzzy set is reduced to an interval-valued type-1 fuzzy set. There are as many type-reduction methods as there are type-1 defuzzification methods. An algorithm developed by Karnik and Mendel (,) now known as the *KM Algorithm* is used for type-reduction. Although this algorithm is iterative, it is very fast.

The second step of Output Processing, which occurs after type-reduction, is still called *defuzzification*. Because a type-reduced set of an interval type-2 fuzzy set is always a finite interval of numbers, the defuzzified value is just the average of the two end-points of this interval.

It is clear from Fig. 3 that there can be two outputs to an interval type-2 FLS—crisp numerical values and the type-reduced set. The latter provides a measure of the uncertainties that have flowed through the interval type-2 FLS, due to the (possibly) uncertain input measurements that have activated rules whose antecedents or consequents or both are uncertain. Just as standard deviation is widely used in probability and statistics to provide a measure of unpredictable uncertainty about a mean value, the type-reduced set can provided a measure of uncertainty about the crisp output of an interval type-2 FLS.

Computing with Words

Another application for fuzzy sets has also been inspired by Prof. Zadeh (–)—*Computing With Words*. Different acronyms have been used for "computing with words," e.g., CW and CWW. According to Zadeh:

> CWW is a methodology in which the objects of computation are words and propositions drawn from a natural language. [It is] inspired by the remarkable human capability to perform a wide variety of physical and mental tasks without any measurements and any computations.

Of course, he did not mean that computers would actually compute using words—single words or phrases—rather than numbers. He meant that computers would be activated by words, which would be converted into a mathematical representation using fuzzy sets and that these fuzzy sets would be mapped by a CWW engine into some other fuzzy set after which the latter would be converted back into a word. A natural question to ask is: Which kind of fuzzy set—type-1 or type-2—should be used as a model for a word? Mendel (,) has argued, on the basis of Karl Popper's concept of *Falsificationism* (,), that using a type-1 fuzzy set as a model for a word is scientifically incorrect. An interval type-2 fuzzy set should be used as a (first-order uncertainty) model for a word. Much research is under way about CWW.

Applications

Type-2 fuzzy sets were applied in image processing, video processing and computer vision.

Defuzzification

Defuzzification is the process of producing a quantifiable result in Crisp logic, given fuzzy sets and corresponding membership degrees. It is typically needed in fuzzy control systems. These will have a number of rules that transform a number of variables into a fuzzy result, that is, the result is described in terms of membership in fuzzy sets. For example, rules designed to decide how much pressure to apply might result in "Decrease Pressure (15%), Maintain Pressure (34%), Increase Pressure (72%)". Defuzzification is interpreting the membership degrees of the fuzzy sets into a specific decision or real value.

The simplest but least useful defuzzification method is to choose the set with the highest membership, in this case, "Increase Pressure" since it has a 72% membership, and ignore the others, and convert this 72% to some number. The problem with this approach is that it loses information. The rules that called for decreasing or maintaining pressure might as well have not been there in this case.

A common and useful defuzzification technique is *center of gravity*. First, the results of the rules must be added together in some way. The most typical fuzzy set membership function has the graph of a triangle. Now, if this triangle were to be cut in a straight horizontal line somewhere between the top and the bottom, and the top portion were to be removed, the remaining portion forms a trapezoid. The first step of defuzzification typically "chops off" parts of the graphs to form trapezoids (or other shapes if the initial shapes were not triangles). For example, if the output has "Decrease Pressure (15%)", then this triangle will be cut 15% the way up from the bottom. In the most common technique, all of these trapezoids are then superimposed one upon another, forming a single geometric shape. Then, the centroid of this shape, called the *fuzzy centroid*, is calculated. The x coordinate of the centroid is the defuzzified value.

Methods

There are many different methods of defuzzification available, including the following:

- AI (adaptive integration)
- BADD (basic defuzzification distributions)
- BOA (bisector of area)
- CDD (constraint decision defuzzification)
- COA (center of area)
- COG (center of gravity)
- ECOA (extended center of area)
- EQM (extended quality method)
- FCD (fuzzy clustering defuzzification)
- FM (fuzzy mean)
- FOM (first of maximum)
- GLSD (generalized level set defuzzification)
- ICOG (indexed center of gravity)
- IV (influence value)
- LOM (last of maximum)
- MeOM (mean of maxima)
- MOM (middle of maximum)
- QM (quality method)
- RCOM (random choice of maximum)
- SLIDE (semi-linear defuzzification)
- WFM (weighted fuzzy mean)

The maxima methods are good candidates for fuzzy reasoning systems. The distribution methods and the area methods exhibit the property of continuity that makes them suitable for fuzzy controllers.

Membership Function (Mathematics)

The membership function of a fuzzy set is a generalization of the indicator function in classical sets. In fuzzy logic, it represents the degree of truth as an extension of valuation. Degrees of truth are often confused with probabilities, although they are conceptually distinct, because fuzzy truth represents membership in vaguely defined sets, not likelihood of some event or condition. Membership functions were introduced by Zadeh in the first paper on fuzzy sets (1965). Zadeh, in his theory of fuzzy sets, proposed using a membership function (with a range covering the interval (0,1)) operating on the domain of all possible values.

Definition

For any set X, a membership function on X is any function from X to the real unit interval [0,1].

Membership functions on X represent fuzzy subsets of X. The membership function which represents a fuzzy set $\tilde{A}$ is usually denoted by μ_A. For an element x of X, the value $\mu_A(x)$ is called the *membership degree* of x in the fuzzy set $\tilde{A}$. The membership degree $\mu_A(x)$ quantifies the grade of membership of the element x to the fuzzy set $\tilde{A}$. The value 0 means that x is not a member of the fuzzy set; the value 1 means that x is fully a member of the fuzzy set. The values between 0 and 1 characterize fuzzy members, which belong to the fuzzy set only partially.

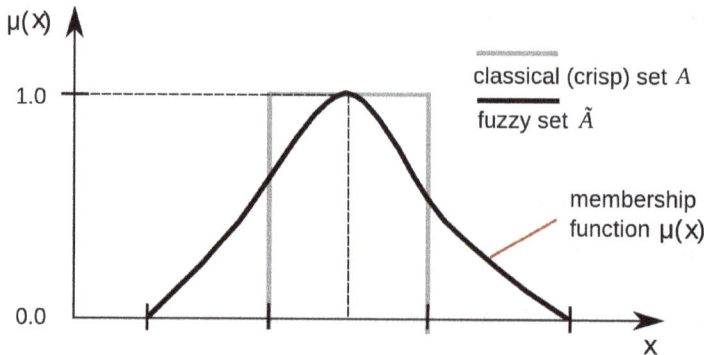

Membership function of a fuzzy set

Sometimes, a more general definition is used, where membership functions take values in an arbitrary fixed algebra or structure L; usually it is required that L be at least a poset or lattice. The usual membership functions with values in [0, 1] are then called [0, 1]-valued membership functions.

Capacity

One application of membership functions is as capacities in decision theory.

In decision theory, a capacity is defined as a function, v from **S**, the set of subsets of some set, into $[0,1]$, such that v is set-wise monotone and is normalized (i.e. $v(\varnothing) = 0, v(\Omega) = 1)$. This is a generalization of the notion of a probability measure, where the probability axiom of countable additivity is weakened. A capacity is used as a subjective measure of the likelihood of an event, and the "expected value" of an outcome given a certain capacity can be found by taking the Choquet integral over the capacity.

Vector Logic

Vector logic is an algebraic model of elementary logic based on matrix algebra. Vector logic assumes that the truth values map on vectors, and that the monadic and dyadic operations are executed by matrix operators.

Overview

Classic binary logic is represented by a small set of mathematical functions depending on one (monadic) or two (dyadic) variables. In the binary set, the value 1 corresponds to *true* and the value 0 to *false*. A two-valued vector logic requires a correspondence between the truth-values *true* (t) and *false* (f), and two q-dimensional normalized column vectors composed by real numbers s and n, hence:

$$t \mapsto s \quad \text{and} \quad f \mapsto n$$

(where $q \geq 2$ is an arbitrary natural number, and "normalized" means that the length of the vector is 1; usually s and n are orthogonal vectors). This correspondence generates a space of vector truth-values: $V_2 = \{s,n\}$. The basic logical operations defined using this set of vectors lead to matrix operators.

The operations of vector logic are based on the scalar product between q-dimensional column vectors: $u^T v = \langle u,v \rangle$: the orthonormality between vectors s and n implies that $\langle u,v \rangle = 1$ if $u = v$, and $\langle u,v \rangle = 0$ if $u \neq v$.

Monadic Operators

The monadic operators result from the application $Mon : V_2 \to V_2$, and the associated matrices have q rows and q columns. The two basic monadic operators for this two-valued vector logic are the identity and the negation:

- Identity: A logical identity ID(p)is represented by matrix $I = ss^T + nn^T$. This matrix operates as follows: $Ip = p, p \in V_2$; due to the orthogonality of s respect to n, we have $Is = ss^T s + nn^T s = s\langle s,s \rangle + n\langle n,s \rangle = s,$ and conversely $In = n..$

- Negation: A logical negation ¬p is represented by matrix $N = ns^T + sn^T$ Consequently, $Ns = n$ and $Nn = s$. The involutory behavior of the logical negation, namely that $\neg(\neg p)$ equals p, corresponds with the fact that $N^2 = I$. Is important to note that this vector logic identity matrix is not generally an identity matrix in the sense of matrix algebra.

Dyadic Operators

The 16 two-valued dyadic operators correspond to functions of the type $Dyad : V_2 \otimes V_2 \rightarrow V_2$; the dyadic matrices have q rows and q^2 columns. The matrices that execute these dyadic operations are based on the properties of the Kronecker product.

Two properties of this product are essential for the formalism of vector logic:

1. The mixed-product property

    If A, B, C and D are matrices of such size that one can form the matrix products AC and BD, then

    $$(A \otimes B)(C \otimes D) = AC \otimes BD$$

2. Distributive transpose The operation of transposition is distributive over the Kronecker product:

    $$(A \otimes B)^T = A^T \otimes B^T.$$

Using these properties, expressions for dyadic logic functions can be obtained:

*   **Conjunction**. The conjunction (p∧q) is executed by a matrix that acts on two vector truth-values: $C(u \otimes v)$ .This matrix reproduces the features of the classical conjunction truth-table in its formulation:

    $$C = s(s \otimes s)^T + n(s \otimes n)^T + n(n \otimes s)^T + n(n \otimes n)^T$$

    and verifies

    $$C(s \otimes s) = s, \text{ and}$$

    $$C(s \otimes n) = C(n \otimes s) = C(n \otimes n) = n.$$

*   Disjunction. The disjunction (p∨q) is executed by the matrix

    $$D = s(s \otimes s)^T + s(s \otimes n)^T + s(n \otimes s)^T + n(n \otimes n)^T, \text{resulting in}$$

    $$D(s \otimes s) = D(s \otimes n) = D(n \otimes s) = s \text{ and}$$

    $$D(n \otimes n) = n.$$

*   Implication. The implication corresponds in classical logic to the expression

$p \to q \equiv \neg p \lor q$. The vector logic version of this equivalence leads to a matrix that represents this implication in vector logic: $L = D(N \otimes I)$. . The explicit expression for this implication is:

$$L = s(s \otimes s)^T + n(s \otimes n)^T + s(n \otimes s)^T + n(n \otimes n)^T,$$

and the properties of classical implication are satisfied:

$$L(s \otimes s) = L(n \otimes s) = L(n \otimes n) = s \text{ and}$$

$$L(s \otimes n) = n.$$

- Equivalence and Exclusive or. In vector logic the equivalence p≡q is represented by the following matrix:

$$E = s(s \otimes s)^T + n(s \otimes n)^T + n(n \otimes s)^T + s(n \otimes n)^T \text{ with}$$

$$E(s \otimes s) = E(n \otimes n) = s \text{ and}$$

$$E(s \otimes n) = E(n \otimes s) = n.$$

The Exclusive or is the negation of the equivalence, ¬(p≡q); it corresponds with the matrix $X = NE$ given by

$$X = n(s \otimes s)^T + s(s \otimes n)^T + s(n \otimes s)^T + n(n \otimes n)^T,$$

with $X(s \otimes s) = X(n \otimes n) = n$ and

$$X(s \otimes n) = X(n \otimes s) = s.$$

- NAND and NOR

The matrices S and P correspond to the Sheffer (NAND) and the Peirce (NOR) operations, respectively:

$$S = NC$$

$$P = ND$$

De Morgan'S Law

In the two-valued logic, the conjunction and the disjunction operations satisfy the De Morgan's law: p∧q≡¬(¬p∨¬q), and its dual: p∨q≡¬(¬p∧¬q)). For the two-valued vector logic this Law is also verified:

$$C(u \otimes v) = ND(Nu \otimes Nv), \text{ where } u \text{ and } v \text{ are two logic vectors.}$$

The Kronecker product implies the following factorization:

$$C(u \otimes v) = ND(N \otimes N)(u \otimes v).$$

Then it can be proved that in the two–dimensional vector logic the De Morgan's law is a law involving operators, and not only a law concerning operations:

$$C = ND(N \otimes N)$$

Law of Contraposition

In the classical propositional calculus, the Law of Contraposition $p \rightarrow q \equiv \neg q \rightarrow \neg p$ is proved because the equivalence holds for all the possible combinations of truth-values of p and q. Instead, in vector logic, the law of contraposition emerges from a chain of equalities within the rules of matrix algebra and Kronecker products, as shown in what follows:

$$L(u \otimes v) = D(N \otimes I)(u \otimes v) = D(Nu \otimes v) = D(Nu \otimes NNv) =$$

$$D(NNv \otimes Nu) = D(N \otimes I)(Nv \otimes Nu) = L(Nv \otimes Nu)$$

This result is based in the fact that D, the disjunction matrix, represents a commutative operation.

Many-valued two-dimensional Logic

Many-valued logic was developed by many researchers, particularly by Jan Łukasiewicz and allows extending logical operations to truth-values that include uncertainties. In the case of two-valued vector logic, uncertainties in the truth values can be introduced using vectors with s and n weighted by probabilities.

Let $f = \epsilon s + \delta n$, with $\epsilon, \delta \in [0,1], \epsilon + \delta = 1$ be this kind of "probabilistic" vectors. Here, the many-valued character of the logic is introduced *a posteriori* via the uncertainties introduced in the inputs.

Scalar Projections of Vector Outputs

The outputs of this many-valued logic can be projected on scalar functions and generate a particular class of probabilistic logic with similarities with the many-valued logic of Reichenbach. Given two vectors $u = \alpha s + \beta n$ and $v = \alpha' s + \beta' n$ and a dyadic logical matrix G, a scalar probabilistic logic is provided by the projection over vector s:

$$Val(\text{scalars}) = s^T G(\text{vectors})$$

Here are the main results of these projections:

$$NOT(\alpha) = s^T N u = 1 - \alpha$$

$$OR(\alpha, \alpha') = s^T D(u \otimes v) = \alpha + \alpha' - \alpha\alpha'$$

$$AND(\alpha, \alpha') = s^T C(u \otimes v) = \alpha\alpha'$$

$$IMPL(\alpha,\alpha') = s^T L(u \otimes v) = 1 - \alpha(1-\alpha')$$

$$XOR(\alpha,\alpha') = s^T X(u \otimes v) = \alpha + \alpha' - 2\alpha\alpha'$$

The associated negations are:

$$NOR(\alpha,\alpha') = 1 - OR(\alpha,\alpha')$$

$$NAND(\alpha,\alpha') = 1 - AND(\alpha,\alpha')$$

$$EQUI(\alpha,\alpha') = 1 - XOR(\alpha,\alpha')$$

If the scalar values belong to the set {0, ½, 1}, this many-valued scalar logic is for many of the operators almost identical to the 3-valued logic of Łukasiewicz. Also, it has been proved that when the monadic or dyadic operators act over probabilistic vectors belonging to this set, the output is also an element of this set.

History

The approach has been inspired in neural network models based on the use of high-dimensional matrices and vectors. Vector logic is a direct translation into a matrix-vector formalism of the classical Boolean polynomials. This kind of formalism has been applied to develop a fuzzy logic in terms of complex numbers. Other matrix and vector approaches to logical calculus have been developed in the framework of quantum physics, computer science and optics. Early attempts to use linear algebra to represent logic operations can be referred to Peirce and Copilowish. The Indian biophysicist G.N. Ramachandran developed a formalism using algebraic matrices and vectors to represent many operations of classical Jain Logic known as Syad and Saptbhangi. Indian logic. It requires independent affirmative evidence for each assertion in a proposition, and does not make the assumption for binary complementation.

Boolean Polynomials

George Boole established the development of logical operations as polynomials. For the case of monadic operators (such as identity or negation), the Boolean polynomials look as follows:

$$f(x) = f(1)x + f(0)(1-x)$$

The four different monadic operations result from the different binary values for the coefficients. Identity operation requires $f(1) = 1$ and $f(0) = 0$, and negation occurs if $f(1) = 0$ and $f(0) = 1$. For the 16 dyadic operators, the Boolean polynomials are of the form:

$$f(x,y) = f(1,1)xy + f(1,0)x(1-y) + f(0,1)(1-x)y + f(0,0)(1-x)(1-y)$$

The dyadic operations can be translated to this polynomial format when the coefficients f take the values indicated in the respective truth tables. For instance: the NAND operation requires that:

$$f(1,1) = 0 \text{ and } f(1,0) = f(0,1) = f(0,0) = 1..$$

These Boolean polynomials can be immediately extended to any number of variables, producing a large potential variety of logical operators. In vector logic, the matrix-vector structure of logical operators is an exact translation to the format of liner algebra of these Boolean polynomials, where the x and $1-x$ correspond to vectors s and n respectively (the same for y and $1-y$). In the example of NAND, $f(1,1)=n$ and $f(1,0)=f(0,1)=f(0,0)=s$ and the matrix version becomes:

$$S = n(s \otimes s)^T + s[(s \otimes n)^T + (n \otimes s)^T + (n \otimes n)^T]$$

Extensions

- Vector logic can be extended to include many truth values since large dimensional vector spaces allow to create many orthogonal truth values and the corresponding logical matrices.

- Logical modalities can be fully represented in this context, with recursive process inspired in neural models

- Some cognitive problems about logical computations can be analyzed using this formalism, in particular recursive decisions. Any logical expression of classical propositional calculus can be naturally represented by a tree structure. This fact is retained by vector logic, and has been partially used in neural models focused in the investigation of the branched structure of natural languages.

- The computation via reversible operations as the Fredkin gate can be implemented in vector logic. This implementations provides explicit expressions for matrix operators that produce the input format and the output filtering necessary for obtaining computations

- Elementary cellular automata can be analyzed using the operator structure of vector logic; this analysis leads to a spectral decomposition of the laws governing its dynamics

- In addition, based on this formalism, a discrete differential and integral calculus has been developed

MV-algebra

In abstract algebra, a branch of pure mathematics, an MV-algebra is an algebraic structure with a binary operation $\oplus$, a unary operation $\neg$, , and the constant 0, satisfying certain axioms. MV-algebras are the algebraic semantics of Łukasiewicz logic; the letters MV refer to the *many-valued* logic of Łukasiewicz. MV-algebras coincide with the class of bounded commutative BCK algebras.

Definitions

An MV-algebra is an algebraic structure $\langle A, \oplus, \neg, 0 \rangle$, consisting of

- a non-empty set A,
- a binary operation $\oplus$ on A,
- a unary operation $\neg$ on A, and
- a constant 0 denoting a fixed element of A,

which satisfies the following identities:

- $(x \oplus y) \oplus z = x \oplus (y \oplus z)$,
- $x \oplus 0 = x$,
- $x \oplus y = y \oplus x$,
- $\neg\neg x = x$,
- $x \oplus \neg 0 = \neg 0$, and
- $\neg(\neg x \oplus y) \oplus y = \neg(\neg y \oplus x) \oplus x$.

By virtue of the first three axioms, $\langle A, \oplus, 0 \rangle$ is a commutative monoid. Being defined by identities, MV-algebras form a variety of algebras. The variety of MV-algebras is a subvariety of the variety of BL-algebras and contains all Boolean algebras.

An MV-algebra can equivalently be defined (Hájek 1998) as a prelinear commutative bounded integral residuated lattice $\langle L, \wedge, \vee, \otimes, \rightarrow, 0, 1 \rangle$ satisfying the additional identity

Examples of MV-algebras

A simple numerical example is $A = [0,1]$, with operations $x \oplus y = \min(x+y, 1)$ and $\neg x = 1 - x$. In mathematical fuzzy logic, this MV-algebra is called the *standard MV-algebra*, as it forms the standard real-valued semantics of Łukasiewicz logic.

The *trivial* MV-algebra has the only element 0 and the operations defined in the only possible way, $0 \oplus 0 = 0$ and $\neg 0 = 0$.

The *two-element* MV-algebra is actually the two-element Boolean algebra $\{0,1\}$, with $\oplus$ coinciding with Boolean disjunction and $\neg$ with Boolean negation. In fact adding the axiom $x \oplus x = x$ to the axioms defining an MV-algebra results in an axiomatization of Boolean algebras.

If instead the axiom added is $x \oplus x \oplus x = x \oplus x,$, then the axioms define the MV_3 algebra corresponding to the three-valued Łukasiewicz logic $Ł_3$. Other finite linearly ordered MV-algebras are obtained by restricting the universe and operations of the standard MV-algebra to the set of n equidistant real numbers between 0 and 1 (both included), that is, the set $\{0, 1/(n-1), 2/(n-1), \ldots, 1\}$, which is closed under the operations $\oplus$ and $\neg$ of the standard MV-algebra; these algebras are usually denoted MV_n.

Another important example is *Chang's MV-algebra*, consisting just of infinitesimals (with the order type ω) and their co-infinitesimals.

Chang also constructed an MV-algebra from an arbitrary totally ordered abelian group G by fixing a positive element u and defining the segment $[0, u]$ as $\{x \in G \mid 0 \leq x \leq u\}$, which becomes an MV-algebra with $x \oplus y = \min(u, x+y)$ and $\neg x = u - x$. Furthermore, Chang showed that every linearly ordered MV-algebra is isomorphic to an MV-algebra constructed from a group in this way.

D. Mundici extended the above construction to abelian lattice-ordered groups. If G is such a group with strong (order) unit u, then the "unit interval" $\{x \in G \mid 0 \leq x \leq u\}$ can be equipped with $\neg x = u - x$, $x \oplus y = u \wedge_G (x+y)$, $x \otimes y = 0 \vee_G (x+y-u)$. This construction establishes a categorical equivalence between lattice-ordered abelian groups with strong unit and MV-algebras.

Relation to Łukasiewicz Logic

C. C. Chang devised MV-algebras to study many-valued logics, introduced by Jan Łukasiewicz in 1920. In particular, MV-algebras form the algebraic semantics of Łukasiewicz logic, as described below.

Given an MV-algebra A, an A-valuation is a homomorphism from the algebra of propositional formulas (in the language consisting of $\oplus, \neg$, and 0) into A. Formulas mapped to 1 (or $\neg 0$) for all A-valuations are called A-tautologies. If the standard MV-algebra over $[0,1]$ is employed, the set of all $[0,1]$-tautologies determines so-called infinite-valued Łukasiewicz logic.

Chang's (1958, 1959) completeness theorem states that any MV-algebra equation holding in the standard MV-algebra over the interval $[0,1]$ will hold in every MV-algebra. Algebraically, this means that the standard MV-algebra generates the variety of all MV-algebras. Equivalently, Chang's completeness theorem says that MV-algebras characterize infinite-valued Łukasiewicz logic, defined as the set of $[0,1]$-tautologies.

The way the $[0,1]$ MV-algebra characterizes all possible MV-algebras parallels the well-known fact that identities holding in the two-element Boolean algebra hold in all possible Boolean algebras. Moreover, MV-algebras characterize infinite-valued Łukasiewicz logic in a manner analogous to the way that Boolean algebras characterize classical bivalent logic.

In 1984, Font, Rodriguez and Torrens introduced the Wajsberg algebra as an alternative model for the infinite-valued Łukasiewicz logic. Wajsberg algebras and MV-algebras are isomorphic.

MV_n-algebras

In the 1940s Grigore Moisil introduced his Łukasiewicz–Moisil algebras (LM_n-algebras) in the hope of giving algebraic semantics for the (finitely) n-valued Łukasiewicz logic. However, in 1956 Alan Rose discovered that for $n \geq 5$, the Łukasiewicz–Moisil algebra does not model the Łukasiewicz n-valued logic. Although C. C. Chang published his MV-algebra in 1958, it is faithful model only for the $\aleph_0$-valued (infinitely-many-valued) Łukasiewicz–Tarski logic. For the axiomatically more complicated (finitely) n-valued Łukasiewicz logics, suitable algebras were published in 1977 by Revaz Grigolia and called MV_n-algebras. MV_n-algebras are a subclass of LM_n-algebras; the inclusion is strict for $n \geq 5$.

The MV_n-algebras are MV-algebras which satisfy some additional axioms, just like the n-valued Łukasiewicz logics have additional axioms added to the $\aleph_0$-valued logic.

In 1982 Roberto Cignoli published some additional constraints that added to LM_n-algebras are proper models for n-valued Łukasiewicz logic; Cignoli called his discovery *proper n-valued Łukasiewicz algebras*. The LM_n-algebras that are also MV_n-algebras are precisely Cignoli's proper n-valued Łukasiewicz algebras.

Relation to Functional Analysis

MV-algebras were related by Daniele Mundici to approximately finite-dimensional C*-algebras by establishing a bijective correspondence between all isomorphism classes of AF C*-algebras with lattice-ordered dimension group and all isomorphism classes of countable MV algebras. Some instances of this correspondence include:

Countable MV algebra	AF C*-algebra
$\{0, 1\}$	$\mathbb{C}$
$\{0, 1/n, ..., 1\}$	$M_n(\mathbb{C})$, i.e. n×n complex matrices
finite	finite-dimensional
boolean	commutative

In Software

There are multiple frameworks implementing fuzzy logic (type II), and most of them implement what has been called a multi-adjoint logic. This is no more than the implementation of a MV-algebra.

Fuzzy Subalgebra

Fuzzy subalgebras theory is a chapter of fuzzy set theory. It is obtained from an interpretation in a multi-valued logic of axioms usually expressing the notion of subalgebra of a given algebraic structure.

Definition

Consider a first order language for algebraic structures with a monadic predicate symbol S. Then a *fuzzy subalgebra* is a fuzzy model of a theory containing, for any n-ary operation h, the axioms

$$\forall x_1, ..., \forall x_n (S(x_1) \wedge ..... \wedge S(x_n) \rightarrow S(h(x_1, ..., x_n)))$$

and, for any constant c, S(c).

The first axiom expresses the closure of S with respect to the operation h, and the second expresses the fact that c is an element in S. As an example, assume that the valuation structure is defined in [0,1] and denote by $\odot$ the operation in [0,1] used to interpret the conjunction. Then a fuzzy subalgebra of an algebraic structure whose domain is D is defined by a fuzzy subset s : D → [0,1] of D such that, for every $d_1, ..., d_n$ in D, if h is the interpretation of the n-ary operation symbol h, then

- $s(d_1) \odot ... \odot s(d_n) \leq s(\mathbf{h}(d_1, ..., d_n))$

Moreover, if c is the interpretation of a constant c such that s(**c**) = 1.

A largely studied class of fuzzy subalgebras is the one in which the operation $\odot$ coincides with the minimum. In such a case it is immediate to prove the following proposition.

Proposition. A fuzzy subset s of an algebraic structure defines a fuzzy subalgebra if and only if for every λ in [0,1], the closed cut {x ∈ D : s(x)≥ λ} of s is a subalgebra.

Fuzzy Subgroups and Submonoids

The fuzzy subgroups and the fuzzy submonoids are particularly interesting classes of fuzzy subalgebras. In such a case a fuzzy subset s of a monoid (M,•,u) is a fuzzy submonoid if and only if

1. $s(\mathbf{u}) = 1$
2. $s(x) \odot s(y) \leq s(x \cdot y)$

where u is the neutral element in A.

Given a group G, a fuzzy subgroup of G is a fuzzy submonoid s of G such that

- $s(x) \leq s(x^{-1}).\backslash$

It is possible to prove that the notion of fuzzy subgroup is strictly related with the notions of fuzzy equivalence. In fact, assume that S is a set, G a group of transformations in S and (G,s) a fuzzy subgroup of G. Then, by setting

- e(x,y) = Sup{s(h) : h is an element in G such that h(x) = y}

we obtain a fuzzy equivalence. Conversely, let e be a fuzzy equivalence in S and, for every transformation h of S, set

- s(h)= Inf{e(x,h(x)): x∈S}.

Then s defines a fuzzy subgroup of transformation in S. In a similar way we can relate the fuzzy submonoids with the fuzzy orders.

De Morgan Algebra

In mathematics, a De Morgan algebra (named after Augustus De Morgan, a British mathematician and logician) is a structure $A = (A, \vee, \wedge, 0, 1, \neg)$ such that:

- $(A, \vee, \wedge, 0, 1)$ is a boundeddistributive lattice, and

- $\neg$ is a De Morgan involution: $\neg(x \wedge y) = \neg x \vee \neg y$ and $\neg\neg x = x$. (i.e. an involution that additionally satisfies De Morgan's laws)

In a De Morgan algebra, the laws

- $\neg x \vee x = 1$ (law of the excluded middle), and

- $\neg x \wedge x = 0$ (law of noncontradiction)

do not always hold. In the presence of the De Morgan laws, either law implies the other, and an algebra which satisfies them becomes a Boolean algebra.

Remark: It follows that $\neg(x \vee y) = \neg x \wedge \neg y$, $\neg 1 = 0$ and $\neg 0 = 1$ (e.g. $\neg 1 = \neg 1 \vee 0 = \neg 1 \vee \neg\neg 0 = \neg(1 \wedge \neg 0) = \neg\neg 0 = 0$). Thus $\neg$ is a dual automorphism.

If the lattice is defined in terms of the order instead, i.e. $(A, \leq)$ is a bounded partial order with a least upper bound and greatest lower bound for every pair of elements, and the meet and join operations so defined satisfy the distributive law, then the complementation can also be defined as an involutive anti-automorphism, that is, a structure $A = (A, \leq, \neg)$ such that:

- $(A, \leq)$ is a boundeddistributive lattice, and

- $\neg\neg x = x$, and

- $x \leq y \rightarrow \neg y \leq \neg x.$

De Morgan algebras were introduced by Grigore Moisil around 1935. although without the restriction of having a 0 and an 1. They were then variously called quasi-boolean algebras in the Polish school, e.g. by Rasiowa and also distributive i-lattices by J. A. Kalman. (i-lattice being an abbreviation for lattice with involution.) They have been further studied in the Argentian algebraic logic school of Antonio Monteiro.

De Morgan algebras are important for the study of the mathematical aspects of fuzzy logic. The standard fuzzy algebra $F = ([0, 1], \max(x, y), \min(x, y), 0, 1, 1 - x)$ is an example of a De Morgan algebra where the laws of excluded middle and noncontradiction do not hold.

Another example is Dunn's 4-valued logic, in which *false<neither-true-nor-false<true* and *false<both-true-and-false<true*, while *neither-true-nor-false* and *both-true-and-false* are not comparable.

Kleene Algebra

If a De Morgan algebra additionally satisfies $x \wedge \neg x \le y \vee \neg y$, it is called a Kleene algebra. This notion has also been called a normal i-lattice by Kalman.

Examples of Kleene algebras in the sense defined above include: lattice-ordered groups, Post algebras and Łukasiewicz algebras.Boolean algebras also meet this definition of Kleene algebra. The simplest Kleene algebra that is not Boolean is Kleene's three-valued logic K_3. K_3 made its first appearance in Kleene's *On notation for ordinal numbers* (1938). The algebra was named after Kleene by Brignole and Monteiro.

Related Notions

De Morgan algebra is not the only plausible way to generalize the Boolean algebra. Another way is to keep $\neg x \wedge x = 0$ (i.e. the law of noncontradiction) but to drop the law of the excluded middle and the law of double negation. This approach (called *semicomplementation*) is well-defined even for a [meet] semilattice; if the set of semicomplements has a greatest element it is usually called pseudocomplement. If the pseudocomplement thus defined satisfies the law of the excluded middle, the resulting algebra is also Boolean. However, if only the weaker law $\neg x \vee \neg\neg x = 1$ is required, this results in Stone algebras. More generally, both De Morgan and Stone algebras are proper subclasses of Ockham algebras.

Residuated Lattice

In abstract algebra, a residuated lattice is an algebraic structure that is simultaneously a lattice $x \le y$ and a monoid $x \bullet y$ which admits operations $x \backslash z$ and z/y loosely analogous to division or implication when $x{\bullet}y$ is viewed as multiplication or conjunction respectively. Called respectively right and left residuals, these operations coincide when the monoid is commutative. The general concept was introduced by Ward and Dilworth in 1939. Examples, some of which existed prior to the general concept, include Boolean algebras, Heyting algebras, residuated Boolean algebras, relation algebras, and MV-algebras. Residuated semilattices omit the meet operation $\wedge$, for example Kleene algebras and action algebras.

Definition

In mathematics, a residuated lattice is an algebraic structure $L = (L, \leq, \bullet, I)$ such that

(i) $(L, \leq)$ is a lattice.

(ii) $(L, \bullet, I)$ is a monoid.

(iii) For all z there exists for every x a greatest y, and for every y a greatest x, such that $x{\bullet}y \leq z$ (the residuation properties).

In (iii), the "greatest y", being a function of z and x, is denoted $x{\backslash}z$ and called the right residual of z by x, thinking of it as what remains of z on the right after "dividing" z on the left by x. Dually the "greatest x" is denoted z/y and called the left residual of z by y. An equivalent more formal statement of (iii) that uses these operations to name these greatest values is

(iii)' for all x, y, z in L, $\quad y \leq x{\backslash}z \iff x{\bullet}y \leq z \iff x \leq z/y$.

As suggested by the notation the residuals are a form of quotient. More precisely, for a given x in L, the unary operations $x{\bullet}$ and $x{\backslash}$ are respectively the lower and upper adjoints of a Galois connection on L, and dually for the two functions ${\bullet}y$ and $/y$. By the same reasoning that applies to any Galois connection, we have yet another definition of the residuals, namely,

$$x{\bullet}(x{\backslash}y) \leq y \leq x{\backslash}(x{\bullet}y), \text{ and}$$

$$(y/x){\bullet}x \leq y \leq (y{\bullet}x)/x,$$

together with the requirement that $x{\bullet}y$ be monotone in x and y. (When axiomatized using (iii) or (iii)' monotonicity becomes a theorem and hence not required in the axiomatization.) These give a sense in which the functions $x{\bullet}$ and $x{\backslash}$ are pseudoinverses or adjoints of each other, and likewise for ${\bullet}x$ and $/x$.

This last definition is purely in terms of inequalities, noting that monotonicity can be axiomatized as $x{\bullet}y \leq (x{\vee}z){\bullet}y$ and similarly for the other operations and their arguments. Moreover, any inequality $x \leq y$ can be expressed equivalently as an equation, either $x{\wedge}y = x$ or $x{\vee}y = y$. This along with the equations axiomatizing lattices and monoids then yields a purely equational definition of residuated lattices, provided the requisite operations are adjoined to the signature $(L, \leq, \bullet, I)$ thereby expanding it to $(L, \wedge, \vee, \bullet, I, /, \backslash)$. When thus organized, residuated lattices form an equational class or variety, whose homomorphisms respect the residuals as well as the lattice and monoid operations. Note that distributivity $x\bullet(y\vee z) = (x\bullet y) \vee (x\bullet z)$ and $x\bullet 0 = 0$ are consequences of these axioms and so do not need to be made part of the definition. This necessary distributivity of $\bullet$ over $\vee$ does not in general entail distributivity of $\wedge$ over $\vee$, that is, a residuated lattice need not be a distributive lattice. However it does do so when $\bullet$ and $\wedge$ are the same operation, a special case of residuated lattices called a Heyting algebra.

Alternative notations for $x\bullet y$ include $x\circ y$, $x;y$ (relation algebra), and $x\otimes y$ (linear logic). Alternatives for I include e and 1'. Alternative notations for the residuals are $x \to y$ for $x\backslash y$ and $y \leftarrow x$ for y/x, suggested by the similarity between residuation and implication in logic, with the multiplication of the monoid understood as a form of conjunction that need not be commutative. When the monoid is commutative the two residuals coincide. When not commutative, the intuitive meaning of the monoid as conjunction and the residuals as implications can be understood as having a temporal quality: $x\bullet y$ means *x and then y*, $x \to y$ means *had x* (in the past) *then y* (now), and $y \leftarrow x$ means *if-ever x* (in the future) *then y* (at that time), as illustrated by the natural language example at the end of the examples.

Examples

One of the original motivations for the study of residuated lattices was the lattice of ideals of a ring. Given a ring R, the ideals of R, denoted $\mathrm{Id}(R)$, forms a complete lattice with set intersection acting as the meet operation and "ideal addition" acting as the join operation. The monoid operation $\bullet$ is given by "ideal multiplication", and the element R of $\mathrm{Id}(R)$ acts as the identity for this operation. Given two ideals A and B in $\mathrm{Id}(R)$, the residuals are given by

$$A / B := \{r \in R \mid rB \subseteq A\}$$
$$A / B := \{r \in R \mid rB \subseteq A\}$$

It is worth noting that $\{0\}/B$ and $B\backslash\{0\}$ are respectively the left and right annihilators of B. This residuation is related to the *conductor* (or *transporter*) in commutative algebra written as $(A:B)=A/B$. One difference in usage is that B need not be an ideal of R: it may just be a subset.

Boolean algebras and Heyting algebras are commutative residuated lattices in which $x\bullet y = x\wedge y$ (whence the unit I is the top element 1 of the algebra) and both residuals $x\backslash y$ and y/x are the same operation, namely implication $x \to y$. The second example is quite general since Heyting algebras include all finite distributive lattices, as well as all chains or total orders forming a complete lattice, for example the unit interval $[0,1]$ in the real line, or the integers and $\pm$.

The structure ($\mathbf{Z}$, *min*, *max*, +, 0, −, −) (the integers with subtraction for both residuals) is a commutative residuated lattice such that the unit of the monoid is not the greatest element (indeed there is no least or greatest integer), and the multiplication of the monoid is not the meet operation of the lattice. In this example the inequalities are equalities because − (subtraction) is not merely the adjoint or pseudoinverse of + but the true inverse. Any totally ordered group under addition such as the rationals or the reals can be substituted for the integers in this example. The nonnegative portion of any of these examples is an example provided *min* and *max* are interchanged and − is replaced by monus, defined (in this case) so that x-y = 0 when $x \leq y$ and otherwise is ordinary subtraction.

A more general class of examples is given by the Boolean algebra of all binary relations on a set X, namely the power set of X^2, made a residuated lattice by taking the monoid multiplication • to be composition of relations and the monoid unit to be the identity relation $\mathbf{I}$ on X consisting of all pairs (x,x) for x in X. Given two relations R and S on X, the right residual $R\backslash S$ of S by R is the binary relation such that $x(R\backslash S)y$ holds just when for all z in X, zRx implies zSy (notice the connection with implication). The left residual is the mirror image of this: $y(S/R)x$ holds just when for all z in X, xRz implies ySz.

This can be illustrated with the binary relations < and > on {0,1} in which 0 < 1 and 1 > 0 are the only relationships that hold. Then $x(>\backslash<)y$ holds just when $x = 1$, while $x(</>)y$ holds just when $y = 0$, showing that residuation of < by > is different depending on whether we residuate on the right or the left. This difference is a consequence of the difference between <•> and >•<, where the only relationships that hold are 0(<•>)0 (since 0<1>0) and 1(>•<)1 (since 1>0<1). Had we chosen ≤ and ≥ instead of < and >, ≥\≤ and ≤/≥ would have been the same because ≤•≥ = ≥•≤, both of which always hold between all x and y (since $x{\leq}1{\geq}y$ and $x{\geq}0{\leq}y$).

The Boolean algebra 2^{Σ^*} of all formal languages over an alphabet (set) Σ forms a residuated lattice whose monoid multiplication is language concatenation LM and whose monoid unit $\mathbf{I}$ is the language {ε} consisting of just the empty string ε. The right residual $M\backslash L$ consists of all words w over Σ such that $Mw \subseteq L$. The left residual L/M is the same with wM in place of Mw.

The residuated lattice of all binary relations on X is finite just when X is finite, and commutative just when X has at most one element. When X is empty the algebra is the degenerate Boolean algebra in which 0 = 1 = $\mathbf{I}$. The residuated lattice of all languages on Σ is commutative just when Σ has at most one letter. It is finite just when Σ is empty, consisting of the two languages 0 (the empty language {}) and the monoid unit $\mathbf{I}$ = {ε} = 1.

The examples forming a Boolean algebra have special properties treated in the article on residuated Boolean algebras.

In natural language residuated lattices formalize the logic of "and" when used with its noncommutative meaning of "and then." Setting x = *bet*, y = *win*, z = *rich*, we can read $x{\bullet}y \leq z$ as "bet and then win entails rich." By the axioms this is equivalent to $y \leq x{\rightarrow}z$

meaning "win entails had bet then rich", and also to $x \le z\!\leftarrow\!y$ meaning "bet entails if-ever win then rich." Humans readily detect such non-sequiturs as "bet entails had win then rich" and "win entails if-ever bet then rich" as both being equivalent to the wishful thinking "win and then bet entails rich." Humans do not so readily detect that Peirce's law $((P\!\rightarrow\!Q)\!\rightarrow\!P)\!\rightarrow\!P$ is a tautology, giving an interesting situation where humans exhibit more proficiency with nonclassical reasoning than classical.

Residuated Semilattice

A residuated semilattice is defined almost identically for residuated lattices, omitting just the meet operation ∧. Thus it is an algebraic structure L = (L, ∨, •, 1, /, \) satisfying all the residuated lattice equations as specified above except those containing an occurrence of the symbol ∧. The option of defining $x \le y$ as $x\!\wedge\!y = x$ is then not available, leaving only the other option $x\!\vee\!y = y$ (or any equivalent thereof).

Any residuated lattice can be made a residuated semilattice simply by omitting ∧. Residuated semilattices arise in connection with action algebras, which are residuated semilattices that are also Kleene algebras, for which ∧ is ordinarily not required.

Residuated Boolean Algebra

In mathematics, a residuated Boolean algebra is a residuated lattice whose lattice structure is that of a Boolean algebra. Examples include Boolean algebras with the monoid taken to be conjunction, the set of all formal languages over a given alphabet Σ under concatenation, the set of all binary relations on a given set X under relational composition, and more generally the power set of any equivalence relation, again under relational composition. The original application was to relation algebras as a finitely axiomatized generalization of the binary relation example, but there exist interesting examples of residuated Boolean algebras that are not relation algebras, such as the language example.

Definition

A residuated Boolean algebra is an algebraic structure $(L, \wedge, \vee, \neg, 0, 1, \bullet, \mathbf{I}, \backslash, /)$ such that

(i) $(L, \wedge, \vee, \bullet, \mathbf{I}, \backslash, /)$ is a residuated lattice, and

(ii) $(L, \wedge, \vee, \neg, 0, 1)$ is a Boolean algebra.

An equivalent signature better suited to the relation algebra application is $(L, \wedge, \vee, \neg, 0, 1, \bullet, \mathbf{I}, \rhd, \lhd)$ where the unary operations $x\backslash$ and $x\rhd$ are intertranslatable in the manner of De Morgan's laws via

$$x\backslash y = \neg(x\rhd\neg y), \quad x\rhd y = \neg(x\backslash\neg y), \quad \text{and dually } /y \text{ and } \lhd y \text{ as}$$

$$x/y = \neg(\neg x \triangleleft y), \quad x \triangleleft y = \neg(\neg x/y),$$

with the residuation axioms in the residuated lattice article reorganized accordingly (replacing z by $\neg z$) to read

$$(x \triangleright z) \wedge y = 0 \iff (x \bullet y) \wedge z = 0 \iff (z \triangleleft y) \wedge x = 0$$

This De Morgan dual reformulation is motivated and discussed in more detail in the section below on conjugacy.

Since residuated lattices and Boolean algebras are each definable with finitely many equations, so are residuated Boolean algebras, whence they form a finitely axiomatizable variety.

Examples

1. Any Boolean algebra, with the monoid multiplication • taken to be conjunction and both residuals taken to be material implication $x \to y$. Of the remaining 15 binary Boolean operations that might be considered in place of conjunction for the monoid multiplication, only five meet the monotonicity requirement, namely 0, 1, x, y, and $x \vee y$. Setting $y = z = 0$ in the residuation axiom $y \le x \backslash z \iff x \bullet y \le z$, we have $0 \le x \backslash 0 \iff x \bullet 0 \le 0$, which is falsified by taking $x = 1$ when $x \bullet y = 1$, x, or $x \vee y$. The dual argument for z/y rules out $x \bullet y = y$. This just leaves $x \bullet y = 0$ (a constant binary operation independent of x and y), which satisfies almost all the axioms when the residuals are both taken to be the constant operation $x/y = x \backslash y = 1$. The axiom it fails is $x \bullet I = x = I \bullet x$, for want of a suitable value for I. Hence conjunction is the only binary Boolean operation making the monoid multiplication that of a residuated Boolean algebra.

2. The power set 2^{X^2} made a Boolean algebra as usual with $\cap$, $\cup$ and complement relative to X^2, and made a monoid with relational composition. The monoid unit I is the identity relation $\{(x,x) \mid x \in X\}$. The right residual $R \backslash S$ is defined by $x(R \backslash S)y$ if and only if for all z in X, zRx implies zSy. Dually the left residual S/R is defined by $y(S/R)x$ if and only if for all z in X, xRz implies ySz.

3. The power set 2^{Σ^*} made a Boolean algebra as for example 2, but with language concatenation for the monoid. Here the set Σ is used as an alphabet while Σ^* denotes the set of all finite (including empty) words over that alphabet. The concatenation LM of languages L and M consists of all words uv such that $u \in L$ and $v \in M$. The monoid unit is the language $\{\epsilon\}$ consisting of just the empty word ϵ. The right residual $M \backslash L$ consists of all words w over Σ such that $Mw \subseteq L$. The left residual L/M is the same with wM in place of Mw.

Conjugacy

The De Morgan duals $\triangleright$ and $\triangleleft$ of residuation arise as follows. Among residuated lattices,

Boolean algebras are special by virtue of having a complementation operation $\neg$. This permits an alternative expression of the three inequalities

$$y \leq x \backslash z \iff x \bullet y \leq z \iff x \leq z/y$$

in the axiomatization of the two residuals in terms of disjointness, via the equivalence $x \leq y \iff x \wedge \neg y = 0$. Abbreviating $x \wedge y = 0$ to $x \# y$ as the expression of their disjointness, and substituting $\neg z$ for z in the axioms, they become with a little Boolean manipulation

$$\neg(x \backslash \neg z) \# y \iff x \bullet y \# z \iff \neg(\neg z/y) \# x$$

Now $\neg(x \backslash \neg z)$ is reminiscent of De Morgan duality, suggesting that $x \backslash$ be thought of as a unary operation f, defined by $f(y) = x \backslash y$, that has a De Morgan dual $\neg f(\neg y)$, analogous to $\forall x \varphi(x) = \neg \exists x \neg \varphi(x)$. Denoting this dual operation as $x \triangleright$, we define $x \triangleright z$ as $\neg(x \backslash \neg z)$. Similarly we define another operation $z \triangleleft y$ as $\neg(\neg z/y)$. By analogy with $x \backslash$ as the residual operation associated with the operation $x \bullet$, we refer to $x \triangleright$ as the conjugate operation, or simply conjugate, of $x \bullet$. Likewise $\triangleleft y$ is the conjugate of $\bullet y$. Unlike residuals, conjugacy is an equivalence relation between operations: if f is the conjugate of g then g is also the conjugate of f, i.e. the conjugate of the conjugate of f is f. Another advantage of conjugacy is that it becomes unnecessary to speak of right and left conjugates, that distinction now being inherited from the difference between $x \bullet$ and $\bullet x$, which have as their respective conjugates $x \triangleright$ and $\triangleleft x$. (But this advantage accrues also to residuals when $x \backslash$ is taken to be the residual operation to $x \bullet$.)

All this yields (along with the Boolean algebra and monoid axioms) the following equivalent axiomatization of a residuated Boolean algebra.

$$y \# x \triangleright z \iff x \bullet y \# z \iff x \# z \triangleleft y$$

With this signature it remains the case that this axiomatization can be expressed as finitely many equations.

Converse

In examples 2 and 3 it can be shown that $x \triangleright I = I \triangleleft x$. In example 2 both sides equal the converse $x^{\smile}$ of x, while in example 3 both sides are I when x contains the empty word and 0 otherwise. In the former case $x^{\smile\smile} = x$. This is impossible for the latter because $x \triangleright I$ retains hardly any information about x. Hence in example 2 we can substitute $x^{\smile}$ for x in $x \triangleright I = x^{\smile} = I \triangleleft x$ and cancel (soundly) to give

$$x^{\smile} \triangleright I = x = I \triangleleft x^{\smile}.$$

$x^{\smile\smile} = x$ can be proved from these two equations. Tarski's notion of a relation algebra can be defined as a residuated Boolean algebra having an operation $x^{\smile}$ satisfying these two equations.

The cancellation step in the above is not possible for example 3, which therefore is not a relation algebra, $x\breve{}$ being uniquely determined as $x \triangleright I$.

Consequences of this axiomatization of converse include $x^{\breve{}\breve{}} = x$, $\neg(x\breve{}) = (\neg x)\breve{}$, $(x \vee y)\breve{} = x\breve{} \vee y\breve{}$, and $(x \bullet y)\breve{} = y\breve{} \bullet x\breve{}$.

References

- Didier Dubois, Henri M. Prade, ed. (2000). Fundamentals of fuzzy sets. The Handbooks of Fuzzy Sets Series. 7. Springer. ISBN 978-0-7923-7732-0.

- Ulrich Höhle, Stephen Ernest Rodabaugh, ed. (1999). Mathematics of fuzzy sets: logic, topology, and measure theory. The Handbooks of Fuzzy Sets Series. 3. Springer. ISBN 978-0-7923-8388-8.

- Kerre, E.E. (2001). B. Reusch; K-H. Temme, eds. "A first view on the alternatives of fuzzy set theory". Computational Intelligence in Theory and Practice. Heidelberg: Physica-Verlag: 55–72. ISBN 3-7908-1357-5.

- Pawlak, Zdzisław (1991). Rough Sets: Theoretical Aspects of Reasoning About Data. Dordrecht: Kluwer Academic Publishing. ISBN 0-7923-1472-7.

- Burgin, M. (2011), Theory of Named Sets, Mathematics Research Developments, Nova Science Pub Inc, ISBN 978-1-61122-788-8

- Lavinia Corina Ciungu (2013). Non-commutative Multiple-Valued Logic Algebras. Springer. pp. vii–viii. ISBN 978-3-319-01589-7.

- Kalle Kaarli; Alden F. Pixley (21 July 2000). Polynomial Completeness in Algebraic Systems. CRC Press. pp. 297–. ISBN 978-1-58488-203-9.

- Nikolaos Galatos, Peter Jipsen, Tomasz Kowalski, and Hiroakira Ono (2007), Residuated Lattices. An Algebraic Glimpse at Substructural Logics, Elsevier, ISBN 978-0-444-52141-5.

Many-valued Logic: An Overview

The logical aspects of fuzzy sets include the principle of bivalence, three-valued logic, canonical form and probilistic logic. Three-valued logic is a logical system in which there are three truth-values, true, false and an unknown value. Three valued logic is in contrast to Boolean logic, which is binary and has only true and false. The aspects explicated are of vital importance, and help in the better comprehension of fuzzy logic.

Many-valued Logic

In logic, a many-valued logic (also multi- or multiple-valued logic) is a propositional calculus in which there are more than two truth values. Traditionally, in Aristotle's logical calculus, there were only two possible values (i.e., "true" and "false") for any proposition. Classical two-valued logic may be extended to n-valued logic for n greater than 2. Those most popular in the literature are three-valued (e.g., Łukasiewicz›s and Kleene's, which accept the values "true", "false", and "unknown"), the finite-valued (finitely-many valued) with more than three values, and the infinite-valued (infinitely-many valued), such as fuzzy logic and probability logic.

History

The first known classical logician who didn't fully accept the law of excluded middle was Aristotle (who, ironically, is also generally considered to be the first classical logician and the "father of logic"). Aristotle admitted that his laws did not all apply to future events (*De Interpretatione*, *ch. IX*), but he didn't create a system of multi-valued logic to explain this isolated remark. Until the coming of the 20th century, later logicians followed Aristotelian logic, which includes or assumes the law of the excluded middle.

The 20th century brought back the idea of multi-valued logic. The Polish logician and philosopher Jan Łukasiewicz began to create systems of many-valued logic in 1920, using a third value, "possible", to deal with Aristotle's paradox of the sea battle. Meanwhile, the American mathematician, Emil L. Post (1921), also introduced the formulation of additional truth degrees with $n \geq 2$, where n are the truth values. Later, Jan Łukasiewicz and Alfred Tarski together formulated a logic on n truth values where $n \geq 2$. In 1932 Hans Reichenbach formulated a logic of many truth values where $n \rightarrow \infty$. Kurt Gödel in 1932 showed that intuitionistic logic is not a finitely-many valued logic, and defined a system of Gödel logics intermediate between classical and intuitionistic logic; such logics are known as intermediate logics.

Kleene (Strong) K_3 and Priest Logic P_3

Kleene's "(strong) logic of indeterminacy" K_3 (sometimes K_3^S)) and Priest's "logic of paradox" add a third "undefined" or "indeterminate" truth value I. The truth functions for negation (¬), conjunction (∧), disjunction (∨), implication (→K), and biconditional (↔K) are given by:

¬	
T	F
I	I
F	T

∧	T	I	F
T	T	I	F
I	I	I	F
F	F	F	F

∨	T	I	F
T	T	T	T
I	T	I	I
F	T	I	F

→K	T	I	F
T	T	I	F
I	T	I	I
F	T	T	T

↔K	T	I	F
T	T	I	F
I	I	I	I
F	F	I	T

The difference between the two logics lies in how tautologies are defined. In K_3 only T is a *designated truth value*, while in P_3 both T and I are (a logical formula is considered a tautology if it evaluates to a designated truth value). In Kleene's logic I can be interpreted as being "underdetermined", being neither true nor false, while in Priest's logic I can be interpreted as being "overdetermined", being both true and false. K_3 does not have any tautologies, while P_3 has the same tautologies as classical two-valued logic.

Bochvar's Internal Three-valued Logic (Also Known as Kleene's Weak Three-valued Logic)

Another logic is Bochvar's "internal" three-valued logic (B_3^I) also called Kleene's weak three-valued logic. Except for negation and biconditional, its truth tables are all different from the above.

∧+	T	I	F
T	T	I	F
I	I	I	I
F	F	I	F

∨+	T	I	F
T	T	I	T
I	I	I	I
F	T	I	F

→+	T	I	F
T	T	I	F
I	I	I	I
F	T	I	T

The intermediate truth value in Bochvar's "internal" logic can be described as "contagious" because it propagates in a formula regardless of the value of any other variable.

Belnap Logic (B_4)

Belnap's logic B_4 combines K_3 and P_3. The overdetermined truth value is here denoted as B and the underdetermined truth value as N.

f	
T	F
B	B
N	N
F	T

f	T	B	N	F
T	T	B	N	F
B	B	B	F	F
N	N	F	N	F
F	F	F	F	F

f	T	B	N	F
T	T	T	T	T
B	T	B	T	B
N	T	T	N	N
F	T	B	N	F

Relation to Classical Logic

Logics are usually systems intended to codify rules for preserving some semantic property of propositions across transformations. In classical logic, this property is "truth." In a valid argument, the truth of the derived proposition is guaranteed if the premises are jointly true, because the application of valid steps preserves the property. However, that property doesn't have to be that of "truth"; instead, it can be some other concept.

Multi-valued logics are intended to preserve the property of designationhood (or being designated). Since there are more than two truth values, rules of inference may be intended to preserve more than just whichever corresponds (in the relevant sense) to truth. For example, in a three-valued logic, sometimes the two greatest truth-values (when they are represented as e.g. positive integers) are designated and the rules of inference preserve these values. Precisely, a valid argument will be such that the value of the premises taken jointly will always be less than or equal to the conclusion.

For example, the preserved property could be *justification*, the foundational concept of intuitionistic logic. Thus, a proposition is not true or false; instead, it is justified or flawed. A key difference between justification and truth, in this case, is that the law of excluded middle doesn't hold: a proposition that is not flawed is not necessarily justified; instead, it's only not proven that it's flawed. The key difference is the determinacy of the preserved property: One may prove that P is justified, that P is flawed, or be unable to prove either. A valid argument preserves justification across transformations, so a proposition derived from justified propositions is still justified. However, there are proofs in classical logic that depend upon the law of excluded middle; since that law is not usable under this scheme, there are propositions that cannot be proven that way.

Applications

Known applications of many-valued logic can be roughly classified into two groups. The first group uses many-valued logic domain to solve binary problems more efficiently. For example, a well-known approach to represent a multiple-output Boolean function is to treat its output part as a single many-valued variable and convert it to a single-output characteristic function. Other applications of many-valued logic include design of Programmable Logic Arrays (PLAs) with input decoders, optimization of finite state machines, testing, and verification.

The second group targets the design of electronic circuits which employ more than two discrete levels of signals, such as many-valued memories, arithmetic circuits, Field Programmable Gate Arrays (FPGA) etc. Many-valued circuits have a number of theoretical advantages over standard binary circuits. For example, the interconnect on and off chip can be reduced if signals in the circuit assume four or more levels rather than only two. In memory design, storing two instead of one bit of information per memory cell doubles the density of the memory in the same die size. Applications using arithmetic circuits often benefit from using alternatives to binary number systems. For example, residue and redundant number systems can reduce or eliminate the ripple-through carries which are involved in normal binary addition or subtraction, resulting in high-speed arithmetic operations. These number systems have a natural implementation using many-valued circuits. However, the practicality of these potential advantages heavily depends on the availability of circuit realizations, which must be compatible or competitive with present-day standard technologies.

Research Venues

An IEEEInternational Symposium on Multiple-Valued Logic (ISMVL) has been held annually since 1970. It mostly caters to applications in digital design and verification.

T-norm

In mathematics, a t-norm (also T-norm or, unabbreviated, triangular norm) is a kind of binary operation used in the framework of probabilistic metric spaces and in multi-valued logic, specifically in fuzzy logic. A t-norm generalizes intersection in a lattice and conjunction in logic. The name *triangular norm* refers to the fact that in the framework of probabilistic metric spaces t-norms are used to generalize triangle inequality of ordinary metric spaces.

Definition

A t-norm is a function T: $[0, 1] \times [0, 1] \to [0, 1]$ which satisfies the following properties:

- Commutativity: $T(a, b) = T(b, a)$
- Monotonicity: $T(a, b) \leq T(c, d)$ if $a \leq c$ and $b \leq d$
- Associativity: $T(a, T(b, c)) = T(T(a, b), c)$
- The number 1 acts as identity element: $T(a, 1) = a$

Since a t-norm is a binary algebraic operation on the interval $[0, 1]$, infix algebraic notation is also common, with the t-norm usually denoted by $*$.

The defining conditions of the t-norm are exactly those of the partially ordered Abelian monoid on the real unit interval $[0, 1]$. *(Cf. ordered group.)* The monoidal operation of any partially ordered Abelian monoid L is therefore by some authors called a *triangular norm on L*.

Motivations and Applications

T-norms are a generalization of the usual two-valued logical conjunction, studied by classical logic, for fuzzy logics. Indeed, the classical Boolean conjunction is both commutative and associative. The monotonicity property ensures that the degree of truth of conjunction does not decrease if the truth values of conjuncts increase. The requirement that 1 be an identity element corresponds to the interpretation of 1 as *true* (and consequently 0 as *false*). Continuity, which is often required from fuzzy conjunction as well, expresses the idea that, roughly speaking, very small changes in truth values of conjuncts should not macroscopically affect the truth value of their conjunction.

T-norms are also used to construct the intersection of fuzzy sets or as a basis for aggregation operators. In probabilistic metric spaces, t-norms are used to generalize triangle inequality of ordinary metric spaces. Individual t-norms may of course frequently occur in further disciplines of mathematics, since the class contains many familiar functions.

Prominent Examples

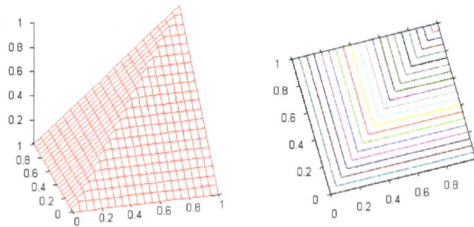

Graph of the minimum t-norm (3D and contours)

- Minimum t-norm $T_{min}(a,b) = \min\{a,b\}$, also called the Gödel t-norm, as it is the standard semantics for conjunction in Gödel fuzzy logic. Besides that, it occurs in most t-norm based fuzzy logics as the standard semantics for weak conjunction. It is the pointwise largest t-norm.

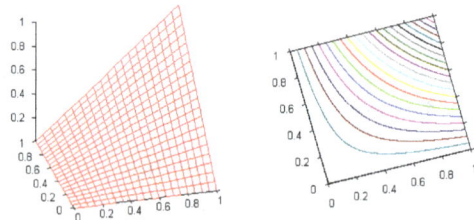

Graph of the product t-norm

- Product t-norm $T_{prod}(a,b) = a \cdot b$ (the ordinary product of real numbers). Besides other uses, the product t-norm is the standard semantics for strong conjunction in product fuzzy logic. It is a strict Archimedean t-norm.

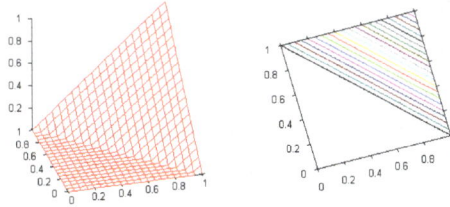

Graph of the Łukasiewicz t-norm

- Łukasiewicz t-norm $\top_{Luk}(a,b) = \max\{0, a+b-1\}$. The name comes from the fact that the t-norm is the standard semantics for strong conjunction in Łukasiewicz fuzzy logic. It is a nilpotent Archimedean t-norm, pointwise smaller than the product t-norm.

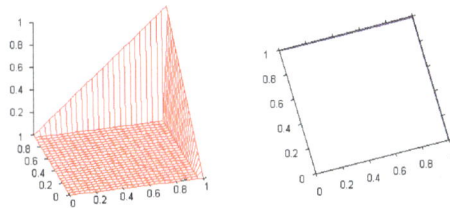

Graph of the drastic t-norm. The function is discontinuous at the lines $0 < x = 1$ and $0 < y = 1$.

- Drastic t-norm

$$\top_D(a,b) = \begin{cases} b & \text{if } a = 1 \\ a & \text{if } b = 1 \\ 0 & \text{otherwise.} \end{cases}$$

The name reflects the fact that the drastic t-norm is the pointwise smallest t-norm. It is a right-continuous Archimedean t-norm.

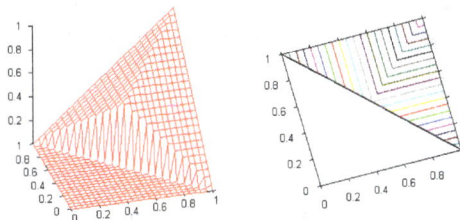

Graph of the nilpotent minimum. The function is discontinuous at the line $0 < x = 1 - y < 1$.

- Nilpotent minimum

$$T_{nM}(a,b) = \begin{cases} \min(a,b) & \text{if } a+b>1 \\ 0 & \text{otherwise} \end{cases}$$

is a standard example of a t-norm which is left-continuous, but not continuous. Despite its name, the nilpotent minimum is not a nilpotent t-norm.

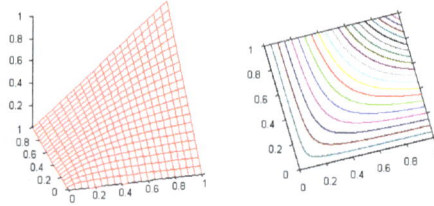

Graph of the Hamacher product

- Hamacher Product

$$T_{H_0}(a,b) = \begin{cases} 0 & \text{if } a=b=0 \\ \dfrac{ab}{a+b-ab} & \text{otherwise} \end{cases}$$

is a strict Archimedean t-norm, and an important representative of the parametric classes of Hamacher t-norms and Schweizer–Sklar t-norms.

Properties of t-norms

The drastic t-norm is the pointwise smallest t-norm and the minimum is the pointwise largest t-norm:

$$T_D(a,b) \leq T(a,b) \leq T_{min}(a,b), \text{for any t-norm } T \text{ and all } a, b \text{ in } [0, 1].$$

For every t-norm T, the number 0 acts as null element: T(a, 0) = 0 for all a in [0, 1].

A t-norm T has zero divisors if and only if it has nilpotent elements; each nilpotent element of T is also a zero divisor of T. The set of all nilpotent elements is an interval [0, a] or [0, a), for some a in [0, 1].

Properties of Continuous t-norms

Although real functions of two variables can be continuous in each variable without being continuous on [0, 1]², this is not the case with t-norms: a t-norm T is continuous if and only if it is continuous in one variable, i.e., if and only if the functions $f_y(x)$ = T(x, y) are continuous for each y in [0, 1]. Analogous theorems hold for left- and right-continuity of a t-norm.

A continuous t-norm is Archimedean if and only if 0 and 1 are its only idempotents.

A continuous Archimedean t-norm is strict if 0 is its only nilpotent element; otherwise it is nilpotent. By definition, moreover, a continuous Archimedean t-norm T is nilpotent if and only if *eachx* < 1 is a nilpotent element of T. Thus with a continuous Archimedean t-norm T, either all or none of the elements of (0, 1) are nilpotent. If it is the case that all elements in (0, 1) are nilpotent, then the t-norm is isomorphic to the Łukasiewicz t-norm; i.e., there is a strictly increasing function f such that

$$\top (x, y) = f^{-1}(\top_{\text{Luk}} (f(x), f(y))).$$

If on the other hand it is the case that there are no nilpotent elements of T, the t-norm is isomorphic to the product t-norm. In other words, all nilpotent t-norms are isomorphic, the Łukasiewicz t-norm being their prototypical representative; and all strict t-norms are isomorphic, with the product t-norm as their prototypical example. The Łukasiewicz t-norm is itself isomorphic to the product t-norm undercut at 0.25, i.e., to the function $p(x, y) = \max(0.25, x \cdot y)$ on $[0.25, 1]^2$.

For each continuous t-norm, the set of its idempotents is a closed subset of $[0, 1]$. Its complement — the set of all elements which are not idempotent — is therefore a union of countably many non-overlapping open intervals. The restriction of the t-norm to any of these intervals (including its endpoints) is Archimedean, and thus isomorphic either to the Łukasiewicz t-norm or the product t-norm. For such x, y that do not fall into the same open interval of non-idempotents, the t-norm evaluates to the minimum of x and y. These conditions actually give a characterization of continuous t-norms, called the Mostert–Shields theorem, since every continuous t-norm can in this way be decomposed, and the described construction always yields a continuous t-norm. The theorem can also be formulated as follows:

> A t-norm is continuous if and only if it is isomorphic to an ordinal sum of the minimum, Łukasiewicz, and product t-norm.

A similar characterization theorem for non-continuous t-norms is not known (not even for left-continuous ones), only some non-exhaustive methods for the construction of t-norms have been found.

Basic Properties of Residua

If $\Rightarrow$ is the residuum of a left-continuous t-norm $\top$,, then

$$(x \Rightarrow y) = \sup\{z \mid \top (z, x) \le y\}.$$

Consequently, for all x, y in the unit interval,

$$(x \Rightarrow y) = 1 \text{ if and only if} (x \Rightarrow y) = 1 \textit{if and only if}$$

and

$$(1 \Rightarrow y) = y.$$

If $*$ is a left-continuous t-norm and $\Rightarrow$ its residuum, then

$$\min(x, y) \quad \geq \qquad\qquad x * (x \Rightarrow y)$$
$$\max(x, y) \quad = \quad \min((x \Rightarrow y) \Rightarrow y, (y \Rightarrow x) \Rightarrow x).$$

If $*$ is continuous, then equality holds in the former.

Residua of Prominent Left-continuous t-norms

If $x \leq y$, then R$(x, y) = 1$ for any residuum R. The following table therefore gives the values of prominent residua only for $x > y$.

Residuum of the	Name	Value for $x>y$	Graph
Minimum t-norm	Standard Gödel implication	y	Standard Gödel implication. The function is discontinuous at the line $y = x < 1$.
Product t-norm	Goguen implication	y / x	Goguen implication. The function is discontinuous at the point $x = y = 0$.
Łukasiewicz t-norm	Standard Łukasiewicz implication	$1 - x + y$	Standard Łukasiewicz implication.

Nilpotent minimum		$\max(1 - x, y)$	
			Residuum of the nilpotent minimum. The function is discontinuous at the line $0 < y = x < 1$.

T-conorms

T-conorms (also called S-norms) are dual to t-norms under the order-reversing operation which assigns $1 - x$ to x on $[0, 1]$. Given a t-norm, the complementary conorm is defined by

$$\perp (a,b) = 1 - \top (1-a, 1-b).$$

This generalizes De Morgan's laws.

It follows that a t-conorm satisfies the following conditions, which can be used for an equivalent axiomatic definition of t-conorms independently of t-norms:

- Commutativity: $\perp(a, b) = \perp(b, a)$

- Monotonicity: $\perp(a, b) \leq \perp(c, d)$ if $a \leq c$ and $b \leq d$

- Associativity: $\perp(a, \perp(b, c)) = \perp(\perp(a, b), c)$

- Identity element: $\perp(a, 0) = a$

T-conorms are used to represent logical disjunction in fuzzy logic and union in fuzzy set theory.

T-norm Fuzzy Logics

T-norm fuzzy logics are a family of non-classical logics, informally delimited by having a semantics which takes the real unit interval $[0, 1]$ for the system of truth values and functions called t-norms for permissible interpretations of conjunction. They are mainly used in applied fuzzy logic and fuzzy set theory as a theoretical basis for approximate reasoning.

T-norm fuzzy logics belong in broader classes of fuzzy logics and many-valued logics. In order to generate a well-behaved implication, the t-norms are usually required to be left-continuous; logics of left-continuous t-norms further belong in the class of substructural logics, among which they are marked with the validity of the *law of prelin-*

earity, $(A \rightarrow B) \vee (B \rightarrow A)$. Both propositional and first-order (or higher-order) t-norm fuzzy logics, as well as their expansions by modal and other operators, are studied. Logics which restrict the t-norm semantics to a subset of the real unit interval (for example, finitely valued Łukasiewicz logics) are usually included in the class as well.

Important examples of t-norm fuzzy logics are monoidal t-norm logic MTL of all left-continuous t-norms, basic logic BL of all continuous t-norms, product fuzzy logic of the product t-norm, or the nilpotent minimum logic of the nilpotent minimum t-norm. Some independently motivated logics belong among t-norm fuzzy logics, too, for example Łukasiewicz logic (which is the logic of the Łukasiewicz t-norm) or Gödel–Dummett logic (which is the logic of the minimum t-norm).

Motivation

As members of the family of fuzzy logics, t-norm fuzzy logics primarily aim at generalizing classical two-valued logic by admitting intermediary truth values between 1 (truth) and 0 (falsity) representing *degrees* of truth of propositions. The degrees are assumed to be real numbers from the unit interval [0, 1]. In propositional t-norm fuzzy logics, propositional connectives are stipulated to be truth-functional, that is, the truth value of a complex proposition formed by a propositional connective from some constituent propositions is a function (called the *truth function* of the connective) of the truth values of the constituent propositions. The truth functions operate on the set of truth degrees (in the standard semantics, on the [0, 1] interval); thus the truth function of an n-ary propositional connective c is a function $F_c \colon [0, 1]^n \rightarrow [0, 1]$. Truth functions generalize truth tables of propositional connectives known from classical logic to operate on the larger system of truth values.

T-norm fuzzy logics impose certain natural constraints on the truth function of conjunction. The truth function $* \colon [0,1]^2 \rightarrow [0,1]$ $x * y = y * x$ of conjunction is assumed to satisfy the following conditions:

- *Commutativity*, that is, $x * y = y * x$ for all x and y in [0, 1]. This expresses the assumption that the order of fuzzy propositions is immaterial in conjunction, even if intermediary truth degrees are admitted.

- *Associativity*, that is, $(x * y) * z = x * (y * z)$ for all x, y, and z in [0, 1]. This expresses the assumption that the order of performing conjunction is immaterial, even if intermediary truth degrees are admitted.

- *Monotony*, that is, if $x \leq y$ then $x * z \leq y * z$ for all x, y, and z in [0, 1]. This expresses the assumption that increasing the truth degree of a conjunct should not decrease the truth degree of the conjunction.

- *Neutrality of 1*, that is, $1 * x = x$ for all x in [0, 1]. This assumption corresponds to regarding the truth degree 1 as full truth, conjunction with which does not decrease the truth value of the other conjunct. Together with the previous conditions this condition ensures that also $0 * x = 0$ for all x in $[0, 1]$, which corresponds to regarding the truth degree 0 as full falsity, conjunction with which is always fully false.

- *Continuity* of the function $*$ (the previous conditions reduce this requirement to the continuity in either argument). Informally this expresses the assumption that microscopic changes of the truth degrees of conjuncts should not result in a macroscopic change of the truth degree of their conjunction. This condition, among other things, ensures a good behavior of (residual) implication derived from conjunction; to ensure the good behavior, however, *left*-continuity (in either argument) of the function $*$ is sufficient. In general t-norm fuzzy logics, therefore, only left-continuity of $*$ is required, which expresses the assumption that a microscopic *decrease* of the truth degree of a conjunct should not macroscopically decrease the truth degree of conjunction.

These assumptions make the truth function of conjunction a left-continuous t-norm, which explains the name of the family of fuzzy logics (*t-norm based*). Particular logics of the family can make further assumptions about the behavior of conjunction (for example, Gödel logic requires its idempotence) or other connectives (for example, the logic IMTL requires the involutiveness of negation).

All left-continuous t-norms $*$ have a unique residuum, that is, a binary function $\Rightarrow$ such that for all x, y, and z in $[0, 1]$,

$$x * y \leq z \text{ if and only if } x \leq y \Rightarrow z.$$

The residuum of a left-continuous t-norm can explicitly be defined as

$$(x \Rightarrow y) = \sup\{z \mid z * x \leq y\}.$$

This ensures that the residuum is the pointwise largest function such that for all x and y,

$$x * (x \Rightarrow y) \leq y.$$

The latter can be interpreted as a fuzzy version of the modus ponens rule of inference. The residuum of a left-continuous t-norm thus can be characterized as the weakest function that makes the fuzzy modus ponens valid, which makes it a suitable truth function for implication in fuzzy logic. Left-continuity of the t-norm is the necessary and sufficient condition for this relationship between a t-norm conjunction and its residual implication to hold.

Truth functions of further propositional connectives can be defined by means of the t-norm and its residuum, for instance the residual negation $\neg x = (x \Rightarrow 0)$ or bi-residual equivalence $x \Leftrightarrow y = (x \Rightarrow y) * (y \Rightarrow x)$. Truth functions of propositional connectives may also be introduced by additional definitions: the most usual ones are the minimum (which plays a role of another conjunctive connective), the maximum (which plays a role of a disjunctive connective), or the Baaz Delta operator, defined in $[0, 1]$ as $\Delta x = 1$ if $x = 1$ and $\Delta x = 0$ otherwise. In this way, a left-continuous t-norm, its residuum, and the truth functions of additional propositional connectives determine the truth values of complex propositional formulae in $[0, 1]$.

Formulae that always evaluate to 1 are called *tautologies* with respect to the given left-continuous t-norm *, or *0 *tautologies*. The set of all *0 tautologies is called the *logic* of the t-norm *, as these formulae represent the laws of fuzzy logic (determined by the t-norm) which hold (to degree 1) regardless of the truth degrees of atomic formulae. Some formulae are tautologies with respect to a larger class of left-continuous t-norms; the set of such formulae is called the logic of the class. Important t-norm logics are the logics of particular t-norms or classes of t-norms, for example:

- Łukasiewicz logic is the logic of the Łukasiewicz t-norm $x * y = \max(x + y - 1, 0)$

- Gödel–Dummett logic is the logic of the minimum t-norm $x * y = \min(x, y)$

- Product fuzzy logic is the logic of the product t-norm $x * y = x \cdot y$

- Monoidal t-norm logic MTL is the logic of (the class of) *all* left-continuous t-norms

- Basic fuzzy logic BL is the logic of (the class of) all *continuous* t-norms

It turns out that many logics of particular t-norms and classes of t-norms are axiomatizable. The completeness theorem of the axiomatic system with respect to the corresponding t-norm semantics on $[0, 1]$ is then called the *standard completeness* of the logic. Besides the standard real-valued semantics on $[0, 1]$, the logics are sound and complete with respect to general algebraic semantics, formed by suitable classes of prelinear commutative bounded integral residuated lattices.

History

Some particular t-norm fuzzy logics have been introduced and investigated long before the family was recognized (even before the notions of fuzzy logic or t-norm emerged):

- Łukasiewicz logic (the logic of the Łukasiewicz t-norm) was originally defined by Jan Łukasiewicz (1920) as a three-valued logic; it was later generalized to n-valued (for all finite n) as well as infinitely-many-valued variants, both propositional and first-order.

- Gödel–Dummett logic (the logic of the minimum t-norm) was implicit in Gödel's 1932 proof of infinite-valuedness of intuitionistic logic. Later (1959) it was explicitly studied by Dummett who proved a completeness theorem for the logic.

A systematic study of particular t-norm fuzzy logics and their classes began with Hájek's (1998) monograph *Metamathematics of Fuzzy Logic*, which presented the notion of the logic of a continuous t-norm, the logics of the three basic continuous t-norms (Łukasiewicz, Gödel, and product), and the 'basic' fuzzy logic BL of all continuous t-norms (all of them both propositional and first-order). The book also started the investigation of fuzzy logics as non-classical logics with Hilbert-style calculi, algebraic semantics, and metamathematical properties known from other logics (completeness theorems, deduction theorems, complexity, etc.).

Since then, a plethora of t-norm fuzzy logics have been introduced and their metamathematical properties have been investigated. Some of the most important t-norm fuzzy logics were introduced in 2001, by Esteva and Godo (MTL, IMTL, SMTL, NM, WNM), Esteva, Godo, and Montagna (propositional ŁΠ), and Cintula (first-order ŁΠ).

Semantics

Algebraic semantics is predominantly used for propositional t-norm fuzzy logics, with three main classes of algebras with respect to which a t-norm fuzzy logic L is complete:

- General semantics, formed of all $L - $-algebras — that is, all algebras for which the logic is sound.

- Linear semantics, formed of all *linear* L-algebras — that is, all L-algebras whose lattice order is linear.

- Standard semantics, formed of all *standard* L-algebras — that is, all L-algebras whose lattice reduct is the real unit interval [0, 1] with the usual order. In standard L-algebras, the interpretation of strong conjunction is a left-continuous t-norm and the interpretation of most propositional connectives is determined by the t-norm (hence the names *t-norm-based logics* and *t-norm L-algebras*, which is also used for L-algebras on the lattice [0, 1]). In t-norm logics with additional connectives, however, the real-valued interpretation of the additional connectives may be restricted by further conditions for the t-norm algebra to be called standard: for example, in standard L-algebras of the logic L with involution, the interpretation of the additional involutive negation $\sim$ is required to be the *standard involution* $f_\sim(x) = 1 - x$,

Monoidal t-norm Logic

Monoidal t-norm based logic (or shortly MTL), the logic of left-continuous t-norms, is one of t-norm fuzzy logics. It belongs to the broader class of substructural logics, or logics of residuated lattices; it extends the logic of commutative bounded integral residuated lattices (known as Höhle's monoidal logic, Ono's FL_{ew}, or intuitionistic logic without contraction) by the axiom of prelinearity.

Motivation

T-norms are binary functions on the real unit interval [0, 1] which are often used to represent a conjunction connective in fuzzy logic. Every *left-continuous* t-norm $*$ has a unique residuum, that is, a function $\Rightarrow$ such that for all x, y, and z,

$$x * y \leq z \text{ if and only if } x \leq (y \Rightarrow z).$$

The residuum of a left-continuous t-norm can explicitly be defined as

$$(x \Rightarrow y) = \sup\{z \mid z * x \leq y\}.$$

This ensures that the residuum is the largest function such that for all x and y,

$$x * (x \Rightarrow y) \leq y.$$

The latter can be interpreted as a fuzzy version of the modus ponens rule of inference. The residuum of a left-continuous t-norm thus can be characterized as the weakest function that makes the fuzzy modus ponens valid, which makes it a suitable truth function for implication in fuzzy logic. Left-continuity of the t-norm is the necessary and sufficient condition for this relationship between a t-norm conjunction and its residual implication to hold.

Construction of t-norms

In mathematics, t-norms are a special kind of binary operations on the real unit interval [0, 1]. Various constructions of t-norms, either by explicit definition or by transformation from previously known functions, provide a plenitude of examples and classes of t-norms. This is important, e.g., for finding counter-examples or supplying t-norms with particular properties for use in engineering applications of fuzzy logic. The main ways of construction of t-norms include using *generators*, defining *parametric classes* of t-norms, *rotations*, or *ordinal sums* of t-norms.

Relevant background can be found in the article on t-norms.

Generators of t-norms

The method of constructing t-norms by generators consists in using a unary function (*generator*) to transform some known binary function (most often, addition or multiplication) into a t-norm.

In order to allow using non-bijective generators, which do not have the inverse function, the following notion of *pseudo-inverse function* is employed:

Let $f: [a, b] \to [c, d]$ be a monotone function between two closed subintervals of extended real line. The *pseudo-inverse function* to f is the function $f^{(-1)}: [c, d] \to [a, b]$ defined as

$$f^{(-1)}(y) = \begin{cases} \sup\{x \in [a,b] \mid f(x) < y\} & \text{for } f \text{ non-decreasing} \\ \sup\{x \in [a,b] \mid f(x) > y\} & \text{for } f \text{ non-increasing.} \end{cases}$$

Additive Generators

The construction of t-norms by additive generators is based on the following theorem:

Let $f: [0, 1] \to [0, +\infty]$ be a strictly decreasing function such that $f(1) = 0$ and $f(x) + f(y)$ is in the range of f or equal to $f(0^+)$ or $+\infty$ for all x, y in $[0, 1]$. Then the function $T: [0, 1]^2 \to [0, 1]$ defined as

$$T(x, y) = f^{(-1)}(f(x) + f(y))$$

is a t-norm.

Alternatively, one may avoid using the notion of pseudo-inverse function by having $T(x, y) = f^{-1}\left(\min\left(f(0^+), f(x) + f(y)\right)\right)$. The corresponding residuum can then be expressed as $(x \Rightarrow y) = f^{-1}\left(\max\left(0, f(y) - f(x)\right)\right)$ $(x \Leftrightarrow y) = f^{-1}\left(|f(x) - f(y)|\right)$. And the biresiduum as $(x \Leftrightarrow y) = f^{-1}\left(|f(x) - f(y)|\right)$.

If a t-norm T results from the latter construction by a function f which is right-continuous in 0, then f is called an *additive generator* of T.

Examples:

- The function $f(x) = 1 - x$ for x in $[0, 1]$ is an additive generator of the Łukasiewicz t-norm.

- The function f defined as $f(x) = -\log(x)$ if $0 < x \le 1$ and $f(0) = +\infty$ is an additive generator of the product t-norm.

- The function f defined as $f(x) = 2 - x$ if $0 \le x < 1$ and $f(1) = 0$ is an additive generator of the drastic t-norm.

Basic properties of additive generators are summarized by the following theorem:

Let $f: [0, 1] \to [0, +\infty]$ be an additive generator of a t-norm T. Then:

- T is an Archimedean t-norm.

- T is continuous if and only if f is continuous.

- T is strictly monotone if and only if $f(0) = +\infty$.

- Each element of (0, 1) is a nilpotent element of T if and only if f(0) < +∞.

- The multiple of f by a positive constant is also an additive generator of T.

- T has no non-trivial idempotents. (Consequently, e.g., the minimum t-norm has no additive generator.)

Multiplicative Generators

The isomorphism between addition on [0, +∞] and multiplication on [0, 1] by the logarithm and the exponential function allow two-way transformations between additive and multiplicative generators of a t-norm. If f is an additive generator of a t-norm T, then the function h: [0, 1] → [0, 1] defined as $h(x) = e^{-f(x)}$ is a *multiplicative generator* of T, that is, a function h such that

- h is strictly increasing

- $h(1) = 1$

- $h(x) \cdot h(y)$ is in the range of h or equal to 0 or $h(0+)$ for all x, y in [0, 1]

- h is right-continuous in 0

- $T(x, y) = h^{(-1)}(h(x) \cdot h(y))$.

Vice versa, if h is a multiplicative generator of T, then f: [0, 1] → [0, +∞] defined by $f(x) = -\log(h(x))$ is an additive generator of T.

Parametric Classes of t-norms

Many families of related t-norms can be defined by an explicit formula depending on a parameter p. This section lists the best known parameterized families of t-norms. The following definitions will be used in the list:

- A family of t-norms T_p parameterized by p is *increasing* if $T_p(x, y) \leq T_q(x, y)$ for all x, y in [0, 1] whenever $p \leq q$ (similarly for *decreasing* and *strictly* increasing or decreasing).

- A family of t-norms T_p is *continuous* with respect to the parameter p if

$$\lim_{p \to p_0} T_p = T_{p_0}$$

for all values p_0 of the parameter.

Schweizer–Sklar t-norms

The family of *Schweizer–Sklar t-norms*, introduced by Berthold Schweizer and Abe Sklar in the early 1960s, is given by the parametric definition

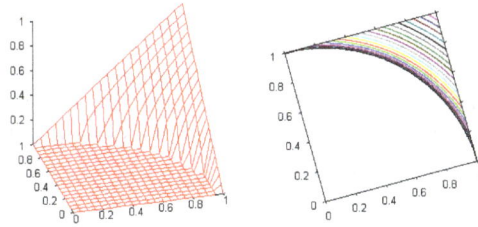

Graph (3D and contours) of the Schweizer–Sklar t-norm with $p = 2$

$$T_p^{SS}(x,y) = \begin{cases} T_{\min}(x,y) & \text{if } p = -\infty \\ (x^p + y^p - 1)^{1/p} & \text{if } -\infty < p < 0 \\ T_{\text{prod}}(x,y) & \text{if } p = 0 \\ (\max(0, x^p + y^p - 1))^{1/p} & \text{if } 0 < p < +\infty \\ T_D(x,y) & \text{if } p = +\infty. \end{cases}$$

A Schweizer–Sklar t-norm T_p^{SS} is

- Archimedean if and only if $p > -\infty$ T_p^{SS}

- Continuous if and only if $p < +\infty$

- Strict if and only if $-\infty < p \leq 0$ (for $p = -1$ it is the Hamacher product)

- Nilpotent if and only if $0 < p < +\infty$ (for $p = 1$ it is the Łukasiewicz t-norm).

The family is strictly decreasing for $p \geq 0$ and continuous with respect to p in $[-\infty, +\infty]$. An additive generator for T_p^{SS} for $-\infty < p < +\infty$ is

$$f_p^{SS}(x) = \begin{cases} -\log x & \text{if } p = 0 \\ \dfrac{1 - x^p}{p} & \text{otherwise.} \end{cases}$$

Hamacher t-norms

The family of *Hamacher t-norms*, introduced by Horst Hamacher in the late 1970s, is given by the following parametric definition for $0 \leq p \leq +\infty$:

$$T_p^{H}(x,y) = \begin{cases} T_D(x,y) & \text{if } p = +\infty \\ 0 & \text{if } p = x = y = 0 \\ \dfrac{xy}{p + (1-p)(x + y - xy)} & \text{otherwise.} \end{cases}$$

The t-norm T_0^H is called the *Hamacher product.*

Hamacher t-norms are the only t-norms which are rational functions. The Hamacher t-norm is strict T_p^H if and only if $p < +\infty$ (for $p = 1$ it is the product t-norm). The family is strictly decreasing and continuous with respect to p. An additive generator of T_p^H for $p < +\infty$ is

Frank t-norms

The family of *Frank t-norms*, introduced by M.J. Frank in the late 1970s, is given by the parametric definition for $0 \le p \le +\infty$ as follows:

$$f_p^H(x) = \begin{cases} \dfrac{1-x}{x} & \text{if } p = 0 \\[2ex] \log \dfrac{p+(1-p)x}{x} & \text{otherwise.} \end{cases}$$

The Frank t-norm $T_p^F(x,y) = \begin{cases} T_{\min}(x,y) & \text{if } p = 0 \\ T_{\text{prod}}(x,y) & \text{if } p = 1 \\ T_{\text{Luk}}(x,y) & \text{if } p = +\infty \\ \log_p\left(1+\dfrac{(p^x-1)(p^y-1)}{p-1}\right) & \text{otherwise.} \end{cases}$ is strict if $p < +\infty$. The

family is strictly decreasing and continuous with respect to p. An additive generator for T_p^F is

$$f_p^F(x) = \begin{cases} -\log x & \text{if } p = 1 \\ 1-x & \text{if } p = +\infty \\ \log \dfrac{p-1}{p^x-1} & \text{otherwise.} \end{cases}$$

Yager t-norms

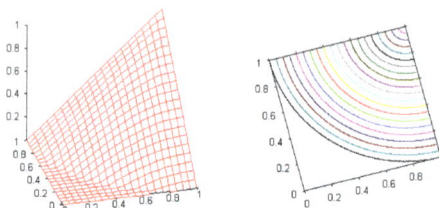

Graph of the Yager t-norm with $p = 2$

The family of *Yager t-norms*, introduced in the early 1980s by Ronald R. Yager, is given for $0 \leq p \leq +\infty$ by

$$T_p^{\mathrm{Y}}(x,y) = \begin{cases} T_{\mathrm{D}}(x,y) & \text{if } p = 0 \\ \max\left(0, 1 - ((1-x)^p + (1-y)^p)^{1/p}\right) & \text{if } 0 < p < +\infty \\ T_{\min}(x,y) & \text{if } p = +\infty \end{cases}$$

The Yager t-norm T_p^{Y} is nilpotent if and only if $0 < p < +\infty$ (for $p = 1$ it is the Łukasiewicz t-norm). The family is strictly increasing and continuous with respect to p. The Yager $p < +\infty$ arises from the Łukasiewicz t-norm by raising its additive generator to the power of p. An additive generator of T_p^{Y} for $0 < p < +\infty$ is

$$f_p^{\mathrm{Y}}(x) = (1-x)^p.$$

Aczél–Alsina t-norms

The family of *Aczél–Alsina t-norms*, introduced in the early 1980s by János Aczél and Claudi Alsina, is given for $0 \leq p \leq +\infty$ by

$$T_p^{\mathrm{AA}}(x,y) = \begin{cases} T_{\mathrm{D}}(x,y) & \text{if } p = 0 \\ e^{-\left(|\log x|^p + |\log y|^p\right)^{1/p}} & \text{if } 0 < p < +\infty \\ T_{\min}(x,y) & \text{if } p = +\infty \end{cases}$$

The Aczél–Alsina t-norm T_p^{AA} is strict if and only if $0 < p < +\infty$ (for $p = 1$ it is the product t-norm). The family is strictly increasing and continuous with respect to p. The Aczél–Alsina t-norm T_p^{AA} for $0 < p < +\infty$ arises from the product t-norm by raising its additive generator to the power of p. An additive generator of T_p^{AA} for $0 < p < +\infty$ is

$$f_p^{\mathrm{AA}}(x) = (-\log x)^p.$$

Dombi t-norms

The family of *Dombi t-norms*, introduced by József Dombi (1982), is given for $0 \leq p \leq +\infty$ by

$$T_p^{\mathrm{D}}(x,y) = \begin{cases} 0 & \text{if } x = 0 \text{ or } y = 0 \\ T_{\mathrm{D}}(x,y) & \text{if } p = 0 \\ T_{\min}(x,y) & \text{if } p = +\infty \\ \dfrac{1}{1 + \left(\left(\dfrac{1-x}{x}\right)^p + \left(\dfrac{1-y}{y}\right)^p\right)^{1/p}} & \text{otherwise.} \end{cases}$$

The Dombi t-norm T_p^D is strict if and only if $0 < p < +\infty$ (for $p = 1$ it is the Hamacher product). The family is strictly increasing and continuous with respect to p. The Dombi t-norm T_p^D for $0 < p < +\infty$ arises from the Hamacher product t-norm by raising its additive generator to the power of p. An additive generator of T_p^D for $0 < p < +\infty$ is

$$f_p^D(x) = \left(\frac{1-x}{x}\right)^p.$$

Sugeno–Weber t-norms

The family of *Sugeno–Weber t-norms* was introduced in the early 1980s by Siegfried Weber; the dual t-conorms were defined already in the early 1970s by Michio Sugeno. It is given for $-1 \le p \le +\infty$ by

$$T_p^{SW}(x,y) = \begin{cases} T_D(x,y) & \text{if } p = -1 \\ \max\left(0, \dfrac{x+y-1+pxy}{1+p}\right) & \text{if } -1 < p < +\infty \\ T_{prod}(x,y) & \text{if } p = +\infty \end{cases}$$

The Sugeno–Weber t-norm T_p^{SW} is nilpotent if and only if $-1 < p < +\infty$ (for $p = 0$ it is the Łukasiewicz t-norm). The family is strictly increasing and continuous with respect to p. An additive generator of T_p^{SW} for $0 < p < +\infty$ [sic] is

$$f_p^{SW}(x) = \begin{cases} 1-x & \text{if } p = 0 \\ 1 - \log_{1+p}(1+px) & \text{otherwise.} \end{cases}$$

Ordinal Sums

The ordinal sum constructs a t-norm from a family of t-norms, by shrinking them into disjoint subintervals of the interval [0, 1] and completing the t-norm by using the minimum on the rest of the unit square. It is based on the following theorem:

Let T_i for i in an index set I be a family of t-norms and (a_i, b_i) a family of pairwise disjoint (non-empty) open subintervals of [0, 1]. Then the function $T: [0, 1]^2 \to [0, 1]$ defined as

$$T(x,y) = \begin{cases} a_i + (b_i - a_i) \cdot T_i\left(\dfrac{x-a_i}{b_i-a_i}, \dfrac{y-a_i}{b_i-a_i}\right) & \text{if } x, y \in [a_i, b_i]^2 \\ \min(x,y) & \text{otherwise} \end{cases}$$

is a t-norm.

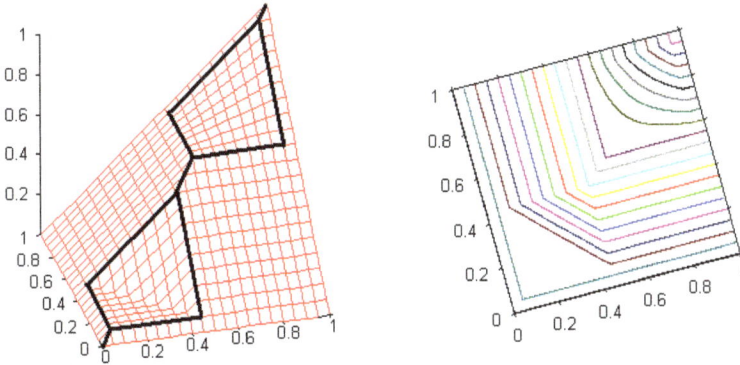

Ordinal sum of the Łukasiewicz t-norm on the interval [0.05, 0.45] and the product t-norm on the interval [0.55, 0.95]

The resulting t-norm is called the *ordinal sum* of the summands (T_i, a_i, b_i) for i in I, denoted by

$$T = \bigoplus_{i \in I}(T_i, a_i, b_i),$$

or $(T_1, a_1, b_1) \oplus \ldots \oplus (T_n, a_n, b_n)$ if I is finite.

Ordinal sums of t-norms enjoy the following properties:

- Each t-norm is a trivial ordinal sum of itself on the whole interval [0, 1].

- The empty ordinal sum (for the empty index set) yields the minimum t-norm $T_{\min}$. Summands with the minimum t-norm can arbitrarily be added or omitted without changing the resulting t-norm.

- It can be assumed without loss of generality that the index set is countable, since the real line can only contain at most countably many disjoint subintervals.

- An ordinal sum of t-norm is continuous if and only if each summand is a continuous t-norm. (Analogously for left-continuity.)

- An ordinal sum is Archimedean if and only if it is a trivial sum of one Archimedean t-norm on the whole unit interval.

- An ordinal sum has zero divisors if and only if for some index i, $a_i = 0$ and T_i has zero divisors. (Analogously for nilpotent elements.)

If $T = \bigoplus_{i \in I}(T_i, a_i, b_i)$ is a left-continuous t-norm, then its residuum R is given as follows:

$$R(x, y) = \begin{cases} 1 & \text{if } x \leq y \\ a_i + (b_i - a_i) \cdot R_i \left(\dfrac{x - a_i}{b_i - a_i}, \dfrac{y - a_i}{b_i - a_i} \right) & \text{if } a_i < y < x \leq b_i \\ y & \text{otherwise.} \end{cases}$$

where R_i is the residuum of T_i, for each i in I.

Ordinal Sums of Continuous t-Norms

The ordinal sum of a family of continuous t-norms is a continuous t-norm. By the Mostert–Shields theorem, every continuous t-norm is expressible as the ordinal sum of Archimedean continuous t-norms. Since the latter are either nilpotent (and then isomorphic to the Łukasiewicz t-norm) or strict (then isomorphic to the product t-norm), each continuous t-norm is isomorphic to the ordinal sum of Łukasiewicz and product t-norms.

Important examples of ordinal sums of continuous t-norms are the following ones:

- Dubois–Prade t-norms, introduced by Didier Dubois and Henri Prade in the early 1980s, are the ordinal sums of the product t-norm on $[0, p]$ for a parameter p in $[0, 1]$ and the (default) minimum t-norm on the rest of the unit interval. The family of Dubois–Prade t-norms is decreasing and continuous with respect to p..

- Mayor–Torrens t-norms, introduced by Gaspar Mayor and Joan Torrens in the early 1990s, are the ordinal sums of the Łukasiewicz t-norm on $[0, p]$ for a parameter p in $[0, 1]$ and the (default) minimum t-norm on the rest of the unit interval. The family of Mayor–Torrens t-norms is decreasing and continuous with respect to p..

Rotations

The construction of t-norms by rotation was introduced by Sándor Jenei (2000). It is based on the following theorem:

Let T be a left-continuous t-norm without zero divisors, $N: [0, 1] \to [0, 1]$ the function that assigns $1 - x$ to x and $t = 0.5$. Let T_1 be the linear transformation of T into $[t, 1]$ and $R_{T_1}(x, y) = \sup \{z \mid T_1(z, x) \leq y\}$. Then the function

$$T_{\text{rot}} = \begin{cases} T_1(x, y) & \text{if } x, y \in (t, 1] \\ N(R_{T_1}(x, N(y))) & \text{if } x \in (t, 1] \text{ and } y \in [0, t] \\ N(R_{T_1}(y, N(x))) & \text{if } x \in [0, t] \text{ and } y \in (t, 1] \\ 0 & \text{if } x, y \in [0, t] \end{cases}$$

is a left-continuous t-norm, called the *rotation* of the t-norm T.

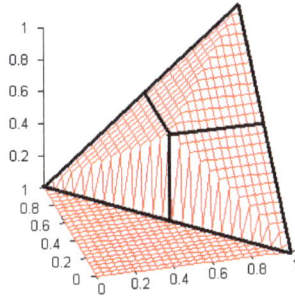

The nilpotent minimum as a rotation of the minimum t-norm

Geometrically, the construction can be described as first shrinking the t-norm T to the interval [0.5, 1] and then rotating it by the angle $2\pi/3$ in both directions around the line connecting the points (0, 0, 1) and (1, 1, 0).

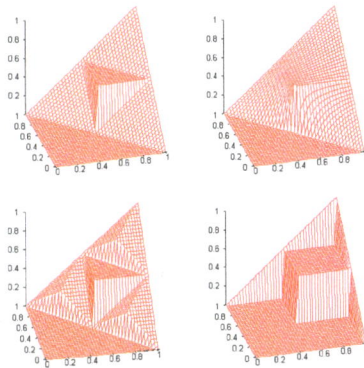

Rotations of the Łukasiewicz, product, nilpotent minimum, and drastic t-norm

The theorem can be generalized by taking for N any *strong negation*, that is, an involutive strictly decreasing continuous function on [0, 1], and for t taking the unique fixed point of N.

The resulting t-norm enjoys the following *rotation invariance* property with respect to N:

$$T(x, y) \leq z \text{ if and only if } T(y, N(z)) \leq N(x) \text{ for all } x, y, z \text{ in } [0, 1].$$

The negation induced by T_{rot} is the function N, that is, $N(x) = R_{rot}(x, 0)$ for all x, where R_{rot} is the residuum of T_{rot}.

Principle of Bivalence

In logic, the semantic principle (or law) of bivalence states that every declarative sentence expressing a proposition (of a theory under inspection) has exactly one truth value, either true or false. A logic satisfying this principle is called a two-valued logic or bivalent logic.

In formal logic, the principle of bivalence becomes a property that a semantics may or may not possess. It is not the same as the law of excluded middle, however, and a semantics may satisfy that law without being bivalent.

The principle of bivalence is studied in philosophical logic to address the question of which natural-language statements have a well-defined truth value. Sentences which predict events in the future, and sentences which seem open to interpretation, are particularly difficult for philosophers who hold that the principle of bivalence applies to all declarative natural-language statements.Many-valued logics formalize ideas that a realistic characterization of the notion of consequence requires the admissibility of premises which, owing to vagueness, temporal or quantum indeterminacy, or reference-failure, cannot be considered classically bivalent. Reference failures can also be addressed by free logics.

Relationship with the Law of the Excluded Middle

The principle of bivalence is related to the law of excluded middle though the latter is a syntactic expression of the language of a logic of the form "P ∨ ¬P". The difference between the principle and the law is important because there are logics which validate the law but which do not validate the principle. For example, the three-valuedLogic of Paradox (LP) validates the law of excluded middle, but not the law of non-contradiction, ¬(P ∧ ¬P), and its intended semantics is not bivalent. In classical two-valued logic both the law of excluded middle and the law of non-contradiction hold.

Many modern logic programming systems replace the law of the excluded middle with the concept of negation as failure. The programmer may wish to add the law of the excluded middle by explicitly asserting it as true; however, it is not assumed *a priori*.

Classical Logic

The intended semantics of classical logic is bivalent, but this is not true of every semantics for classical logic. In Boolean-valued semantics (for classical propositional logic), the truth values are the elements of an arbitrary Boolean algebra, "true" corresponds to the maximal element of the algebra, and "false" corresponds to the minimal element. Intermediate elements of the algebra correspond to truth values other than "true" and "false". The principle of bivalence holds only when the Boolean algebra is taken to be the two-element algebra, which has no intermediate elements.

Assigning Boolean semantics to classical predicate calculus requires that the model be a complete Boolean algebra because the universal quantifier maps to the infimum operation, and the existential quantifier maps to the supremum; this is called a Boolean-valued model. All finite Boolean algebras are complete.

Suszko's Thesis

In order to justify his claim that true and false are the only logical values, Suszko (1977) observes that every structural Tarskian many-valued propositional logic can be provided with a bivalent semantics.

Criticisms

Future Contingents

A famous example is the *contingent sea battle* case found in Aristotle's work, *De Interpretatione*, chapter 9:

> Imagine P refers to the statement "There will be a sea battle tomorrow."

The principle of bivalence here asserts:

> Either it is true that there will be a sea battle tomorrow, or it is false that there will be a sea battle tomorrow.

Aristotle hesitated to embrace bivalence for such future contingents;Chrysippus, the Stoic logician, did embrace bivalence for this and all other propositions. The controversy continues to be of central importance in both the philosophy of time and the philosophy of logic.

One of the early motivations for the study of many-valued logics has been precisely this issue. In the early 20th century, the Polish formal logician Jan Łukasiewicz proposed three truth-values: the true, the false and the *as-yet-undetermined*. This approach was later developed by Arend Heyting and L. E. J. Brouwer.

Issues such as this have also been addressed in various temporal logics, where one can assert that "*Eventually*, either there will be a sea battle tomorrow, or there won't be." (Which is true if "tomorrow" eventually occurs.)

Vagueness

Such puzzles as the Sorites paradox and the related continuum fallacy have raised doubt as to the applicability of classical logic and the principle of bivalence to concepts that may be vague in their application. Fuzzy logic and some other multi-valued logics have been proposed as alternatives that handle vague concepts better. Truth (and falsity) in fuzzy logic, for example, comes in varying degrees. Consider the following statement in the circumstance of sorting apples on a moving belt:

> This apple is red.

Upon observation, the apple is an undetermined color between yellow and red, or it is motled both colors. Thus the color falls into neither category " red " nor " yellow ", but these are the only categories available to us as we sort the apples. We might say it is "50% red". This could be rephrased: it is 50% true that the apple is red. Therefore, P is 50% true, and 50% false. Now consider:

This apple is red and it is not-red.

In other words, P and not-P. This violates the law of noncontradiction and, by extension, bivalence. However, this is only a partial rejection of these laws because P is only partially true. If P were 100% true, not-P would be 100% false, and there is no contradiction because P and not-P no longer holds.

However, the law of the excluded middle is retained, because P and not-P implies P or not-P, since "or" is inclusive. The only two cases where P and not-P is false (when P is 100% true or false) are the same cases considered by two-valued logic, and the same rules apply.

Example of a 3-valued logic applied to vague (undetermined) cases: Kleene 1952 (§64, pp. 332–340) offers a 3-valued logic for the cases when algorithms involving partial recursive functions may not return values, but rather end up with circumstances «u» = undecided. He lets «t» = «true», «f» = «false», «u» = «undecided» and redesigns all the propositional connectives. He observes that:

We were justified intuitionistically in using the classical 2-valued logic, when we were using the connectives in building primitive and general recursive predicates, since there is a decision procedure for each general recursive predicate; i.e. the law of the excluded middle is proved intuitionistically to apply to general recursive predicates.

Now if Q(x) is a partial recursive predicate, there is a decision procedure for Q(x) on its range of definition, so the law of the excluded middle or excluded "third" (saying that, Q(x) is either t or f) applies intuitionistically on the range of definition. But there may be no algorithm for deciding, given x, whether Q(x) is defined or not. [...] Hence it is only classically and not intuitionistically that we have a law of the excluded fourth (saying that, for each x, Q(x) is either t, f, or u).

The third "truth value" u is thus not on par with the other two t and f in our theory. Consideration of its status will show that we are limited to a special kind of truth table".

The following are his "strong tables":

~Q			QVR	R	t	f	u	Q&R	R	t	f	u	Q⊃R	R	t	f	u	Q=R	R	t	f	u
Q	t	f	Q	t	t	t	t	Q	t	t	f	u	Q	t	t	f	u	Q	t	t	f	u
	f	t		f	t	f	u		f	f	f	f		f	t	t	t		f	f	t	u
	u	u		u	t	u	u		u	u	f	u		u	t	u	u		u	u	u	u

For example, if a determination cannot be made as to whether an apple is red or not-red, then the truth value of the assertion Q: " This apple is red " is " u ". Likewise, the truth value of the assertion R " This apple is not-red " is " u ". Thus the AND of these into the assertion Q AND R, i.e. " This apple is red AND this apple is not-red " will, per the tables, yield " u ". And, the assertion Q OR R, i.e. " This apple is red OR this apple is not-red " will likewise yield " u ".

Three-valued Logic

In logic, a three-valued logic (also trinary logic, trivalent, ternary, or trilean, sometimes abbreviated 3VL) is any of several many-valued logic systems in which there are three truth values indicating *true*, *false* and some indeterminate third value. This is contrasted with the more commonly known bivalent logics (such as classical sentential or Boolean logic) which provide only for *true* and *false*. Conceptual form and basic ideas were initially created by Jan Łukasiewicz and C. I. Lewis. These were then re-formulated by Grigore Moisil in an axiomatic algebraic form, and also extended to n-valued logics in 1945.

Representation of Values

As with bivalent logic, truth values in ternary logic may be represented numerically using various representations of the ternary numeral system. A few of the more common examples are:

- in balanced ternary, each digit has one of 3 values: −1, 0, or +1; these values may also be simplified to −, 0, +, respectively;

- in the redundant binary representation, each digit can have a value of −1, 0, 0/1 (the value 0/1 has two different representations);

- in the ternary numeral system, each digit is a *trit* (trinary digit) having a value of: 0, 1, or 2;

- in the skew binary number system, only most-significant non-zero digit has a value 2, and the remaining digits have a value of 0 or 1;

- 1 for *true*, 2 for *false*, and 0 for *unknown, unknowable/undecidable, irrelevant,* or *both*;

- 0 for *false*, 1 for *true*, and a third non-integer "maybe" symbol such as ?, #, ½, or xy.

Inside a ternary computer, ternary values are represented by ternary signals.

This article mainly illustrates a system of ternary propositional logic using the truth values {false, unknown, true}, and extends conventional Boolean connectives to a trivalent context. Ternary predicate logics exist as well; these may have readings of the quantifier different from classical (binary) predicate logic and may include alternative quantifiers as well.

Logics

Where Boolean logic has $2^2 = 4$ unary operators, the addition of a third value in ternary logic leads to a total of $3^3 = 27$ distinct operators on a single input value. Similarly, where

Boolean logic has $2^{2^2} = 16$ distinct binary operators (operators with 2 inputs), ternary logic has $3^{3^2} = 19{,}683$ such operators. Where we can easily name a significant fraction of the Boolean operators (not, and, or, nand, nor, exclusive or, equivalence, implication), it is unreasonable to attempt to name all but a small fraction of the possible ternary operators.

Kleene and Priest Logics

Below is a set of truth tables showing the logic operations for Kleene's "strong logic of indeterminacy" and Priest's "logic of paradox".

(F, false; U, unknown; T, true)

NOT(A):

A	¬A
F	T
U	U
T	F

AND(A, B), $A \wedge B$:

$A \wedge B$	B = F	U	T
A = F	F	F	F
U	F	U	U
T	F	U	T

OR(A, B), $A \vee B$:

$A \vee B$	B = F	U	T
A = F	F	U	T
U	U	U	T
T	T	T	T

(−1, false; 0, unknown; +1, true)

NEG(A):

A	¬A
−1	+1
0	0
+1	−1

MIN(A, B), $A \wedge B$:

$A \wedge B$	B = −1	0	+1
A = −1	−1	−1	−1
0	−1	0	0
+1	−1	0	+1

MAX(A, B), $A \vee B$:

$A \vee B$	B = −1	0	+1
A = −1	−1	0	+1
0	0	0	+1
+1	+1	+1	+1

In these truth tables, the *unknown* state can be thought of as neither true nor false in Kleene logic, or thought of as both true and false in Priest logic. The difference lies in the definition of tautologies. Where Kleene logic's only designated truth value is T, Priest logic's designated truth values are both T and U. In Kleene logic, the knowledge of whether any particular *unknown* state secretly represents *true* or *false* at any moment in time is not available. However, certain logical operations can yield an unambiguous result, even if they involve at least one *unknown* operand. For example, since *true* OR *true* equals *true*, and *true* OR *false* also equals *true*, one can infer that *true* OR *unknown* equals *true*, as well. In this example, since either bivalent state could be underlying the *unknown* state, but either state also yields the same result, a definitive *true* results in all three cases.

If numeric values, e.g. balanced ternary values, are assigned to *false*, *unknown* and *true* such that *false* is less than *unknown* and *unknown* is less than *true*, then A AND B AND C... = MIN(A, B, C ...) and A OR B OR C ... = MAX(A, B, C...).

Material implication for Kleene logic can be defined as:

$$A \to B \overset{def}{=} \text{NOT}(A) \text{ OR } B,$$

, and its truth table is

$\text{IMP}_K(A, B), \text{OR}(\neg A, B)$:

A → B	B = T	U	F
A = T	T	U	F
A = U	T	U	U
A = F	T	T	T

$\text{IMP}_K(A, B), \text{MAX}(-A, B)$:

A → B	B = 0	−1	+1*
A = +1	+1	0	−1
A = 0	+1	0	0
A = −1	+1	+1	+1

which differs from that for Łukasiewicz logic (described below).

Kleene logic has no tautologies (valid formulas) because whenever all of the atomic components of a well-formed formula are assigned the value Unknown, the formula itself must also have the value Unknown. (And the only *designated* truth value for Kleene logic is True.) However, the lack of valid formulas does not mean that it lacks valid arguments and/or inference rules. An argument is semantically valid in Kleene logic if, whenever (for any interpretation/model) all of its premises are True, the conclusion must also be True. (Note that the Logic of Paradox (LP) has the same truth tables as Kleene logic, but it has two *designated* truth values instead of one; these are: True and Both (the analogue of Unknown), so that LP does have tautologies but it has fewer valid inference rules.)

Łukasiewicz Logic

The Łukasiewicz Ł3 has the same tables for AND, OR, and NOT as the Kleene logic given above, but differs in its definition of implication in that "unknown implies unknown" is true.

$IMP_Ł(A, B)$			
$A \rightarrow B$		B	
T		U	F
	T	T	U	F
A	U	T	T	U
	F	T	T	T

$IMP_Ł(A, B)$			
$A \rightarrow B$		B	
+1	0	−1	
	+1	+1	0	−1
A	0	+1	+1	0
	−1	+1	+1	+1

In fact, using Łukasiewicz's implication and negation, the other usual connectives may be derived as:

- $A \lor B = (A \rightarrow B) \rightarrow B$

- $A \land B = \neg(\neg A \lor \neg B)$

- $A \leftrightarrow B = (A \rightarrow B) \land (B \rightarrow A)$

It's also possible to derive a few other useful unary operators (first derived by Tarski in 1921):

- $MA = \neg A \rightarrow A$

- $LA = \neg M \neg A$

- $IA = MA \land \neg LA$

They have the following truth tables:

A	MA
F	F
U	T
T	T

A	LA
F	F
U	F
T	T

A	IA
F	F
U	T
T	F

M is read as "it is not false that…" or in the (unsuccessful) Tarski–Łukasiewicz attempt to axiomatize modal logic using a three-valued logic, "it is possible that…" L is read "it is true that…" or "it is necessary that…" Finally I is read "it is unknown that…" or "it is contingent that…"

In Łukasiewicz's Ł3 the designated value is True, meaning that only a proposition having this value everywhere is considered a tautology. For example, $A \to A$ and $A \leftrightarrow A$ are tautologies in Ł3 and also in classical logic. Not all tautologies of classical logic lift to Ł3 "as is". For example, the law of excluded middle, $A \lor \neg A$, and the law of non-contradiction, $\neg(A \land \neg A)$ are not tautologies in Ł3. However, using the operator **I** defined above, it is possible to state tautologies that are their analogues:

- $A \lor IA \lor \neg A$ (law of excluded fourth)

- $\neg(A \land \neg IA \land \neg A)$ (extended contradiction principle).

Modular Algebras

Some 3VL modular algebras have been introduced more recently, motivated by circuit problems rather than philosophical issues:

- Cohn algebra

- Pradhan algebra

- Dubrova and Muzio algebra

Application in SQL

The database structural query language SQL implements ternary logic as a means of handling comparisons with NULL field content. The original intent of NULL in SQL was to represent missing data in a database, i.e. the assumption that an actual value exists, but that the value is not currently recorded in the database. SQL uses a common fragment of the Kleene K3 logic, restricted to AND, OR, and NOT tables. In SQL, the intermediate value is intended to be interpreted as UNKNOWN. Explicit comparisons with NULL, including that of another NULL yields UNKNOWN. However this choice of semantics is abandoned for some set operations, e.g. UNION or INTERSECT, where NULLs are treated as equal with each other. Critics assert that this inconsistency deprives SQL of intuitive semantics in its treatment of NULLs. The SQL standard defines an optional feature called F571, which adds some unary operators, among which IS UNKNOWN corresponding to the Łukasiewicz I in this article. The addition of IS UN-

KNOWN to the other operators of SQL's three-valued logic makes the SQL three-valued logic functionally complete, meaning its logical operators can express (in combination) any conceivable three-valued logical function.

Łukasiewicz Logic

In mathematics, Łukasiewicz logic is a non-classical, many valued logic. It was originally defined in the early 20th century by Jan Łukasiewicz as a three-valued logic; it was later generalized to n-valued (for all finite n) as well as infinitely-many-valued ($\aleph_0$-valued) variants, both propositional and first-order. The $\aleph_0$-valued version was published in 1930 by Łukasiewicz and Alfred Tarski; consequently it is sometimes called the Łukasiewicz-Tarski logic. It belongs to the classes of t-norm fuzzy logics and substructural logics.

This article presents the Łukasiewicz[-Tarski] logic in its full generality, i.e. as an infinite-valued logic. For an elementary introduction to the three-valued instantiation $Ł_3$.

Language

The propositional connectives of Łukasiewicz logic are *implication* $\rightarrow$, *negation* $\neg$, *equivalence* $\leftrightarrow$, *weak conjunction* $\wedge$, *strong conjunction* $\otimes$, *weak disjunction* $\vee$, *strong disjunction* $\oplus$, and propositional constants $\overline{0}$ and $\overline{1}$. The presence of conjunction and disjunction is a common feature of substructural logics without the rule of contraction, to which Łukasiewicz logic belongs.

Axioms

The original system of axioms for propositional infinite-valued Łukasiewicz logic used implication and negation as the primitive connectives:

$$A \rightarrow (B \rightarrow A)$$

$$(A \rightarrow B) \rightarrow ((B \rightarrow C) \rightarrow (A \rightarrow C))$$

$$((A \rightarrow B) \rightarrow B) \rightarrow ((B \rightarrow A) \rightarrow A)$$

$$(\neg B \rightarrow \neg A) \rightarrow (A \rightarrow B).$$

Propositional infinite-valued Łukasiewicz logic can also be axiomatized by adding the following axioms to the axiomatic system of monoidal t-norm logic:

- *Divisibility:* $(A \wedge B) \rightarrow (A \otimes (A \rightarrow B))$

- *Double negation:* $\neg\neg A \rightarrow A.$

That is, infinite-valued Łukasiewicz logic arises by adding the axiom of double negation to basic t-norm logic BL, or by adding the axiom of divisibility to the logic IMTL.

Finite-valued Łukasiewicz logics require additional axioms.

Finite-valued and Countable-valued Semantics

Using exactly the same valuation formulas as for real-valued semantics Łukasiewicz (1922) also defined (up to isomorphism) semantics over

- any finite set of cardinality $n \geq 2$ by choosing the domain as $\{\, 0, 1/(n-1), 2/(n-1), ..., 1 \,\}$

- any countable set by choosing the domain as $\{\, p/q \mid 0 \leq p \leq q$ where p is a non-negative integer and q is a positive integer $\}$.

General Algebraic Semantics

The standard real-valued semantics determined by the Łukasiewicz t-norm is not the only possible semantics of Łukasiewicz logic. General algebraic semantics of propositional infinite-valued Łukasiewicz logic is formed by the class of all MV-algebras. The standard real-valued semantics is a special MV-algebra, called the *standard MV-algebra*.

Like other t-norm fuzzy logics, propositional infinite-valued Łukasiewicz logic enjoys completeness with respect to the class of all algebras for which the logic is sound (that is, MV-algebras) as well as with respect to only linear ones. This is expressed by the general, linear, and standard completeness theorems:

> The following conditions are equivalent:
>
> - A is provable in propositional infinite-valued Łukasiewicz logic
>
> - A is valid in all MV-algebras (*general completeness*)
>
> - A is valid in all linearly ordered MV-algebras (*linear completeness*)
>
> - A is valid in the standard MV-algebra (*standard completeness*).

Font, Rodriguez and Torrens introduced in 1984 the Wajsberg algebra as an alternative model for the infinite-valued Łukasiewicz logic.

A 1940s attempt by Grigore Moisil to provide algebraic semantics for the n-valued Łukasiewicz logic by means of his Łukasiewicz–Moisil (LM) algebra (which Moisil called *Łukasiewicz algebras*) turned out to be an incorrect model for $n \geq 5$. This issue was made public by Alan Rose in 1956. C. C. Chang's MV-algebra, which is a model for the $\aleph_0$-valued (infinitely-many-valued) Łukasiewicz-Tarski logic, was published in 1958.

For the axiomatically more complicated (finite) n-valued Łukasiewicz logics, suitable algebras were published in 1977 by Revaz Grigolia and called MV_n-algebras. MV_n-algebras are a subclass of LM_n-algebras, and the inclusion is strict for $n \geq 5$. In 1982 Roberto Cignoli published some additional constraints that added to LM_n-algebras produce proper models for n-valued Łukasiewicz logic; Cignoli called his discovery *proper Łukasiewicz algebras*.

Canonical Form

In mathematics and computer science, a canonical, normal, or standardform of a mathematical object is a standard way of presenting that object as a mathematical expression. The distinction between "canonical" and "normal" forms varies by subfield. In most fields, a canonical form specifies a *unique* representation for every object, while a normal form simply specifies its form, without the requirement of uniqueness.

The canonical form of a positive integer in decimal representation is a finite sequence of digits that does not begin with zero.

More generally, for a class of objects on which an equivalence relation is defined, a canonical form consists in the choice of a specific object in each class. For example, Jordan normal form is a canonical form for matrix similarity, and the row echelon form is a canonical form, when one considers as equivalent a matrix and its left product by an invertible matrix.

In computer science, and more specifically in computer algebra, when representing mathematical objects in a computer, there are usually many different ways to represent the same object. In this context, a canonical form is a representation such that every object has a unique representation. Thus, the equality of two objects can easily be tested by testing the equality of their canonical forms. However canonical forms frequently depend on arbitrary choices (like ordering the variables), and this introduces difficulties for testing the equality of two objects resulting on independent computations. Therefore, in computer algebra, *normal form* is a weaker notion: A normal form is a representation such that zero is uniquely represented. This allows testing for equality by putting the difference of two objects in normal form.

Canonical form can also mean a differential form that is defined in a natural (canonical) way.

In computer science, data that has more than one possible representation can often be canonicalized into a completely unique representation called its **canonical form**. Putting something into canonical form is canonicalization.

Definition

Suppose we have some set S of objects, with an equivalence relation. A canonical form is given by designating some objects of S to be "in canonical form", such that every object under consideration is equivalent to exactly one object in canonical form. In other words, the canonical forms in S represent the equivalence classes, once and only once. To test whether two objects are equivalent, it then suffices to test their canonical forms for equality. A canonical form thus provides a classification theorem and more, in that it not just classifies every class, but gives a distinguished (canonical) representative.

In practical terms, one wants to be able to recognize the canonical forms. There is also a practical, algorithmic question to consider: how to pass from a given object s in S to its canonical form $s*$? Canonical forms are generally used to make operating with equivalence classes more effective. For example, in modular arithmetic, the canonical form for a residue class is usually taken as the least non-negative integer in it. Operations on classes are carried out by combining these representatives and then reducing the result to its least non-negative residue. The uniqueness requirement is sometimes relaxed, allowing the forms to be unique up to some finer equivalence relation, like allowing reordering of terms (if there is no natural ordering on terms).

A canonical form may simply be a convention, or a deep theorem.

For example, polynomials are conventionally written with the terms in descending powers: it is more usual to write $x^2 + x + 30$ than $x + 30 + x^2$, although the two forms define the same polynomial. By contrast, the existence of Jordan canonical form for a matrix is a deep theorem.

Examples

Note: in this section, "up to" some equivalence relation E means that the canonical form is not unique in general, but that if one object has two different canonical forms, they are E-equivalent.

Classical Logic

- Negation normal form
- Conjunctive normal form
- Disjunctive normal form
- Algebraic normal form
- Prenex normal form
- Skolem normal form

- Blake canonical form, also known as the complete sum of prime implicants, the complete sum, or the disjunctive prime form

Number Theory

- canonical representation of a positive integer

- canonical form of a continued fraction

Algebra

Objects	A is equivalent to B if:	Normal form
Finitely generated R-modules with R a principal ideal domain	A and B are isomorphic as R-modules	Primary decomposition (up to reordering) or invariant factor decomposition

Geometry

- The equation of a line: $Ax + By = C$, with $A^2 + B^2 = 1$ and $C \geq 0$

- The equation of a circle: $(x-h)^2 + (y-k)^2 = r^2$

By contrast, there are alternative forms for writing equations. For example, the equation of a line may be written as a linear equation in point-slope and slope-intercept form.

Mathematical Notation

Standard form is used by many mathematicians and scientists to write extremely large numbers in a more concise and understandable way.

Set Theory

- Cantor normal form of an ordinal number

Game Theory

- Normal form game

Proof Theory

- Normal form (natural deduction)

Rewriting Systems

- In an abstract rewriting system a normal form is an irreducible object.

Lambda Calculus

- Beta normal form if no beta reduction is possible; Lambda calculus is a particular case of an abstract rewriting system.

Dynamical Systems

- Normal form of a bifurcation

Graph Theory

In graph theory, a branch of mathematics, graph canonization is the problem finding a canonical form of a given graph G. A canonical form is a labeled graph Canon(G) that is isomorphic to G, such that every graph that is isomorphic to G has the same canonical form as G. Thus, from a solution to the graph canonization problem, one could also solve the problem of graph isomorphism: to test whether two graphs G and H are isomorphic, compute their canonical forms Canon(G) and Canon(H), and test whether these two canonical forms are identical.

Differential Forms

Canonical differential forms include the canonical one-form and canonical symplectic form, important in the study of Hamiltonian mechanics and symplectic manifolds.

Computing

In computing, the reduction of data to any kind of canonical form is commonly called *data normalization*.

For instance, Database normalization is the process of organizing the fields and tables of a relational database to minimize redundancy and dependency. In the field of software security, a common vulnerability is unchecked malicious input. The mitigation for this problem is proper input validation. Before input validation may be performed, the input must be normalized, i.e., eliminating encoding (for instance HTML encoding) and reducing the input data to a single common character set.

Other forms of data, typically associated with signal processing (including audio and imaging) or machine learning, can be normalized in order to provide a limited range of values.

Probabilistic Logic

The aim of a probabilistic logic (also probability logic and probabilistic reasoning) is to combine the capacity of probability theory to handle uncertainty with the capacity of deductive logic to exploit structure of formal argument. The result is a richer and more expressive formalism with a broad range of possible application areas. Probabilistic logics attempt to find a natural extension of traditional logic truth tables: the results they define are derived through probabilistic expressions instead. A difficulty with probabilistic logics is that they tend to multiply the computational complexities of their probabilistic and logical components. Other difficulties include the possibility of counter-intuitive results, such as those of Dempster-Shafer theory. The need to deal with a broad variety of contexts and issues has led to many different proposals.

Historical Context

There are numerous proposals for probabilistic logics. Very roughly, they can be categorized into two different classes: those logics that attempt to make a probabilistic extension to logical entailment, such as Markov logic networks, and those that attempt to address the problems of uncertainty and lack of evidence (evidentiary logics).

That probability and uncertainty are not quite the same thing may be understood by noting that, despite the mathematization of probability in the Enlightenment, mathematical probability theory remains, to this very day, entirely unused in criminal courtrooms, when evaluating the "probability" of the guilt of a suspected criminal.

More precisely, in evidentiary logic, there is a need to distinguish the truth of a statement from the confidence in its truth: thus, being uncertain of a suspect's guilt is not the same as assigning a numerical probability to the commission of the crime. A single suspect may be guilty or not guilty, just as a coin may be flipped heads or tails. Given a large collection of suspects, a certain percentage may be guilty, just as the probability of flipping "heads" is one-half. However, it is incorrect to take this law of averages with regard to a single criminal (or single coin-flip): the criminal is no more "a little bit guilty" than a single coin flip is "a little bit heads and a little bit tails": we are merely uncertain as to which it is. Conflating probability and uncertainty may be acceptable when making scientific measurements of physical quantities, but it is an error, in the context of "common sense" reasoning and logic. Just as in courtroom reasoning, the goal of employing uncertain inference is to gather evidence to strengthen the confidence of a proposition, as opposed to performing some sort of probabilistic entailment.

Historically, attempts to quantify probabilistic reasoning date back to antiquity. There was a particularly strong interest starting in the 12th century, with the work of the Scholastics, with the invention of the half-proof (so that two half-proofs are sufficient to prove guilt), the elucidation of moral certainty (sufficient certainty to act upon, but short of absolute certainty), the development of Catholic probabilism (the idea that it is always safe to follow the established rules of doctrine or the opinion of experts, even when they are less probable), the case-based reasoning of casuistry, and the scandal of Laxism (whereby probabilism was used to give support to almost any statement at all, it being possible to find an expert opinion in support of almost any proposition.).

Modern Proposals

Below is a list of proposals for probabilistic and evidentiary extensions to classical and predicate logic.

- The term "*probabilistic logic*" was first used in a paper by Nils Nilsson published in 1986, where the truth values of sentences are probabilities. The proposed semantical generalization induces a probabilistic logical entailment, which reduces to ordinary logical entailment when the probabilities of all sentences are either 0 or 1. This generalization applies to any logical system for which the consistency of a finite set of sentences can be established.

- The central concept in the theory of subjective logic are *opinions* about some of the propositional variables involved in the given logical sentences. A binomial opinion applies to a single proposition and is represented as a 3-dimensional extension of a single probability value to express various degrees of ignorance about the truth of the proposition. For the computation of derived opinions based on a structure of argument opinions, the theory proposes respective operators for various logical connectives, such as e.g. multiplication (AND), co-multiplication (OR), division (UN-AND) and co-division (UN-OR) of opinions as well as conditional deduction (MP) and abduction (MT).

- Approximate reasoning formalism proposed by fuzzy logic can be used to obtain a logic in which the models are the probability distributions and the theories are the lower envelopes. In such a logic the question of the consistency of the available information is strictly related with the one of the coherence of partial probabilistic assignment and therefore with Dutch book phenomenon.

- Markov logic networks implement a form of uncertain inference based on the maximum entropy principle—the idea that probabilities should be assigned in such a way as to maximize entropy, in analogy with the way that Markov chains assign probabilities to finite state machine transitions.

- Systems such as Pei Wang's Non-Axiomatic Reasoning System (NARS) or Ben Goertzel's Probabilistic Logic Networks (PLN) add an explicit confidence ranking, as well as a probability to atoms and sentences. The rules of deduction and induction incorporate this uncertainty, thus side-stepping difficulties in purely Bayesian approaches to logic (including Markov logic), while also avoiding the paradoxes of Dempster-Shafer theory. The implementation of PLN attempts to use and generalize algorithms from logic programming, subject to these extensions.

- In the theory of probabilistic argumentation, probabilities are not directly attached to logical sentences. Instead it is assumed that a particular subset W of the variables V involved in the sentences defines a probability space over the corresponding sub-σ-algebra. This induces two distinct probability measures with respect to V, which are called *degree of support* and *degree of possibility*, respectively. Degrees of support can be regarded as non-additive *probabilities of provability*, which generalizes the concepts of ordinary logical entailment (for $V = \{\}$) and classical posterior probabilities (for $V = W$). Mathematically, this view is compatible with the Dempster-Shafer theory.

- The theory of evidential reasoning also defines non-additive *probabilities of probability* (or *epistemic probabilities*) as a general notion for both logical entailment (provability) and probability. The idea is to augment standard propositional logic by considering an epistemic operator **K** that represents the state of knowledge that a rational agent has about the world. Probabilities are then defined over the resulting *epistemic universe* **K***p* of all propositional sentences *p*, and it is argued that this is the best information available to an analyst. From this view, Dempster-Shafer theory appears to be a generalized form of probabilistic reasoning.

Possible Application Areas

- Argumentation theory
- Artificial intelligence
- Artificial general intelligence
- Bioinformatics
- Formal epistemology
- Game theory
- Philosophy of science
- Psychology
- Statistics

Sugeno Integral

In mathematics, the Sugeno integral, named after M. Sugeno, is a type of integral with respect to a fuzzy measure.

Let (X, Ω) be a measurable space and let $h : X \to [0,1]$ be an Ω-measurable function.

The Sugeno integral over the crisp set $A \subseteq X$ of the function h with respect to the fuzzy measure g is defined by:

$$\int_A h(x)°g = \sup_{E \subseteq X}\left[\min\left(\min_{x \in E} h(x), g(A \cap E)\right)\right] = \sup_{\alpha \in [0,1]}\left[\min\left(\alpha, g(A \cap F_\alpha)\right)\right]$$

where $F_\alpha = \{x \mid h(x) \geq \alpha\}$.

The Sugeno integral over the fuzzy set $\tilde{A}$ of the function h with respect to the fuzzy measure g is defined by:

$$\int_A h(x)°g = \int_X \left[h_A(x) \wedge h(x)\right]°g$$

where $h_A(x)$ is the membership function of the fuzzy set $\tilde{A}$.

References

- Lou Goble (2001). The Blackwell guide to philosophical logic. Wiley-Blackwell. p. 309. ISBN 978-0-631-20693-4.

- Mark Hürlimann (2009). Dealing with Real-World Complexity: Limits, Enhancements and New Approaches for Policy Makers. Gabler Verlag. p. 42. ISBN 978-3-8349-1493-4.

- Dov M. Gabbay; John Woods (2007). The Many Valued and Nonmonotonic Turn in Logic. The handbook of the history of logic. 8. Elsevier. p. vii. ISBN 978-0-444-51623-7. Retrieved 4 April 2011.

- Graham Priest (2008). An introduction to non-classical logic: from if to is. Cambridge University Press. pp. 124–125. ISBN 978-0-521-85433-7.

- Morten Heine Sørensen; Paweł Urzyczyn (2006). Lectures on the Curry-Howard isomorphism. Elsevier. pp. 206–207. ISBN 978-0-444-52077-7.

- Stephen C. Kleene 1952 Introduction to Metamathematics, 6th Reprint 1971, North-Holland Publishing Company, Amsterdam NY, ISBN 0-7294-2130-9.

- Miller, D. Michael; Thornton, Mitchell A. (2008). Multiple valued logic: concepts and representations. Synthesis lectures on digital circuits and systems. 12. Morgan & Claypool Publishers. pp. 41–42. ISBN 978-1-59829-190-2.

- Gunther Schmidt Relational Mathematics, Encyclopedia of Mathematics and its Applications, vol. 132, Cambridge University Press, 2011, ISBN 978-0-521-76268-7

- Hansen, Vagn Lundsgaard (2006), Functional Analysis: Entering Hilbert Space, World Scientific Publishing, ISBN 981-256-563-9.

- James Franklin, The Science of Conjecture: Evidence and Probability before Pascal, 2001 The Johns Hopkins Press, ISBN 0-8018-7109-3

- Klement, Erich Peter; Mesiar, Radko; and Pap, Endre (2000), Triangular Norms. Dordrecht: Kluwer. ISBN 0-7923-6416-3.

6

Understanding Logic

Logic helps in analyzing and structuring arguments. A valid argument is considered to be an argument that has a specific relation to the arguments and to its conclusion. This section is an overview of the subject matter incorporating all the major aspects to explain logic.

Logic

Logic is generally held to consist of the systematic study of the form of arguments. A valid argument is one where there is a specific relation of logical support between the assumptions of the argument and its conclusion. (In ordinary discourse, the conclusion of such an argument may be signified by words like *therefore*, *hence*, *ergo* and so on.)

There is no universal agreement as to the exact scope and subject matter of logic, but it has traditionally included the classification of arguments, the systematic exposition of the 'logical form' common to all valid arguments, the study of inference, including fallacies, and the study of semantics, including paradoxes. Historically, logic has been studied in philosophy (since ancient times) and mathematics (since the mid-1800s), and recently logic has been studied in computer science, linguistics, psychology, and other fields.

Concepts

> " Upon this first, and in one sense this sole, rule of reason, that in order to learn you must desire to learn, and in so desiring not be satisfied with what you already incline to think, there follows one corollary which itself deserves to be inscribed upon every wall of the city of philosophy: Do not block the way of inquiry. "
>
> — *Charles Sanders Peirce*, "First Rule of Logic"

The concept of logical form is central to logic. The validity of an argument is determined by its logical form, not by its content. Traditional Aristotelian syllogistic logic and modern symbolic logic are examples of formal logic.

- Informal logic is the study of natural language arguments. The study of fallacies is an important branch of informal logic. Since much informal argument is not strictly speaking deductive, on some conceptions of logic, informal logic is not logic at all.

- Formal logic is the study of inference with purely formal content. An inference possesses a *purely formal content* if it can be expressed as a particular application of a wholly abstract rule, that is, a rule that is not about any particular thing or property. The works of Aristotle contain the earliest known formal study of logic. Modern formal logic follows and expands on Aristotle. In many definitions of logic, logical inference and inference with purely formal content are the same. This does not render the notion of informal logic vacuous, because no formal logic captures all of the nuances of natural language.

- Symbolic logic is the study of symbolic abstractions that capture the formal features of logical inference. Symbolic logic is often divided into two main branches: propositional logic and predicate logic.

- Mathematical logic is an extension of symbolic logic into other areas, in particular to the study of model theory, proof theory, set theory, and recursion theory.

However, agreement on what logic is has remained elusive, and although the field of universal logic has studied the common structure of logics, in 2007 Mossakowski et al. commented that "it is embarrassing that there is no widely acceptable formal definition of 'a logic'".

Logical Form

Logic is generally considered formal when it analyzes and represents the *form* of any valid argument type. The form of an argument is displayed by representing its sentences in the formal grammar and symbolism of a logical language to make its content usable in formal inference. Simply put, formalising simply means translating English sentences into the language of logic.

This is called showing the *logical form* of the argument. It is necessary because indicative sentences of ordinary language show a considerable variety of form and complexity that makes their use in inference impractical. It requires, first, ignoring those grammatical features irrelevant to logic (such as gender and declension, if the argument is in Latin), replacing conjunctions irrelevant to logic (such as "but") with logical conjunctions like "and" and replacing ambiguous, or alternative logical expressions ("any", "every", etc.) with expressions of a standard type (such as "all", or the universal quantifier $\forall$).

Second, certain parts of the sentence must be replaced with schematic letters. Thus, for example, the expression "all Ps are Qs" shows the logical form common to the sentences "all men are mortals", "all cats are carnivores", "all Greeks are philosophers", and so on. The schema can further be condensed into the formula $A(P,Q)$, where the letter A indicates the judgement 'all - are -'.

The importance of form was recognised from ancient times. Aristotle uses variable letters to represent valid inferences in *Prior Analytics*, leading Jan Łukasiewicz to say that the introduction of variables was "one of Aristotle's greatest inventions". According to the followers of Aristotle (such as Ammonius), only the logical principles stated in schematic terms belong to logic, not those given in concrete terms. The concrete terms "man", "mortal", etc., are analogous to the substitution values of the schematic placeholders P, Q, R, which were called the "matter" (Greek *hyle*) of the inference.

There is a big difference between the kinds of formulas seen in traditional term logic and the predicate calculus that is the fundamental advance of modern logic. The formula $A(P,Q)$ (all Ps are Qs) of traditional logic corresponds to the more complex formula $\forall x.(P(x) \rightarrow Q(x))$ in predicate logic, involving the logical connectives for universal quantification and implication rather than just the predicate letter A and using variable arguments $P(x)$ where traditional logic uses just the term letter P. With the complexity comes power, and the advent of the predicate calculus inaugurated revolutionary growth of the subject.

Semantics

The validity of an argument depends upon the meaning or *semantics* of the sentences that make it up.

Aristotle's Organon, especially On Interpretation, gives a cursory outline of semantics which the scholastic logicians, particularly in the thirteenth and fourteenth century, developed into a complex and sophisticated theory, called Supposition Theory. This showed how the truth of simple sentences, expressed schematically, depend on how the terms 'supposit' or *stand for* certain extra-linguistic items. For example, in part II of his Summa Logicae, William of Ockham presents a comprehensive account of the necessary and sufficient conditions for the truth of simple sentences, in order to show which arguments are valid and which are not. Thus 'every A is B' is true if and only if there is something for which 'A' stands for, and there is nothing for which 'A' stands for, which 'B' does not also stand for.

Early modern logic defined semantics purely as a relation between ideas. Antoine Arnauld in the Port Royal Logic, says that 'after conceiving things by our ideas, we compare these ideas, and, finding that some belong together and some do not, we unite or separate them. This is called *affirming* or *denying*, and in general *judging*'. Thus truth and falsity are no more than the agreement or disagreement of ideas. This suggests obvious difficulties, leading Locke to distinguish between 'real' truth, when our ideas have 'real existence' and 'imaginary' or 'verbal' truth, where ideas like harpies or centaurs exist only in the mind. This view (psychologism) was taken to the extreme in the nineteenth century, and is generally held by modern logicians to signify a low point in the decline of logic before the twentieth century.

Modern semantics is in some ways closer to the medieval view, in rejecting such psychological truth-conditions. However, the introduction of quantification, needed to solve the problem of multiple generality, rendered impossible the kind of subject-predicate analysis that underlies medieval semantics. The main modern approach is *model-theoretic semantics*, based on Alfred Tarski's semantic theory of truth. The approach assumes that the meaning of the various parts of the propositions are given by the possible ways we can give a recursively specified group of interpretation functions from them to some predefined domain of discourse: an interpretation of first-order predicate logic is given by a mapping from terms to a universe of individuals, and a mapping from propositions to the truth values "true" and "false". Model-theoretic semantics is one of the fundamental concepts of model theory. Modern semantics also admits rival approaches, such as the proof-theoretic semantics that associates the meaning of propositions with the roles that they can play in inferences, an approach that ultimately derives from the work of Gerhard Gentzen on structural proof theory and is heavily influenced by Ludwig Wittgenstein's later philosophy, especially his aphorism "meaning is use".

Inference

An implication is a sentence of the form 'If p then q', and can be true or false. The Stoic logician Philo of Megara was the first to define the truth conditions of such an implication: false only when the antecedent p is true and the consequent q is false, in all other cases true. An inference, on the other hand, consists of two separately asserted propositions of the form 'p therefore q'. An inference is not true or false, but valid or invalid. However, there is a connection between implication and inference, as follows: if the implication 'if p then q' is *true*, the inference 'p therefore q' is *valid*. This was given an apparently paradoxical formulation by Philo, who said that the implication 'if it is day, it is night' is true only at night, so the inference 'it is day, therefore it is night' is valid in the night, but not in the day.

The theory of inference (or 'consequences') was systematically developed in medieval times by logicians such as William of Ockham and Walter Burley. It is uniquely medieval, though it has its origins in Aristotle's Topics and Boethius' *De Syllogismis hypotheticis*. This is why many terms in logic are Latin. For example, the rule that licenses the move from the implication 'if p then q' plus the assertion of its antecedent p, to the assertion of the consequent q is known as modus ponens (or 'mode of positing'). Its Latin formulation is 'Posito antecedente ponitur consequens'. The Latin formulations of many other rules such as 'ex falso quodlibet' (anything follows from a falsehood), 'reductio ad absurdum' (disproof by showing the consequence is absurd) also date from this period.

However, the theory of consequences, or of the so-called 'hypothetical syllogism' was never fully integrated into the theory of the 'categorical syllogism'. This was partly because of the resistance to reducing the categorical judgment 'Every S is P' to the so-called hypothetical judgment 'if anything is S, it is P'. The first was thought to imply 'some S is P', the second was not, and as late as 1911 in the Encyclopedia Britannica article on Logic, we find the Oxford logician T.H. Case arguing against Sigwart's and Brentano's modern analysis of the universal proposition. Cf. problem of existential import

Logical Systems

A formal system is an organisation of terms used for the analysis of deduction. It consists of an alphabet, a language over the alphabet to construct sentences, and a rule for deriving sentences. Among the important properties that logical systems can have are:

- Consistency, which means that no theorem of the system contradicts another.

- Validity, which means that the system's rules of proof never allow a false inference from true premises.

- Completeness, which means that if a formula is true, it can be proven, i.e. is a *theorem* of the system.

- Soundness, meaning that if any formula is a theorem of the system, it is true. This is the converse of completeness. (Note that in a distinct philosophical use of the term, an argument is sound when it is both valid and its premises are true).

Some logical systems do not have all four properties. As an example, Kurt Gödel's incompleteness theorems show that sufficiently complex formal systems of arithmetic cannot be consistent and complete; however, first-order predicate logics not extended by specific axioms to be arithmetic formal systems with equality can be complete and consistent.

Logic and Rationality

As the study of argument is of clear importance to the reasons that we hold things to be true, logic is of essential importance to rationality. Here we have defined logic to be "the systematic study of the form of arguments"; the reasoning behind argument is of several sorts, but only some of these arguments fall under the aegis of logic proper.

Deductive reasoning concerns the logical consequence of given premises and is the form of reasoning most closely connected to logic. On a narrow conception of logic logic concerns just deductive reasoning, although such a narrow conception controversially excludes most of what is called informal logic from the discipline.

There are other forms of reasoning that are rational but that are generally not taken to be part of logic. These include inductive reasoning, which covers forms of inference that move from collections of particular judgements to universal judgements, and abductive reasoning, which is a form of inference that goes from observation to a hypothesis that accounts for the reliable data (observation) and seeks to explain relevant evidence. The

American philosopher Charles Sanders Peirce (1839–1914) first introduced the term as "guessing". Peirce said that to *abduce* a hypothetical explanation *a* from an observed surprising circumstance *b* is to surmise that may be true because then *b* would be a matter of course. Thus, to abduce *a* from *b* involves determining that *a* is sufficient (or nearly sufficient), but not necessary, for *b* .

While inductive and abductive inference are not part of logic proper, the methodology of logic has been applied to them with some degree of success. For example, the notion of deductive validity (where an inference is deductively valid if and only if there is no possible situation in which all the premises are true but the conclusion false) exists in an analogy to the notion of inductive validity, or "strength", where an inference is inductively strong if and only if its premises give some degree of probability to its conclusion. Whereas the notion of deductive validity can be rigorously stated for systems of formal logic in terms of the well-understood notions of semantics, inductive validity requires us to define a reliable generalization of some set of observations. The task of providing this definition may be approached in various ways, some less formal than others; some of these definitions may use logical association rule induction, while others may use mathematical models of probability such as decision trees.

Rival Conceptions

Logic arose from a concern with correctness of argumentation. Modern logicians usually wish to ensure that logic studies just those arguments that arise from appropriately general forms of inference. For example, Thomas Hofweber writes in the *Stanford Encyclopedia of Philosophy* that logic "does not, however, cover good reasoning as a whole. That is the job of the theory of rationality. Rather it deals with inferences whose validity can be traced back to the formal features of the representations that are involved in that inference, be they linguistic, mental, or other representations."

By contrast, Immanuel Kant (1724–1804) argued for seeing logic as the science of judgement, an idea taken up in the logical and philosophical work of Gottlob Frege (1848–1925). But Frege's work is ambiguous in the sense that it is both concerned with the "laws of thought" as well as with the "laws of truth", i.e. it both treats logic in the context of a theory of the mind, and treats logic as the study of abstract formal structures.

Logic has been defined as "the study of arguments correct in virtue of their form". This has not been the definition taken in this article, but the idea that logic treats special forms of argument, deductive argument, rather than argument in general, has a history in logic that dates back at least to logicism in mathematics (19th and 20th centuries) and the advent of the influence of mathematical logic on philosophy. A consequence of taking logic to treat special kinds of argument is that it leads to identification of special kinds of truth, the logical truths (with logic equivalently being the study of logical truth), and excludes many of the original objects of study of logic that are treated as informal logic. Robert Brandom has argued against the idea that logic is the study of a special kind of logical truth, arguing that instead one can talk of the logic of material inference (in the terminology of Wilfred Sellars), with logic making explicit the commitments that were originally implicit in informal inference.

History

Aristotle, 384–322 BCE.

In Europe, logic was first developed by Aristotle. Aristotelian logic became widely accepted in science and mathematics and remained in wide use in the West until the early 19th century. Aristotle's system of logic was responsible for the introduction of hypothetical syllogism, temporal modal logic, and inductive logic, as well as influential terms such as terms, predicables, syllogisms and propositions. In Europe during the later medieval period, major efforts were made to show that Aristotle's ideas were compatible with Christian faith. During the High Middle Ages, logic became a main focus of philosophers, who would engage in critical logical analyses of philosophical arguments, often using variations of the methodology of scholasticism. In 1323, William of Ockham's influential *Summa Logicae* was released. By the 18th century, the structured approach to arguments had degenerated and fallen out of favour, as depicted in Holberg's satirical play *Erasmus Montanus*.

The Chinese logical philosopher Gongsun Long (c. 325–250 BCE) proposed the paradox "One and one cannot become two, since neither becomes two." In China, the tradition of scholarly investigation into logic, however, was repressed by the Qin dynasty following the legalist philosophy of Han Feizi.

In India, innovations in the scholastic school, called Nyaya, continued from ancient times into the early 18th century with the Navya-Nyaya school. By the 16th century, it developed theories resembling modern logic, such as Gottlob Frege's "distinction between sense and reference of proper names" and his "definition of number", as well as the theory of "restrictive conditions for universals" anticipating some of the developments in modern set theory. Since 1824, Indian logic attracted the attention of many Western scholars, and has had an influence on important 19th-century logicians such as Charles Babbage, Augustus De Morgan, and George Boole. In the 20th century, Western philosophers like Stanislaw Schayer and Klaus Glashoff have explored Indian logic more extensively.

The syllogistic logic developed by Aristotle predominated in the West until the mid-19th century, when interest in the foundations of mathematics stimulated the development of symbolic logic (now called mathematical logic). In 1854, George Boole published *An Investigation of the Laws of Thought on Which are Founded the Mathematical Theories of Logic and Probabilities*, introducing symbolic logic and the principles of what is now known as Boolean logic. In 1879, Gottlob Frege published *Begriffsschrift*, which inaugurated modern logic with the invention of quantifier notation. From 1910 to 1913, Alfred North Whitehead and Bertrand Russell published *Principia Mathematica* on the foundations of mathematics, attempting to derive mathematical truths from axioms and inference rules in symbolic logic. In 1931, Gödel raised serious problems with the foundationalist program and logic ceased to focus on such issues.

The development of logic since Frege, Russell, and Wittgenstein had a profound influence on the practice of philosophy and the perceived nature of philosophical problems, and Philosophy of mathematics. Logic, especially sentential logic, is implemented in computer logic circuits and is fundamental to computer science. Logic is commonly taught by university philosophy departments, often as a compulsory discipline.

Types

Syllogistic Logic

A depiction from the 15th century of the square of opposition,
which expresses the fundamental dualities of syllogistic.

The *Organon* was Aristotle's body of work on logic, with the *Prior Analytics* constituting the first explicit work in formal logic, introducing the syllogistic. The parts of syllogistic logic, also known by the name term logic, are the analysis of the judgements into propositions consisting of two terms that are related by one of a fixed number of relations, and the expression of inferences by means of syllogisms that consist of two propositions sharing a common term as premise, and a conclusion that is a proposition involving the two unrelated terms from the premises.

Aristotle's work was regarded in classical times and from medieval times in Europe and the Middle East as the very picture of a fully worked out system. However, it was not alone: the Stoics proposed a system of propositional logic that was studied by medieval logicians. Also, the problem of multiple generality was recognized in medieval times. Nonetheless, problems with syllogistic logic were not seen as being in need of revolutionary solutions.

Today, some academics claim that Aristotle's system is generally seen as having little more than historical value (though there is some current interest in extending term logics), regarded as made obsolete by the advent of propositional logic and the predicate calculus. Others use Aristotle in argumentation theory to help develop and critically question argumentation schemes that are used in artificial intelligence and legal arguments.

Propositional Logic

A propositional calculus or logic (also a sentential calculus) is a formal system in which formulae representing propositions can be formed by combining atomic propositions using logical connectives, and in which a system of formal proof rules establishes certain formulae as "theorems". An example of a theorem of propositional logic is $A \rightarrow B \rightarrow A$, which says that if A holds, then B implies A.

Predicate Logic

$$\exists x\ F(x)$$

Predicate logic is the generic term for symbolic formal systems such as first-order logic, second-order logic, many-sorted logic, and infinitary logic. It provides an account of quantifiers general enough to express a wide set of arguments occurring in natural language. For example, Bertrand Russell's famous barber paradox, "there is a man who shaves all and only men who do not shave themselves" can be formalised by the sentence $(\exists x)(man(x) \wedge (\forall y)(man(y) \rightarrow (shaves(x, y) \leftrightarrow \neg shaves(y, y))))$, using the non-logical predicate $man(x)$ to indicate that x is a man, and the non-logical relation shaves(x, y) to indicate that x shaves y; all other symbols of the formulae are logical, expressing the universal and existential quantifiers, conjunction, implication, negation and biconditional.

Whilst Aristotelian syllogistic logic specifies a small number of forms that the relevant part of the involved judgements may take, predicate logic allows sentences to be analysed into subject and argument in several additional ways—allowing predicate logic to solve the problem of multiple generality that had perplexed medieval logicians.

The development of predicate logic is usually attributed to Gottlob Frege, who is also credited as one of the founders of analytical philosophy, but the formulation of predicate logic most often used today is the first-order logic presented in Principles of Mathematical Logic by David Hilbert and Wilhelm Ackermann in 1928. The analytical generality of predicate logic allowed the formalization of mathematics, drove the investigation of set theory, and allowed the development of Alfred Tarski's approach to model theory. It provides the foundation of modern mathematical logic.

Frege's original system of predicate logic was second-order, rather than first-order. Second-order logic is most prominently defended (against the criticism of Willard Van Orman Quine and others) by George Boolos and Stewart Shapiro.

Modal Logic

In languages, modality deals with the phenomenon that sub-parts of a sentence may have their semantics modified by special verbs or modal particles. For example, "*We go to the games*" can be modified to give "*We should go to the games*", and "*We can go to the games*" and perhaps "*We will go to the games*". More abstractly, we might say that modality affects the circumstances in which we take an assertion to be satisfied.

Aristotle's logic is in large parts concerned with the theory of non-modalized logic. Although, there are passages in his work, such as the famous sea-battle argument in *De Interpretatione* § 9, that are now seen as anticipations of modal logic and its connection with potentiality and time, the earliest formal system of modal logic was developed by Avicenna, whom ultimately developed a theory of "temporally modalized" syllogistic.

While the study of necessity and possibility remained important to philosophers, little logical innovation happened until the landmark investigations of Clarence Irving Lewis in 1918, who formulated a family of rival axiomatizations of the alethic modalities. His work unleashed a torrent of new work on the topic, expanding the kinds of modality treated to include deontic logic and epistemic logic. The seminal work of Arthur Prior applied the same formal language to treat temporal logic and paved the way for the marriage of the two subjects. Saul Kripke discovered (contemporaneously with rivals) his theory of frame semantics, which revolutionized the formal technology available to modal logicians and gave a new graph-theoretic way of looking at modality that has driven many applications in computational linguistics and computer science, such as dynamic logic.

Informal Reasoning and Dialectic

The motivation for the study of logic in ancient times was clear: it is so that one may learn to distinguish good arguments from bad arguments, and so become more effective in argument and oratory, and perhaps also to become a better person. Half of the works of Aristotle's Organon treat inference as it occurs in an informal setting, side by side with the development of the syllogistic, and in the Aristotelian school, these informal works on logic were seen as complementary to Aristotle's treatment of rhetoric.

This ancient motivation is still alive, although it no longer takes centre stage in the picture of logic; typically dialectical logic forms the heart of a course in critical thinking, a compulsory course at many universities. Dialectic has been linked to logic since ancient times, but it has not been until recent decades that European and American logicians have attempted to provide mathematical foundations for logic and dialectic by formalising dialectical logic. Dialectical logic is also the name given to the special treatment of dialectic in Hegelian and Marxist thought. There have been pre-formal treatises on argument and dialectic, from authors such as Stephen Toulmin (*The Uses of Argument*), Nicholas Rescher (*Dialectics*), and van Eemeren and Grootendorst (Pragma-dialectics). Theories of defeasible reasoning can provide a foundation for the formalisation of dialectical logic and dialectic itself can be formalised as moves in a game, where an advocate for the truth of a proposition and an opponent argue. Such games can provide a formal game semantics for many logics.

Argumentation theory is the study and research of informal logic, fallacies, and critical questions as they relate to every day and practical situations. Specific types of dialogue can be analyzed and questioned to reveal premises, conclusions, and fallacies. Argumentation theory is now applied in artificial intelligence and law.

Mathematical Logic

Mathematical logic comprises two distinct areas of research: the first is the application of the techniques of formal logic to mathematics and mathematical reasoning, and the second, in the other direction, the application of mathematical techniques to the representation and analysis of formal logic.

The earliest use of mathematics and geometry in relation to logic and philosophy goes back to the ancient Greeks such as Euclid, Plato, and Aristotle. Many other ancient and medieval philosophers applied mathematical ideas and methods to their philosophical claims.

One of the boldest attempts to apply logic to mathematics was the logicism pioneered by philosopher-logicians such as Gottlob Frege and Bertrand Russell. Mathematical theories were supposed to be logical tautologies, and the programme was to show this by means to a reduction of mathematics to logic. The various attempts to carry this out met with failure, from the crippling of Frege's project in his *Grundgesetze* by Russell's paradox, to the defeat of Hilbert's program by Gödel's incompleteness theorems.

Both the statement of Hilbert's program and its refutation by Gödel depended upon their work establishing the second area of mathematical logic, the application of mathematics to logic in the form of proof theory. Despite the negative nature of the incompleteness theorems, Gödel's completeness theorem, a result in model theory and another application of mathematics to logic, can be understood as showing how close logicism came to being true: every rigorously defined mathematical theory can be exactly captured by a first-order logical theory; Frege's proof calculus is enough to *describe* the whole of mathematics, though not *equivalent* to it.

If proof theory and model theory have been the foundation of mathematical logic, they have been but two of the four pillars of the subject. Set theory originated in the study of the infinite by Georg Cantor, and it has been the source of many of the most challenging and important issues in mathematical logic, from Cantor's theorem, through the status of the Axiom of Choice and the question of the independence of the continuum hypothesis, to the modern debate on large cardinal axioms.

Recursion theory captures the idea of computation in logical and arithmetic terms; its most classical achievements are the undecidability of the Entscheidungsproblem by Alan Turing, and his presentation of the Church–Turing thesis. Today recursion theory is mostly concerned with the more refined problem of complexity classes—when is a problem efficiently solvable?—and the classification of degrees of unsolvability.

Philosophical Logic

Philosophical logic deals with formal descriptions of ordinary, non-specialist ("natural") language. Most philosophers assume that the bulk of everyday reasoning can be captured in logic if a method or methods to translate ordinary language into that logic can be found. Philosophical logic is essentially a continuation of the traditional discipline called "logic" before the invention of mathematical logic. Philosophical logic has a much greater concern with the connection between natural language and logic. As a result, philosophical logicians have contributed a great deal to the development of non-standard logics (e.g. free logics, tense logics) as well as various extensions of classical logic (e.g. modal logics) and non-standard semantics for such logics (e.g. Kripke's supervaluationism in the semantics of logic).

Logic and the philosophy of language are closely related. Philosophy of language has to do with the study of how our language engages and interacts with our thinking. Logic has an immediate impact on other areas of study. Studying logic and the relationship between logic and ordinary speech can help a person better structure his own arguments and critique the arguments of others. Many popular arguments are filled with errors because so many people are untrained in logic and unaware of how to formulate an argument correctly.

Computational Logic

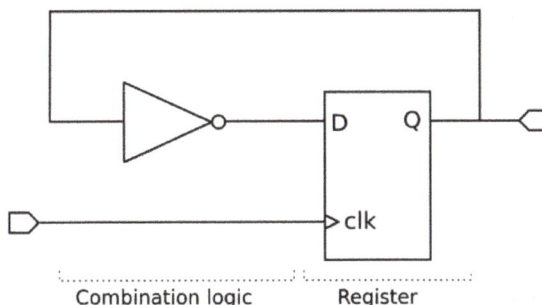

A simple toggling circuit is expressed using a logic gate and a synchronous register.

Logic cut to the heart of computer science as it emerged as a discipline: Alan Turing's work on the *Entscheidungsproblem* followed from Kurt Gödel's work on the incompleteness theorems. The notion of the general purpose computer that came from this work was of fundamental importance to the designers of the computer machinery in the 1940s.

In the 1950s and 1960s, researchers predicted that when human knowledge could be expressed using logic with mathematical notation, it would be possible to create a machine that reasons, or artificial intelligence. This was more difficult than expected because of the complexity of human reasoning. In logic programming, a program consists of a set of axioms and rules. Logic programming systems such as Prolog compute the consequences of the axioms and rules in order to answer a query.

Today, logic is extensively applied in the fields of Artificial Intelligence and Computer Science, and these fields provide a rich source of problems in formal and informal logic. Argumentation theory is one good example of how logic is being applied to artificial intelligence. The ACM Computing Classification System in particular regards:

- Section F.3 on Logics and meanings of programs and F.4 on Mathematical logic and formal languages as part of the theory of computer science: this work covers formal semantics of programming languages, as well as work of formal methods such as Hoare logic;

- Boolean logic as fundamental to computer hardware: particularly, the system's section B.2 on Arithmetic and logic structures, relating to operatives AND, NOT, and OR;

- Many fundamental logical formalisms are essential to section I.2 on artificial intelligence, for example modal logic and default logic in Knowledge representation formalisms and methods, Horn clauses in logic programming, and description logic.

Furthermore, computers can be used as tools for logicians. For example, in symbolic logic and mathematical logic, proofs by humans can be computer-assisted. Using automated theorem proving, the machines can find and check proofs, as well as work with proofs too lengthy to write out by hand.

Non-classical Logics

The logics discussed above are all "bivalent" or "two-valued"; that is, they are most naturally understood as dividing propositions into true and false propositions. Non-classical logics are those systems that reject various rules of Classical logic.

Hegel developed his own dialectic logic that extended Kant's transcendental logic but also brought it back to ground by assuring us that "neither in heaven nor in earth, neither in the world of mind nor of nature, is there anywhere such an abstract 'either–or' as the understanding maintains. Whatever exists is concrete, with difference and opposition in itself".

In 1910, Nicolai A. Vasiliev extended the law of excluded middle and the law of contradiction and proposed the law of excluded fourth and logic tolerant to contradiction. In the early 20th century Jan Łukasiewicz investigated the extension of the traditional true/false values to include a third value, "possible", so inventing ternary logic, the first multi-valued logic.

Logics such as fuzzy logic have since been devised with an infinite number of "degrees of truth", represented by a real number between 0 and 1.

Intuitionistic logic was proposed by L.E.J. Brouwer as the correct logic for reasoning about mathematics, based upon his rejection of the law of the excluded middle as part of his intuitionism. Brouwer rejected formalization in mathematics, but his student Arend Heyting studied intuitionistic logic formally, as did Gerhard Gentzen. Intuitionistic logic is of great interest to computer scientists, as it is a constructive logic and can be applied for extracting verified programs from proofs.

Modal logic is not truth conditional, and so it has often been proposed as a non-classical logic. However, modal logic is normally formalized with the principle of the excluded middle, and its relational semantics is bivalent, so this inclusion is disputable.

Controversies

"Is Logic Empirical?"

What is the epistemological status of the laws of logic? What sort of argument is appropriate for criticizing purported principles of logic? In an influential paper entitled "Is

logic empirical?" Hilary Putnam, building on a suggestion of W. V. Quine, argued that in general the facts of propositional logic have a similar epistemological status as facts about the physical universe, for example as the laws of mechanics or of general relativity, and in particular that what physicists have learned about quantum mechanics provides a compelling case for abandoning certain familiar principles of classical logic: if we want to be realists about the physical phenomena described by quantum theory, then we should abandon the principle of distributivity, substituting for classical logic the quantum logic proposed by Garrett Birkhoff and John von Neumann.

Another paper of the same name by Michael Dummett argues that Putnam's desire for realism mandates the law of distributivity. Distributivity of logic is essential for the realist's understanding of how propositions are true of the world in just the same way as he has argued the principle of bivalence is. In this way, the question, "Is logic empirical?" can be seen to lead naturally into the fundamental controversy in metaphysics on realism versus anti-realism.

Implication: Strict or Material

The notion of implication formalized in classical logic does not comfortably translate into natural language by means of "if ... then ...", due to a number of problems called the paradoxes of material implication.

The first class of paradoxes involves counterfactuals, such as *If the moon is made of green cheese, then 2+2=5*, which are puzzling because natural language does not support the principle of explosion. Eliminating this class of paradoxes was the reason for C. I. Lewis's formulation of strict implication, which eventually led to more radically revisionist logics such as relevance logic.

The second class of paradoxes involves redundant premises, falsely suggesting that we know the succedent because of the antecedent: thus "if that man gets elected, granny will die" is materially true since granny is mortal, regardless of the man's election prospects. Such sentences violate the Gricean maxim of relevance, and can be modelled by logics that reject the principle of monotonicity of entailment, such as relevance logic.

Tolerating the Impossible

Hegel was deeply critical of any simplified notion of the Law of Non-Contradiction. It was based on Leibniz's idea that this law of logic also requires a sufficient ground to specify from what point of view (or time) one says that something cannot contradict itself. A building, for example, both moves and does not move; the ground for the first is our solar system and for the second the earth. In Hegelian dialectic, the law of non-contradiction, of identity, itself relies upon difference and so is not independently assertable.

Closely related to questions arising from the paradoxes of implication comes the suggestion that logic ought to tolerate inconsistency. Relevance logic and paraconsistent logic are the most important approaches here, though the concerns are different: a key consequence of classical logic and some of its rivals, such as intuitionistic logic, is that they respect the principle of explosion, which means that the logic collapses if it is capable of deriving a contradiction. Graham Priest, the main proponent of dialetheism, has argued for paraconsistency on the grounds that there are in fact, true contradictions.

Rejection of Logical Truth

The philosophical vein of various kinds of skepticism contains many kinds of doubt and rejection of the various bases on which logic rests, such as the idea of logical form, correct inference, or meaning, typically leading to the conclusion that there are no logical truths. Observe that this is opposite to the usual views in philosophical skepticism, where logic directs skeptical enquiry to doubt received wisdoms, as in the work of Sextus Empiricus.

Friedrich Nietzsche provides a strong example of the rejection of the usual basis of logic: his radical rejection of idealization led him to reject truth as a "... mobile army of metaphors, metonyms, and anthropomorphisms—in short ... metaphors which are worn out and without sensuous power; coins which have lost their pictures and now matter only as metal, no longer as coins." His rejection of truth did not lead him to reject the idea of either inference or logic completely, but rather suggested that "logic [came] into existence in man's head [out] of illogic, whose realm originally must have been immense. Innumerable beings who made inferences in a way different from ours perished". Thus there is the idea that logical inference has a use as a tool for human survival, but that its existence does not support the existence of truth, nor does it have a reality beyond the instrumental: "Logic, too, also rests on assumptions that do not correspond to anything in the real world".

This position held by Nietzsche however, has come under extreme scrutiny for several reasons. Some philosophers, such as Jürgen Habermas, claim his position is self-refuting—and accuse Nietzsche of not even having a coherent perspective, let alone a theory of knowledge. Georg Lukács, in his book *The Destruction of Reason*, asserts that, "Were we to study Nietzsche's statements in this area from a logico-philosophical angle, we would be confronted by a dizzy chaos of the most lurid assertions, arbitrary and violently incompatible." Bertrand Russell described Nietzsche's irrational claims with "He is fond of expressing himself paradoxically and with a view to shocking conventional readers" in his book *A History of Western Philosophy*.

Logical Form

The logical form of a sentence (or proposition or statement or truthbearer) or set of sentences is the form obtained by abstracting from the subject matter of its content terms or by regarding the content terms as mere placeholders or blanks on a form. In an ideal logical language, the logical form can be determined from syntax alone; formal languages used in formal sciences are examples of such languages. Logical form, however, should not be confused with the mere syntax used to represent it; there may be more than one string that represents the same logical form in a given language.

The logical form of an argument is called the argument form or *test form* of the argument.

History

The importance of the concept of form to logic was already recognized in ancient times. Aristotle, in the Prior Analytics, was probably the first to employ variable letters to represent valid inferences. Therefore, Łukasiewicz claims that the introduction of variables was 'one of Aristotle's greatest inventions'.

According to the followers of Aristotle like Ammonius, only the logical principles stated in schematic terms belong to logic, and not those given in concrete terms. The concrete terms *man*, *mortal*, etc., are analogous to the substitution values of the schematic placeholders 'A', 'B', 'C', which were called the 'matter' (Greek *hyle*, Latin *materia*) of the argument.

The term "logical form" itself was introduced by Bertrand Russell in 1914, in the context of his program to formalize natural language and reasoning, which he called philosophical logic. Russell wrote: "Some kind of knowledge of logical forms, though with most people it is not explicit, is involved in all understanding of discourse. It is the business of philosophical logic to extract this knowledge from its concrete integuments, and to render it explicit and pure."

Example of Argument Form

To demonstrate the important notion of the *form* of an argument, substitute letters for similar items throughout the sentences in the original argument.

Original argument

    All humans are mortal.

    Socrates is human.

    Therefore, Socrates is mortal.

Argument form

    All H are M.

    S is H.

Therefore, S is M.

All we have done in the *Argument form* is to put 'H' for 'human' and 'humans', 'M' for 'mortal', and 'S' for 'Socrates'; what results is the *form* of the original argument. Moreover, each individual sentence of the *Argument form* is the *sentence form* of its respective sentence in the original argument.

Importance of Argument Form

Attention is given to argument and sentence form, because *form is what makes an argument valid or cogent*. Some examples of valid argument forms are modus ponens, modus tollens, disjunctive syllogism, hypothetical syllogism and dilemma. Two invalid argument forms are affirming the consequent and denying the antecedent.

A logical argument, seen as an ordered set of sentences, has a logical form that derives from the form of its constituent sentences; the logical form of an argument is sometimes called argument form. Some authors only define logical form with respect to whole arguments, as the schemata or inferential structure of the argument. In argumentation theory or informal logic, an argument form is sometimes seen as a broader notion than the logical form.

It consists of stripping out all spurious grammatical features from the sentence (such as gender, and passive forms), and replacing all the expressions specific to *the subject matter* of the argument by schematic variables. Thus, for example, the expression 'all A's are B's' shows the logical form which is common to the sentences 'all men are mortals', 'all cats are carnivores', 'all Greeks are philosophers' and so on.

Logical form in Modern Logic

The fundamental difference between modern formal logic and traditional, or Aristotelian logic, lies in their differing analysis of the logical form of the sentences they treat:

- On the traditional view, the form of the sentence consists of (1) a subject (e.g., "man") plus a sign of quantity ("all" or "some" or "no"); (2) the copula, which is of the form "is" or "is not"; (3) a predicate (e.g., "mortal"). Thus: 'all men are mortal'. The logical constants such as "all", "no" and so on, plus sentential connectives such as "and" and "or" were called syncategorematic terms (from the Greek *kategorei* – to predicate, and *syn* – together with). This is a fixed scheme, where each judgment has a specific quantity and copula, determining the logical form of the sentence.

- The modern view is more complex, since a single judgement of Aristotle's system involves two or more logical connectives. For example, the sentence "All men are mortal" involves, in term logic, two non-logical terms "is a man" (here M) and "is mortal" (here D): the sentence is given by the judgement $A(M,D)$. In predicate logic, the sentence involves the same two non-logical concepts, here analyzed as $m(x)$ and $d(x)$, and the sentence is given by $\forall x.(m(x) \rightarrow d(x))$, involving the logical connectives for universal quantification and implication.

The more complex modern view comes with more power. On the modern view, the fundamental form of a simple sentence is given by a recursive schema, like natural language and involving logical connectives, which are joined by juxtaposition to other sentences, which in turn may have logical structure. Medieval logicians recognized the problem of multiple generality, where Aristotelian logic is unable to satisfactorily render such sentences as "Some guys have all the luck", because both quantities "all" and "some" may be relevant in an inference, but the fixed scheme that Aristotle used allows only one to govern the inference. Just as linguists recognize recursive structure in natural languages, it appears that logic needs recursive structure.

Semantics of Logic

In logic, the semantics of logic is the study of the semantics, or interpretations, of formal and (idealizations of) natural languages usually trying to capture the pre-theoretic notion of entailment.

Overview

The truth conditions of various sentences we may encounter in arguments will depend upon their meaning, and so logicians cannot completely avoid the need to provide some treatment of the meaning of these sentences. The semantics of logic refers to the approaches that logicians have introduced to understand and determine that part of meaning in which they are interested; the logician traditionally is not interested in the sentence as uttered but in the proposition, an idealised sentence suitable for logical manipulation.

Until the advent of modern logic, Aristotle's *Organon*, especially *De Interpretatione*, provided the basis for understanding the significance of logic. The introduction of quantification, needed to solve the problem of multiple generality, rendered impossible the kind of subject-predicate analysis that governed Aristotle's account, although there is a renewed interest in term logic, attempting to find calculi in the spirit of Aristotle's syllogistic but with the generality of modern logics based on the quantifier.

The main modern approaches to semantics for formal languages are the following:

- Model-theoretic semantics is the archetype of Alfred Tarski's semantic theory of truth, based on his T-schema, and is one of the founding concepts of model theory. This is the most widespread approach, and is based on the idea that the meaning of the various parts of the propositions are given by the possible ways we can give a recursively specified group of interpretation functions from them to some predefined mathematical domains: an interpretation of first-order predicate logic is given by a mapping from terms to a universe of individuals, and a mapping from propositions to the truth values "true" and "false". Model-theoretic semantics provides the foundations for an approach to the theory of meaning known as Truth-conditional semantics, which was pioneered by Donald Davidson. Kripke semantics introduces innovations, but is broadly in the Tarskian mold.

- Proof-theoretic semantics associates the meaning of propositions with the roles that they can play in inferences. Gerhard Gentzen, Dag Prawitz and Michael Dummett are generally seen as the founders of this approach; it is heavily influenced by Ludwig Wittgenstein's later philosophy, especially his aphorism "meaning is use".

- Truth-value semantics (also commonly referred to as *substitutional quantification*) was advocated by Ruth Barcan Marcus for modal logics in the early 1960s and later championed by Dunn, Belnap, and Leblanc for standard first-order logic. James Garson has given some results in the areas of adequacy for intensional logics outfitted with such a semantics. The truth conditions for quantified formulas are given purely in terms of truth with no appeal to domains whatsoever (and hence its name *truth-value semantics*).

- Game-theoretical semantics has made a resurgence lately mainly due to Jaakko Hintikka for logics of (finite) partially ordered quantification which were originally investigated by Leon Henkin, who studied Henkin quantifiers.

- Probabilistic semantics originated from H. Field and has been shown equivalent to and a natural generalization of truth-value semantics. Like truth-value semantics, it is also non-referential in nature.

Formal System

A formal system is broadly defined as any well-defined system of abstract thought based on the model of mathematics. The entailment of the system by its logical foundation is what distinguishes a formal system from others which may have some basis in an abstract model. Often the formal system will be the basis for or even identified with a larger theory or field (e.g. Euclidean geometry) consistent with the usage in modern mathematics such as model theory. A formal system need not be mathematical as such; for example, Spinoza's *Ethics* imitates the form of Euclid's *Elements*.

Each formal system has a formal language, which is composed by primitive symbols. These symbols act on certain rules of formation and are developed by inference from a set of axioms. The system thus consists of any number of formulas built up through finite combinations of the primitive symbols—combinations that are formed from the axioms in accordance with the stated rules.

Formal systems in mathematics consist of the following elements:

1. A finite set of symbols (i.e. the alphabet), that can be used for constructing formulas (i.e. finite strings of symbols).

2. A grammar, which tells how well-formed formulas (abbreviated *wff*) are constructed out of the symbols in the alphabet. It is usually required that there be a decision procedure for deciding whether a formula is well formed or not.

3. A set of axioms or axiom schemata: each axiom must be a wff.

4. A set of inference rules.

A formal system is said to be recursive (i.e. effective) or recursively enumerable if the set of axioms and the set of inference rules are decidable sets or semidecidable sets, respectively.

Some theorists use the term *formalism* as a rough synonym for *formal system*, but the term is also used to refer to a particular style of *notation*, for example, Paul Dirac's bra–ket notation.

History

Panini is credited with the creation of the world's first formal system well before the innovations of Gottlob Frege and the subsequent development of mathematical logic in the 19th century of europe.

Related Subjects

Logical System

A *logical system* or, for short, *logic*, is a formal system together with a form of semantics, usually in the form of model-theoretic interpretation, which assigns truth values to sentences of the formal language, that is, formulae that contain no free variables. A logic is sound if all sentences that can be derived are true in the interpretation, and complete if, conversely, all true sentences can be derived.

Deductive System

A *deductive system* (also called a deductive apparatus of a formal system) consists of the axioms (or axiom schemata) and rules of inference that can be used to derive the theorems of the system.

Such a deductive system is intended to preserve deductive qualities in the formulas that are expressed in the system. Usually the quality we are concerned with is truth as opposed to falsehood. However, other modalities, such as justification or belief may be preserved instead.

In order to sustain its deductive integrity, a *deductive apparatus* must be definable without reference to any intended interpretation of the language. The aim is to ensure that each line of a derivation is merely a syntactic consequence of the lines that precede it. There should be no element of any interpretation of the language that gets involved with the deductive nature of the system.

Formal Proofs

Formal proofs are sequences of well-formed formulas (or wff for short). For a wff to qualify as part of a proof, it might either be an axiom or be the product of applying an inference rule on previous wffs in the proof sequence. The last wff in the sequence is recognized as a theorem.

The point of view that generating formal proofs is all there is to mathematics is often called *formalism*. David Hilbert founded metamathematics as a discipline for discussing formal systems. Any language that one uses to talk about a formal system is called a *metalanguage*. The metalanguage may be a natural language, or it may be partially formalized itself, but it is generally less completely formalized than the formal language component of the formal system under examination, which is then called the *object language*, that is, the object of the discussion in question.

Once a formal system is given, one can define the set of theorems which can be proved inside the formal system. This set consists of all wffs for which there is a proof. Thus all axioms are considered theorems. Unlike the grammar for wffs, there is no guarantee that there will be a decision procedure for deciding whether a given wff is a theorem or not. The notion of *theorem* just defined should not be confused with *theorems about the formal system*, which, in order to avoid confusion, are usually called metatheorems.

Formal Language

In mathematics, logic, and computer science, a formal language is a language that is defined by precise mathematical or machine processable formulas. Like languages in linguistics, formal languages generally have two aspects:

- the syntax of a language is what the language looks like (more formally: the set of possible expressions that are valid utterances in the language)

- the semantics of a language are what the utterances of the language mean (which is formalized in various ways, depending on the type of language in question)

A special branch of mathematics and computer science exists that is devoted exclusively to the theory of language syntax: formal language theory. In formal language theory, a language is nothing more than its syntax; questions of semantics are not included in this specialty.

Formal Grammar

In computer science and linguistics a formal grammar is a precise description of a formal language: a set of strings. The two main categories of formal grammar are that of generative grammars, which are sets of rules for how strings in a language can be generated, and that of analytic grammars (or reductive grammar, which are sets of rules for how a string can be analyzed to determine whether it is a member of the language. In short, an analytic grammar describes how to *recognize* when strings are members in the set, whereas a generative grammar describes how to *write* only those strings in the set.

Rival Conceptions of Logic

The history of logic as a subject has been characterised by many disputes over what the topic deals with, and the main article 'Logic' has as a result been hesitant to commit to a particular definition of logic. This article surveys various definitions of the subject that have appeared over the centuries through to modern times, and puts them in context as reflecting rival conceptions of the subject.

Rival Conceptions of Logic

In the periodic of scholastic philosophy, logic was predominantly Aristotelian. Following the decline of scholasticism, logic was thought of as an affair of ideas by early modern philosophers such as Locke and Hume. Immanuel Kant took this one step further. He begins with the assumption of the empiricist philosophers, that all knowledge whatsoever is internal to the mind, and that we have no genuine knowledge of 'things in themselves'. Furthermore, (an idea he seemed to have got from Hume) the material of knowledge is a succession of separate ideas which have no intrinsic connection and thus no real unity. In order that these disparate sensations be brought into some sort of order and coherence, there must be an internal mechanism in the mind which provides the *forms* by which we think, perceive and reason.

Kant calls these forms *Categories* (in a somewhat different sense than employed by the Aristotelian logicians), of which he claims there are twelve:

- Quantity (Singular, Particular, Universal)

- Quality (Affirmative, Negative, Infinite)

- Relation (Categorical, Hypothetical, Disjunctive)

- Modality (Problematic, Assertoric, Apodictic)

However, this seems to be an arbitrary arrangement, driven by the desire to present a harmonious appearance than from any underlying method or system. For example, the triple nature of each division forced him to add artificial categories such as the infinite judgment.

This conception of logic eventually developed into an extreme form of psychologism espoused in the nineteenth by Benno Erdmann and others. The view of historians of logic is that Kant's influence was negative.

Another view of logic espoused by Hegel and others of his school (such as Lotze, Bradley, Bosanquet and others), was the 'Logic of the Pure Idea'. The central feature of this view is the identification of Logic and Metaphysics. The Universe has its origin in the categories of thought. Thought in its fullest development becomes the Absolute Idea, a divine mind evolving itself in the development of the Universe.

In the modern period, W. V. Quine (1940, pp. 2–3) defined logic in terms of a logical vocabulary, which in turn is identified by an argument that the many particular vocabularies —Quine mentions geological vocabulary— are used in their particular discourses together with a common, topic-independent kernel of terms. These terms, then, constitute the logical vocabulary, and the logical truths are those truths common to all particular topics.

Hofweber (2004) lists several definitions of logic, and goes on to claim that all definitions of logic are of one of four sorts. These are that logic is the study of: (i) artificial formal structures, (ii) sound inference (e.g., Poinsot), (iii) tautologies (e.g., Watts), or (iv) general features of thought (e.g., Frege). He argues then that these definitions are related to each other, but do not exhaust each other, and that an examination of formal ontology shows that these mismatches between rival definitions are due to tricky issues in ontology.

Informal and Colloquial Definitions

Arranged in approximate chronological order.

- The tool for distinguishing between the true and the false (Averroes).

- The science of reasoning, teaching the way of investigating unknown truth in connection with a thesis (Robert Kilwardby).

- The art whose function is to direct the reason lest it err in the manner of inferring or knowing (John Poinsot).

- The art of conducting reason well in knowing things (Antoine Arnauld).

- The right use of reason in the inquiry after truth (Isaac Watts).

- The Science, as well as the Art, of reasoning (Richard Whately).

- The science of the operations of the understanding which are subservient to the estimation of evidence (John Stuart Mill).

- The science of the laws of discursive thought (James McCosh).

- The science of the most general laws of truth (Gottlob Frege).

History of Logic

The history of logic deals with the study of the development of the science of valid inference (logic). Formal logics developed in ancient times in China, India, and Greece. Greek methods, particularly Aristotelian logic (or term logic) as found in the *Organon*, found wide application and acceptance in Western science and mathematics for millennia. The Stoics, especially Chrysippus, began the development of predicate logic.

Christian and Islamic philosophers such as Boethius (died 524) and William of Ockham (died 1347) further developed Aristotle's logic in the Middle Ages, reaching a high point in the mid-fourteenth century. The period between the fourteenth century and the beginning of the nineteenth century saw largely decline and neglect, and at least one historian of logic regards this time as barren. Empirical methods ruled the day, as evidenced by Sir Francis Bacon's *Novum Organon* of 1620.

Logic revived in the mid-nineteenth century, at the beginning of a revolutionary period when the subject developed into a rigorous and formal discipline which took as its exemplar the exact method of proof used in mathematics, a hearkening back to the Greek tradition. The development of the modern "symbolic" or "mathematical" logic during this period by the likes of Boole, Frege, Russell, and Peano is the most significant in the two-thousand-year history of logic, and is arguably one of the most important and remarkable events in human intellectual history.

Progress in mathematical logic in the first few decades of the twentieth century, particularly arising from the work of Gödel and Tarski, had a significant impact on analytic philosophy and philosophical logic, particularly from the 1950s onwards, in subjects such as modal logic, temporal logic, deontic logic, and relevance logic.

Logic in the West

Prehistory of Logic

Valid reasoning has been employed in all periods of human history. However, logic studies the *principles* of valid reasoning, inference and demonstration. It is probable that the idea of demonstrating a conclusion first arose in connection with geometry, which originally meant the same as "land measurement". The ancient Egyptians discovered geometry, including the formula for the volume of a truncated pyramid. Ancient Babylon was also skilled in mathematics. Esagil-kin-apli's medical *Diagnostic Handbook* in the 11th century BC was based on a logical set of axioms and assumptions, while Babylonian astronomers in the 8th and 7th centuries BC employed an internal logic within their predictive planetary systems, an important contribution to the philosophy of science.

Ancient Greece before Aristotle

While the ancient Egyptians empirically discovered some truths of geometry, the great achievement of the ancient Greeks was to replace empirical methods by demonstrative proof. Both Thales and Pythagoras of the Pre-Socratic philosophers seem aware of geometry's methods.

Fragments of early proofs are preserved in the works of Plato and Aristotle, and the idea of a deductive system was probably known in the Pythagorean school and the Platonic Academy. The proofs of Euclid of Alexandria are a paradigm of Greek geometry. The three basic principles of geometry are as follows:

- Certain propositions must be accepted as true without demonstration; such a proposition is known as an axiom of geometry.

- Every proposition that is not an axiom of geometry must be demonstrated as following from the axioms of geometry; such a demonstration is known as a proof or a "derivation" of the proposition.

- The proof must be *formal*; that is, the derivation of the proposition must be independent of the particular subject matter in question.

Further evidence that early Greek thinkers were concerned with the principles of reasoning is found in the fragment called *dissoi logoi*, probably written at the beginning of the fourth century BC. This is part of a protracted debate about truth and falsity. In the case of the classical Greek city-states, interest in argumentation was also stimulated by the activities of the Rhetoricians or Orators and the Sophists, who used arguments to defend or attack a thesis, both in legal and political contexts.

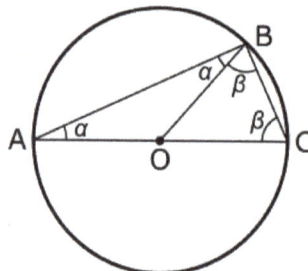

Thales Theorem

Thales

It is said Thales, most widely regarded as the first philosopher in the Greek tradition, measured the height of the pyramids by their shadows at the moment when his own shadow was equal to his height. Thales was said to have had a sacrifice in celebration of discovering Thales' Theorem just as Pythagoras had the Pythagorean Theorem.

Thales is the first known individual to use deductive reasoning applied to geometry, by deriving four corollaries to his theorem, and the first known individual to whom a mathematical discovery has been attributed. Indian and Babylonian mathematicians knew his theorem for special cases before he proved it. It is believed that Thales learned that an angle inscribed in a semicircle is a right angle during his travels to Babylon.

Pythagoras

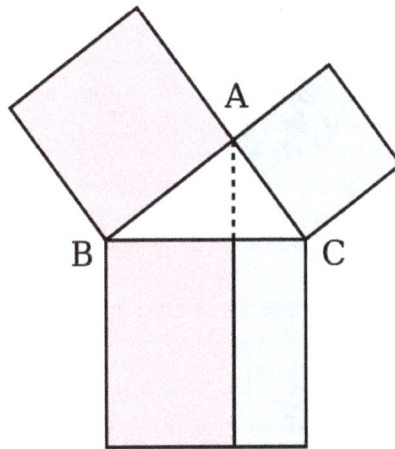

Proof of the Pythagorean Theorem in Euclid's *Elements*

Before 520 BC, on one of his visits to Egypt or Greece, Pythagoras might have met the c. 54 years older Thales. The systematic study of proof seems to have begun with the school of Pythagoras (i. e. the Pythagoreans) in the late sixth century BC. Indeed, the Pythagoreans, believing all was number, are the first philosophers to emphasize *form* rather than *matter*.

Heraclitus and Parmenides

The writing of Heraclitus (c. 535 – c. 475 BC) was the first place where the word *logos* was given special attention in ancient Greek philosophy, Heraclitus held that everything changes and all was fire and conflicting opposites, seemingly unified only by this *Logos*. He is known for his obscure sayings.

This *logos* holds always but humans always prove unable to understand it, both before hearing it and when they have first heard it. For though all things come to be in accordance with this *logos*, humans are like the inexperienced when they experience such words and deeds as I set out, distinguishing each in accordance with its nature and saying how it is. But other people fail to notice what they do when awake, just as they forget what they do while asleep.

—*Diels-Kranz, 22B1*

Parmenides has been called the discoverer of logic.

In contrast to Heraclitus, Parmenides held that all is one and nothing changes. He may have been a dissident Pythagorean, disagreeing that One (a number) produced the many. "X is not" must always be false or meaningless. What exists can in no way not exist. Our sense perceptions with its noticing of generation and destruction are in grievous error. Instead of sense perception, Parmenides advocated *logos* as the means to Truth. He has been called the discoverer of logic,

> For this view, that That Which Is Not exists, can never predominate. You must debar your thought from this way of search, nor let ordinary experience in its variety force you along this way, (namely, that of allowing) the eye, sightless as it is, and the ear, full of sound, and the tongue, to rule; but (you must) judge by means of the Reason (Logos) the much-contested proof which is expounded by me. (B 7.1–8.2)

Zeno of Elea, a pupil of Parmenides, had the idea of a standard argument pattern found in the method of proof known as *reductio ad absurdum*. This is the technique of drawing an obviously false (that is, "absurd") conclusion from an assumption, thus demonstrating that the assumption is false. Therefore, Zeno and his teacher are seen as the first to apply the art of logic. Plato's dialogue Parmenides portrays Zeno as claiming to have written a book defending the monism of Parmenides by demonstrating the absurd consequence of assuming that there is plurality. Zeno famously used this method to develop his paradoxes in his arguments against motion. Such *dialectic* reasoning later became popular. The members of this school were called "dialecticians" (from a Greek word meaning "to discuss").

Plato

Let no one ignorant of geometry enter here.

—Inscribed over the entrance to Plato's Academy.

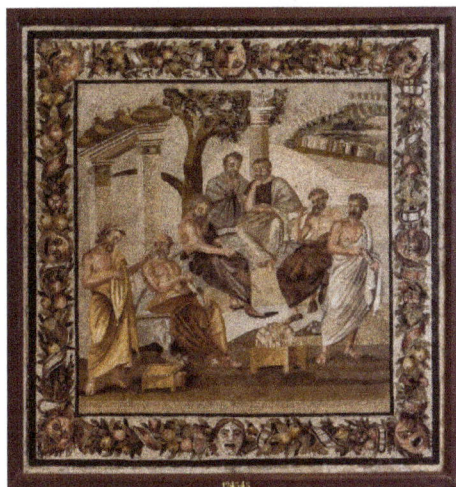

Plato's academy

None of the surviving works of the great fourth-century philosopher Plato (428–347 BC) include any formal logic, but they include important contributions to the field of philosophical logic. Plato raises three questions:

- What is it that can properly be called true or false?

- What is the nature of the connection between the assumptions of a valid argument and its conclusion?

- What is the nature of definition?

The first question arises in the dialogue *Theaetetus*, where Plato identifies thought or opinion with talk or discourse (*logos*). The second question is a result of Plato's theory of Forms. Forms are not things in the ordinary sense, nor strictly ideas in the mind, but they correspond to what philosophers later called universals, namely an abstract entity common to each set of things that have the same name. In both the *Republic* and the *Sophist*, Plato suggests that the necessary connection between the assumptions of a valid argument and its conclusion corresponds to a necessary connection between "forms". The third question is about definition. Many of Plato's dialogues concern the search for a definition of some important concept (justice, truth, the Good), and it is likely that Plato was impressed by the importance of definition in mathematics. What underlies every definition is a Platonic Form, the common nature present in different particular things. Thus, a definition reflects the ultimate object of understanding, and is the foundation of all valid inference. This had a great influence on Plato's student Aristotle, in particular Aristotle's notion of the essence of a thing.

Aristotle

The logic of Aristotle, and particularly his theory of the syllogism, has had an enormous influence in Western thought. His logical works, called the *Organon*, are the earliest formal study of logic that have come down to modern times. Though it is difficult to determine the dates, the probable order of writing of Aristotle's logical works is:

Aristotle

- *The Categories*, a study of the ten kinds of primitive term.

- *The Topics* (with an appendix called *On Sophistical Refutations*), a discussion of dialectics.

- *On Interpretation*, an analysis of simple categorical propositions into simple terms, negation, and signs of quantity. It also contains a comprehensive treatment of the notions of opposition and conversion; chapter 7 is at the origin of the square of opposition (or logical square); chapter 9 contains the beginning of modal logic.

- *The Prior Analytics*, a formal analysis of what makes a syllogism (a valid argument, according to Aristotle).

- *The Posterior Analytics*, a study of scientific demonstration, containing Aristotle's mature views on logic.

Aristotle's logic was still influential in the Renaissance

These works are of outstanding importance in the history of logic. Aristotle was the first logician to attempt a systematic analysis of logical syntax, of noun (or *term*), and of verb. He was the first *formal logician*, in that he demonstrated the principles of reasoning by employing variables to show the underlying logical form of an argument. He was looking for relations of dependence which characterize necessary inference, and distinguished the validity of these relations, from the truth of the premises (the soundness of the argument). He was the first to deal with the principles of contradiction and excluded middle in a systematic way.

In the *Categories*, he attempts to discern all the possible things to which a term can refer; this idea underpins his philosophical work *Metaphysics*, which itself had a profound influence on Western thought. The *Prior Analytics* contains his exposition of the "syllogism", where three important principles are applied for the first time in history: the use of variables, a purely formal treatment, and the use of an axiomatic system. He also developed a theory of non-formal logic (*i.e.*, the theory of fallacies), which is presented in *Topics* and *Sophistical Refutations*.

Stoics

The other great school of Greek logic is that of the Stoics. Stoic logic traces its roots back to the late 5th century BC philosopher Euclid of Megara, a pupil of Socrates and slightly older contemporary of Plato, probably following in the tradition of Parmenides and Zeno. His pupils and successors were called "Megarians", or "Eristics", and later the "Dialecticians". The two most important dialecticians of the Megarian school were Diodorus Cronus and Philo, who were active in the late 4th century BC. The Stoics adopted the Megarian logic and systemized it. The most important member of the school was Chrysippus (c. 278–c. 206 BC), who was its third head, and who formalized much of Stoic doctrine. He is supposed to have written over 700 works, including at least 300 on logic, almost none of which survive. Unlike with Aristotle, we have no complete works by the Megarians or the early Stoics, and have to rely mostly on accounts (sometimes hostile) by later sources, including prominently Diogenes Laertius, Sextus Empiricus, Galen, Aulus Gellius, Alexander of Aphrodisias, and Cicero.

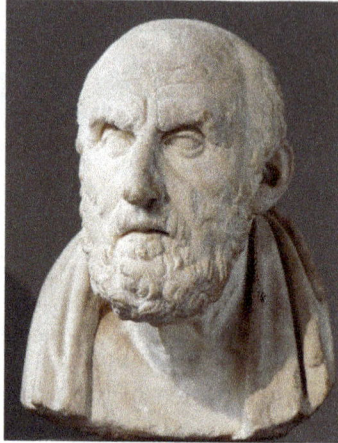

Chrysippus of Soli

Three significant contributions of the Stoic school were (i) their account of modality, (ii) their theory of the Material conditional, and (iii) their account of meaning and truth.

- *Modality.* According to Aristotle, the Megarians of his day claimed there was no distinction between potentiality and actuality. Diodorus Cronus defined the possible as that which either is or will be, the impossible as what will not be true, and the contingent as that which either is already, or will be false. Diodorus is also famous for what is known as his Master argument, which states that each pair of the following 3 propositions contradicts the third proposition:

  - Everything that is past is true and necessary.

  - The impossible does not follow from the possible.

  - What neither is nor will be is possible.

  Diodorus used the plausibility of the first two to prove that nothing is possible if it neither is nor will be true. Chrysippus, by contrast, denied the second premise and said that the impossible could follow from the possible.

- *Conditional statements.* The first logicians to debate conditional statements were Diodorus and his pupil Philo of Megara. Sextus Empiricus refers three times to a debate between Diodorus and Philo. Philo regarded a conditional as true unless it has both a true antecedent and a false consequent. Precisely, let T_o and T_1 be true statements, and let F_o and F_1 be false statements; then, according to Philo, each of the following conditionals is a true statement, because it is not the case that the consequent is false while the antecedent is true (it is not the case that a false statement is asserted to follow from a true statement):

- If T_o, then T_1

- If F_o, then T_o

- If F_o, then F_1

The following conditional does not meet this requirement, and is therefore a false statement according to Philo:

- If T_o, *then* F_o

Indeed, Sextus says "According to [Philo], there are three ways in which a conditional may be true, and one in which it may be false." Philo's criterion of truth is what would now be called a truth-functional definition of "if ... then"; it is the definition used in modern logic.

In contrast, Diodorus allowed the validity of conditionals only when the antecedent clause could never lead to an untrue conclusion. A century later, the Stoic philosopher Chrysippus attacked the assumptions of both Philo and Diodorus.

- *Meaning and truth.* The most important and striking difference between Megarian-Stoic logic and Aristotelian logic is that Megarian-Stoic logic concerns propositions, not terms, and is thus closer to modern propositional logic. The Stoics distinguished between utterance (*phone*), which may be noise, speech (*lexis*), which is articulate but which may be meaningless, and discourse (*logos*), which is meaningful utterance. The most original part of their theory is the idea that what is expressed by a sentence, called a *lekton*, is something real; this corresponds to what is now called a *proposition*. Sextus says that according to the Stoics, three things are linked together: that which signifies, that which is signified, and the object; for example, that which signifies is the word *Dion*, and that which is signified is what Greeks understand but barbarians do not, and the object is Dion himself.

Medieval logic

Logic in the Middle East

The works of Al-Kindi, Al-Farabi, Avicenna, Al-Ghazali, Averroes and other Muslim logicians were based on Aristotelian logic and were important in communicating the ideas of the ancient world to the medieval West. Al-Farabi (Alfarabi) (873–950) was an Aristotelian logician who discussed the topics of future contingents, the number and relation of the categories, the relation between logic and grammar, and non-Aristotelian forms of inference. Al-Farabi also considered the theories of conditional syllogisms and analogical inference, which were part of the Stoic tradition of logic rather than the Aristotelian.

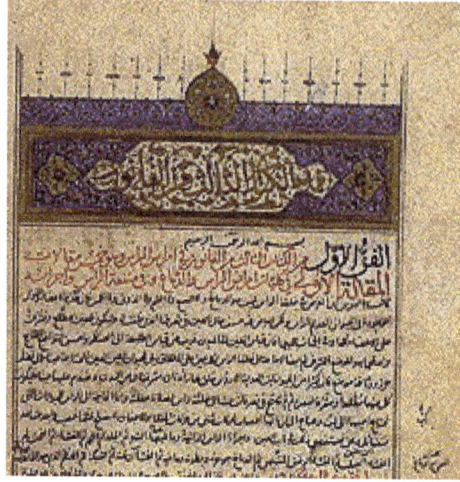

A text by Avicenna, founder of Avicennian logic

Ibn Sina (Avicenna) (980–1037) was the founder of Avicennian logic, which replaced Aristotelian logic as the dominant system of logic in the Islamic world, and also had an important influence on Western medieval writers such as Albertus Magnus. Avicenna wrote on the hypothetical syllogism and on the propositional calculus, which were both part of the Stoic logical tradition. He developed an original "temporally modalized" syllogistic theory, involving temporal logic and modal logic. He also made use of inductive logic, such as the methods of agreement, difference, and concomitant variation which are critical to the scientific method. One of Avicenna's ideas had a particularly important influence on Western logicians such as William of Ockham: Avicenna's word for a meaning or notion (*ma'na*), was translated by the scholastic logicians as the Latin *intentio*; in medieval logic and epistemology, this is a sign in the mind that naturally represents a thing. This was crucial to the development of Ockham's conceptualism: A universal term (*e.g.*, "man") does not signify a thing existing in reality, but rather a sign in the mind (*intentio in intellectu*) which represents many things in reality; Ockham cites Avicenna's commentary on *Metaphysics* V in support of this view.

Fakhr al-Din al-Razi (b. 1149) criticised Aristotle's "first figure" and formulated an early system of inductive logic, foreshadowing the system of inductive logic developed by John Stuart Mill (1806–1873). Al-Razi's work was seen by later Islamic scholars as marking a new direction for Islamic logic, towards a Post-Avicennian logic. This was further elaborated by his student Afdaladdîn al-Khûnajî (d. 1249), who developed a form of logic revolving around the subject matter of conceptions and assents. In response to this tradition, Nasir al-Din al-Tusi (1201–1274) began a tradition of Neo-Avicennian logic which remained faithful to Avicenna's work and existed as an alternative to the more dominant Post-Avicennian school over the following centuries.

The Illuminationist school was founded by Shahab al-Din Suhrawardi (1155–1191), who developed the idea of "decisive necessity", which refers to the reduction of all modalities (necessity, possibility, contingency and impossibility) to the single mode of necessity. Ibn al-Nafis (1213–1288) wrote a book on Avicennian logic, which was a commentary of Avicenna's *Al-Isharat* (*The Signs*) and *Al-Hidayah* (*The Guidance*). Ibn Taymiyyah (1263–1328), wrote the *Ar-Radd 'ala al-Mantiqiyyin*, where he argued against the usefulness, though not the validity, of the syllogism and in favour of inductive reasoning. Ibn Taymiyyah also argued against the certainty of syllogistic arguments and in favour of analogy; his argument is that concepts founded on induction are themselves not certain but only probable, and thus a syllogism based on such concepts is no more certain than an argument based on analogy. He further claimed that induction itself is founded on a process of analogy. His model of analogical reasoning was based on that of juridical arguments. This model of analogy has been used in the recent work of John F. Sowa.

The *Sharh al-takmil fi'l-mantiq* written by Muhammad ibn Fayd Allah ibn Muhammad Amin al-Sharwani in the 15th century is the last major Arabic work on logic that has been studied. However, "thousands upon thousands of pages" on logic were written between the 14th and 19th centuries, though only a fraction of the texts written during this period have been studied by historians, hence little is known about the original work on Islamic logic produced during this later period.

Logic in Medieval Europe

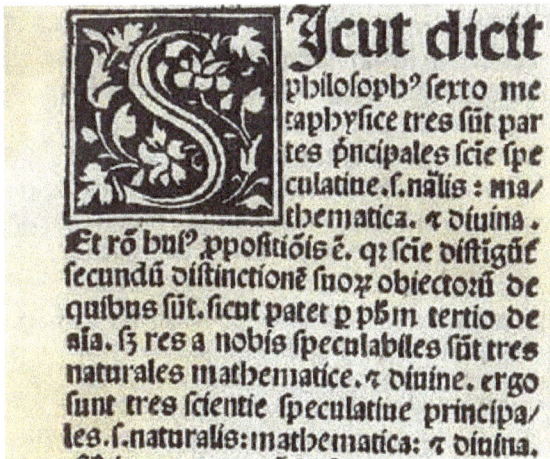

Brito's questions on the *Old Logic*

"Medieval logic" (also known as "Scholastic logic") generally means the form of Aristotelian logic developed in medieval Europe throughout roughly the period 1200–1600. For centuries after Stoic logic had been formulated, it was the dominant system of logic in the classical world. When the study of logic resumed after the Dark Ages, the main source was the work of the Christian philosopher Boethius, who was familiar with some of Aristotle's logic, but almost none of the work of the Stoics. Until the twelfth century, the only works of Aristotle available in the West were the *Categories, On Interpretation,* and Boethius's translation of the Isagoge of Porphyry (a commentary on the Categories). These works were known as the "Old Logic" (*Logica Vetus* or *Ars Vetus*). An important work in this tradition was the *Logica Ingredientibus* of Peter Abelard (1079–1142). His direct influence was small, but his influence through pupils such as John of Salisbury was great, and his method of applying rigorous logical analysis to theology shaped the way that theological criticism developed in the period that followed.

By the early thirteenth century, the remaining works of Aristotle's *Organon* (including the *Prior Analytics, Posterior Analytics,* and the *Sophistical Refutations*) had been recovered in the West. Logical work until then was mostly paraphrasis or commentary on the work of Aristotle. The period from the middle of the thirteenth to the middle of the fourteenth century was one of significant developments in logic, particularly in three areas which were original, with little foundation in the Aristotelian tradition that came before. These were:

- The theory of supposition. Supposition theory deals with the way that predicates (*e.g.,* 'man') range over a domain of individuals (*e.g.,* all men). In the proposition 'every man is an animal', does the term 'man' range over or 'supposit for' men existing just in the present, or does the range include past and future men? Can a term supposit for a non-existing individual? Some medievalists have argued that this idea is a precursor of modern first-order logic. "The theory of supposition with the associated theories of *copulatio* (sign-capacity of adjectival terms), *ampliatio* (widening of referential domain), and *distributio* constitute one of the most original achievements of Western medieval logic".

- The theory of syncategoremata. Syncategoremata are terms which are necessary for logic, but which, unlike *categorematic* terms, do not signify on their own behalf, but 'co-signify' with other words. Examples of syncategoremata are 'and', 'not', 'every', 'if', and so on.

- The theory of consequences. A consequence is a hypothetical, conditional proposition: two propositions joined by the terms 'if … then'. For example, 'if a man runs, then God exists' (*Si homo currit, Deus est*). A fully developed theory of consequences is given in Book III of William of Ockham's work Summa Logicae. There, Ockham distinguishes between 'material' and 'formal' consequences, which are roughly equivalent to the modern material implication and logical implication respectively. Similar accounts are given by Jean Buridan and Albert of Saxony.

The last great works in this tradition are the *Logic* of John Poinsot (1589–1644), known as John of St Thomas), the *Metaphysical Disputations* of Francisco Suarez (1548–1617), and the *Logica Demonstrativa* of Giovanni Girolamo Saccheri (1667–1733).

Traditional Logic

The Textbook Tradition

Dudley Fenner's *Art of Logic* (1584)

Traditional logic generally means the textbook tradition that begins with Antoine Arnauld's and Pierre Nicole's *Logic, or the Art of Thinking*, better known as the *Port-Royal Logic*. Published in 1662, it was the most influential work on logic after Aristotle until the nineteenth century. The book presents a loosely Cartesian doctrine (that the proposition is a combining of ideas rather than terms, for example) within a framework that is broadly derived from Aristotelian and medieval term logic. Between 1664 and 1700, there were eight editions, and the book had considerable influence after that. The Port-Royal introduces the concepts of extension and intension. The account of propositions that Locke gives in the *Essay* is essentially that of the Port-Royal: "Verbal propositions, which are words, [are] the signs of our ideas, put together or separated in affirmative or negative sentences. So that proposition consists in the putting together or separating these signs, according as the things which they stand for agree or disagree."

Dudley Fenner helped popularize Ramist logic, a reaction against Aristotle. Another influential work was the *Novum Organum* by Francis Bacon, published in 1620. The title translates as "new instrument". This is a reference to Aristotle's work known as the *Organon*. In this work, Bacon rejects the syllogistic method of Aristotle in favor of an alternative procedure "which by slow and faithful toil gathers information from things and brings it into understanding". This method is known as inductive reasoning, a method which starts from empirical observation and proceeds to lower axioms or propositions; from these lower axioms, more general ones can be induced. For example, in finding the cause of a *phenomenal nature* such as heat, 3 lists should be constructed:

- The presence list: a list of every situation where heat is found.

- The absence list: a list of every situation that is similar to at least one of those of the presence list, except for the lack of heat.

- The variability list: a list of every situation where heat can vary.

Then, the *form nature* (or cause) of heat may be defined as that which is common to every situation of the presence list, and which is lacking from every situation of the absence list, and which varies by degree in every situation of the variability list.

Other works in the textbook tradition include Isaac Watts's *Logick: Or, the Right Use of Reason* (1725), Richard Whately's *Logic* (1826), and John Stuart Mill's *A System of Logic* (1843). Although the latter was one of the last great works in the tradition, Mill's view that the foundations of logic lie in introspection influenced the view that logic is best understood as a branch of psychology, a view which dominated the next fifty years of its development, especially in Germany.

Logic in Hegel's Philosophy

Georg Wilhelm Friedrich Hegel

G.W.F. Hegel indicated the importance of logic to his philosophical system when he condensed his extensive *Science of Logic* into a shorter work published in 1817 as the first volume of his *Encyclopaedia of the Philosophical Sciences*. The "Shorter" or "Encyclopaedia" *Logic*, as it is often known, lays out a series of transitions which leads from the most empty and abstract of categories—Hegel begins with "Pure Being" and "Pure Nothing"—to the "Absolute, the category which contains and resolves all the categories which preceded it. Despite the title, Hegel's *Logic* is not really a contribution to the science of valid inference. Rather than deriving conclusions about concepts through valid inference from premises, Hegel seeks to show that thinking about one concept compels thinking about another concept (one cannot, he argues, possess the concept of "Quality" without the concept of "Quantity"); this compulsion is, supposedly, not a matter of individual psychology, because it arises almost organically from the content of the concepts themselves. His purpose is to show the rational structure of the "Absolute"—indeed of rationality itself. The method by which thought is driven from one concept to its contrary, and then to further concepts, is known as the Hegelian dialectic.

Although Hegel's *Logic* has had little impact on mainstream logical studies, its influence can be seen elsewhere:

- Carl von Prantl's *Geschichte der Logik in Abendland* (1855–1867).

- The work of the British Idealists, such as F.H. Bradley's *Principles of Logic* (1883).

- The economic, political, and philosophical studies of Karl Marx, and in the various schools of Marxism.

Logic and Psychology

Between the work of Mill and Frege stretched half a century during which logic was widely treated as a descriptive science, an empirical study of the structure of reasoning, and thus essentially as a branch of psychology. The German psychologist Wilhelm Wundt, for example, discussed deriving "the logical from the psychological laws of thought", emphasizing that "psychological thinking is always the more comprehensive form of thinking." This view was widespread among German philosophers of the period:

- Theodor Lipps described logic as "a specific discipline of psychology".

- Christoph von Sigwart understood logical necessity as grounded in the individual's compulsion to think in a certain way.

- Benno Erdmann argued that "logical laws only hold within the limits of our thinking".

Such was the dominant view of logic in the years following Mill's work. This psychological approach to logic was rejected by Gottlob Frege. It was also subjected to an extended and destructive critique by Edmund Husserl in the first volume of his *Logical Investigations* (1900), an assault which has been described as "overwhelming". Husserl argued forcefully that grounding logic in psychological observations implied that all logical truths remained unproven, and that skepticism and relativism were unavoidable consequences.

Such criticisms did not immediately extirpate what is called "psychologism". For example, the American philosopher Josiah Royce, while acknowledging the force of Husserl's critique, remained "unable to doubt" that progress in psychology would be accompanied by progress in logic, and vice versa.

Rise of Modern Logic

The period between the fourteenth century and the beginning of the nineteenth century had been largely one of decline and neglect, and is generally regarded as barren by historians of logic. The revival of logic occurred in the mid-nineteenth century, at the beginning of a revolutionary period where the subject developed into a rigorous and formalistic discipline whose exemplar was the exact method of proof used in mathematics. The development of the modern "symbolic" or "mathematical" logic during this period is the most significant in the 2000-year history of logic, and is arguably one of the most important and remarkable events in human intellectual history.

A number of features distinguish modern logic from the old Aristotelian or traditional logic, the most important of which are as follows: Modern logic is fundamentally a *calculus* whose rules of operation are determined only by the *shape* and not by the *meaning* of the symbols it employs, as in mathematics. Many logicians were impressed by the "success" of mathematics, in that there had been no prolonged dispute about any truly mathematical result. C.S. Peirce noted that even though a mistake in the evaluation of a definite integral by Laplace led to an error concerning the moon's orbit that persisted for nearly 50 years, the mistake, once spotted, was corrected without any serious dispute. Peirce contrasted this with the disputation and uncertainty surrounding traditional logic, and especially reasoning in metaphysics. He argued that a truly "exact" logic would depend upon mathematical, i.e., "diagrammatic" or "iconic" thought. "Those who follow such methods will ... escape all error except such as will be speedily corrected after it is once suspected". Modern logic is also "constructive" rather than "abstractive"; i.e., rather than abstracting and formalising theorems derived from ordinary language (or from psychological intuitions about validity), it constructs theorems by formal methods, then looks for an interpretation in ordinary language. It is entirely symbolic, meaning that even the logical constants (which the medieval logicians called "syncategoremata") and the categoric terms are expressed in symbols.

Modern Logic

The development of modern logic falls into roughly five periods:

- The embryonic period from Leibniz to 1847, when the notion of a logical calculus was discussed and developed, particularly by Leibniz, but no schools were formed, and isolated periodic attempts were abandoned or went unnoticed.

- The algebraic period from Boole's Analysis to Schröder's *Vorlesungen*. In this period, there were more practitioners, and a greater continuity of development.

- The logicist period from the Begriffsschrift of Frege to the *Principia Mathematica* of Russell and Whitehead. The aim of the "logicist school" was to incorporate the logic of all mathematical and scientific discourse in a single unified system which, taking as a fundamental principle that all mathematical truths are logical, did not accept any non-logical terminology. The major logicists were Frege, Russell, and the early Wittgenstein. It culminates with the *Principia*, an important work which includes a thorough examination and attempted solution of the antinomies which had been an obstacle to earlier progress.

- The metamathematical period from 1910 to the 1930s, which saw the development of metalogic, in the finitist system of Hilbert, and the non-finitist system of Löwenheim and Skolem, the combination of logic and metalogic in the work of Gödel and Tarski. Gödel's incompleteness theorem of 1931 was one of the greatest achievements in the history of logic. Later in the 1930s, Gödel developed the notion of set-theoretic constructibility.

- The period after World War II, when mathematical logic branched into four inter-related but separate areas of research: model theory, proof theory, computability theory, and set theory, and its ideas and methods began to influence philosophy.

Algebraic Period

George Boole

Modern logic begins with what is known as the "algebraic school", originating with Boole and including Peirce, Jevons, Schröder, and Venn. Their objective was to develop a calculus to formalise reasoning in the area of classes, propositions, and probabilities. The school begins with Boole's seminal work *Mathematical Analysis of Logic* which appeared in 1847, although De Morgan (1847) is its immediate precursor. The fundamental idea of Boole's system is that algebraic formulae can be used to express logical relations. This idea occurred to Boole in his teenage years, working as an usher in a private school in Lincoln, Lincolnshire. For example, let x and y stand for classes let the symbol = signify that the classes have the same members, xy stand for the class containing all and only the members of x and y and so on. Boole calls these *elective symbols*, i.e. symbols which select certain objects for consideration. An expression in which elective symbols are used is called an *elective function*, and an equation of which the members are elective functions, is an *elective equation*. The theory of elective functions and their "development" is essentially the modern idea of truth-functions and their expression in disjunctive normal form.

Boole's system admits of two interpretations, in class logic, and propositional logic. Boole distinguished between "primary propositions" which are the subject of syllogistic theory, and "secondary propositions", which are the subject of propositional logic, and showed how under different "interpretations" the same algebraic system could represent both. An example of a primary proposition is "All inhabitants are either Europeans or Asiatics." An example of a secondary proposition is "Either all inhabitants are Europeans or they are all Asiatics." These are easily distinguished in modern propositional calculus, where it is also possible to show that the first follows from the second, but it is a significant disadvantage that there is no way of representing this in the Boolean system.

In his *Symbolic Logic* (1881), John Venn used diagrams of overlapping areas to express Boolean relations between classes or truth-conditions of propositions. In 1869 Jevons realised that Boole's methods could be mechanised, and constructed a "logical machine" which he showed to the Royal Society the following year. In 1885 Allan Marquand proposed an electrical version of the machine that is still extant (picture at the Firestone Library).

Charles Sanders Peirce

The defects in Boole's system (such as the use of the letter v for existential propositions) were all remedied by his followers. Jevons published *Pure Logic, or the Logic of Quality apart from Quantity* in 1864, where he suggested a symbol to signify exclusive or, which allowed Boole's system to be greatly simplified. This was usefully exploited by Schröder when he set out theorems in parallel columns in his *Vorlesungen* (1890–1905). Peirce (1880) showed how all the Boolean elective functions could be expressed by the use of a single primitive binary operation, "neither ... nor ..." and equally well "not both ... and ...", however, like many of Peirce's innovations, this remained unknown or unnoticed until Sheffer rediscovered it in 1913. Boole's early work also lacks the idea of the logical sum which originates in Peirce (1867), Schröder (1877) and Jevons (1890), and the concept of inclusion, first suggested by Gergonne (1816) and clearly articulated by Peirce (1870).

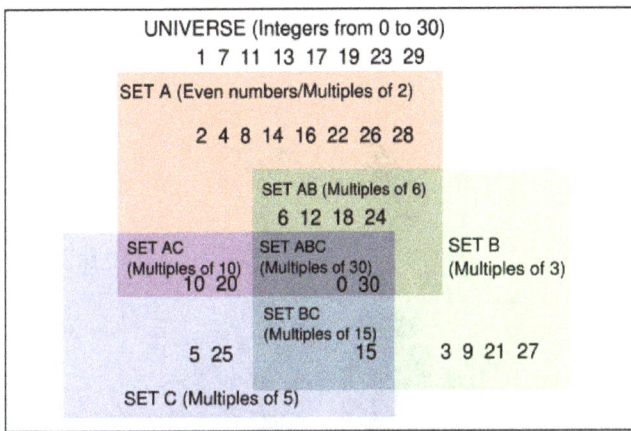

Boolean multiples

The success of Boole's algebraic system suggested that all logic must be capable of algebraic representation, and there were attempts to express a logic of relations in such form, of which the most ambitious was Schröder's monumental *Vorlesungen über die Algebra der Logik* ("Lectures on the Algebra of Logic", vol iii 1895), although the original idea was again anticipated by Peirce.

Boole's unwavering acceptance of Aristotle's logic is emphasized by the historian of logic John Corcoran in an accessible introduction to *Laws of Thought* Corcoran also wrote a point-by-point comparison of *Prior Analytics* and *Laws of Thought*. According to Corcoran, Boole fully accepted and endorsed Aristotle's logic. Boole's goals were "to go under, over, and beyond" Aristotle's logic by 1) providing it with mathematical foundations involving equations, 2) extending the class of problems it could treat — from assessing validity to solving equations — and 3) expanding the range of applications it could handle — e.g. from propositions having only two terms to those having arbitrarily many.

More specifically, Boole agreed with what Aristotle said; Boole's 'disagreements', if they might be called that, concern what Aristotle did not say. First, in the realm of foundations, Boole reduced the four propositional forms of Aristotelian logic to formulas in the form of equations — by itself a revolutionary idea. Second, in the realm of logic's problems, Boole's addition of equation solving to logic — another revolutionary idea — involved Boole's doctrine that Aristotle's rules of inference (the "perfect syllogisms") must be supplemented by rules for equation solving. Third, in the realm of applications, Boole's system could handle multi-term propositions and arguments whereas Aristotle could handle only two-termed subject-predicate propositions and arguments. For example, Aristotle's system could not deduce "No quadrangle that is a square is a rectangle that is a rhombus" from "No square that is a quadrangle is a rhombus that is a rectangle" or from "No rhombus that is a rectangle is a square that is a quadrangle".

Logicist Period

Gottlob Frege.

After Boole, the next great advances were made by the German mathematician Gottlob Frege. Frege's objective was the program of Logicism, i.e. demonstrating that arithmetic is identical with logic. Frege went much further than any of his predecessors in his rigorous and formal approach to logic, and his calculus or Begriffsschrift is important. Frege also tried to show that the concept of number can be defined by purely logical means, so that (if he was right) logic includes arithmetic and all branches of mathematics that are reducible to arithmetic. He was not the first writer to suggest this. In his pioneering work *Die Grundlagen der Arithmetik* (The Foundations of Arithmetic), sections 15–17, he acknowledges the efforts of Leibniz, J.S. Mill as well as Jevons, citing the latter's claim that "algebra is a highly developed logic, and number but logical discrimination."

Frege's first work, the *Begriffsschrift* ("concept script") is a rigorously axiomatised system of propositional logic, relying on just two connectives (negational and conditional), two rules of inference (*modus ponens* and substitution), and six axioms. Frege referred to the "completeness" of this system, but was unable to prove this. The most significant innovation, however, was his explanation of the quantifier in terms of mathematical functions. Traditional logic regards the sentence "Caesar is a man" as of fundamentally the same form as "all men are mortal." Sentences with a proper name subject were regarded as universal in character, interpretable as "every Caesar is a man". At the outset Frege abandons the traditional "concepts *subject* and *predicate*", replacing them with *argument* and *function* respectively, which he believes "will stand the test of time. It is easy to see how regarding a content as a function of an argument leads to the formation of concepts. Furthermore, the demonstration of the connection between the meanings of the words *if, and, not, or, there is, some, all,* and so forth, deserves attention". Frege argued that the quantifier expression "all men" does not have the same logical or semantic form as "all men", and that the universal proposition "every A is B" is a complex proposition involving two *functions*, namely ' – is A' and ' – is B' such that whatever satisfies the first, also satisfies the second. In modern notation, this would be expressed as

$$\forall x(A(x) \rightarrow B(x))$$

In English, "for all x, if Ax then Bx". Thus only singular propositions are of subject-predicate form, and they are irreducibly singular, i.e. not reducible to a general proposition. Universal and particular propositions, by contrast, are not of simple subject-predicate form at all. If "all mammals" were the logical subject of the sentence "all mammals are land-dwellers", then to negate the whole sentence we would have to negate the predicate to give "all mammals are *not* land-dwellers". But this is not the case. This functional analysis of ordinary-language sentences later had a great impact on philosophy and linguistics.

This means that in Frege's calculus, Boole's "primary" propositions can be represented in a different way from "secondary" propositions. "All inhabitants are either men or women" is

Frege's "Concept Script"

$$\forall x(I(x) \rightarrow (M(x) \vee W(x)))$$

whereas "All the inhabitants are men or all the inhabitants are women" is

$$\forall x(I(x) \rightarrow M(x)) \vee \forall x(I(x) \rightarrow W(x))$$

As Frege remarked in a critique of Boole's calculus:

"The real difference is that I avoid [the Boolean] division into two parts ... and give a homogeneous presentation of the lot. In Boole the two parts run alongside one another, so that one is like the mirror image of the other, but for that very reason stands in no organic relation to it'

As well as providing a unified and comprehensive system of logic, Frege's calculus also resolved the ancient problem of multiple generality. The ambiguity of "every girl kissed a boy" is difficult to express in traditional logic, but Frege's logic resolves this through the different scope of the quantifiers. Thus

$$\forall\, x(G(x) \to \exists\, y(B(y) \land K(x,y)))$$

Peano

means that to every girl there corresponds some boy (any one will do) who the girl kissed. But

$$\exists\, x(B(x) \land \forall\, y(G(y) \to K(y,x)))$$

means that there is some particular boy whom every girl kissed. Without this device, the project of logicism would have been doubtful or impossible. Using it, Frege provided a definition of the ancestral relation, of the many-to-one relation, and of mathematical induction.

This period overlaps with the work of what is known as the "mathematical school", which included Dedekind, Pasch, Peano, Hilbert, Zermelo, Huntington, Veblen and Heyting. Their objective was the axiomatisation of branches of mathematics like geometry, arithmetic, analysis and set theory. Most notable was Hilbert's Program, which sought to ground all of mathematics to a finite set of axioms, proving its consistency by "finitistic" means and providing a procedure which would decide the truth or falsity of any mathematical statement. The standard axiomatization of the natural numbers is named the Peano axioms in his honor. Peano maintained a clear distinction between mathematical and logical symbols. While unaware of Frege's work, he independently recreated his logical apparatus based on the work of Boole and Schröder.

Ernst Zermelo

The logicist project received a near-fatal setback with the discovery of a paradox in 1901 by Bertrand Russell. This proved Frege's naive set theory led to a contradiction. Frege's theory contained the axiom that for any formal criterion, there is a set of all objects that meet the criterion. Russell showed that a set containing exactly the sets that are not members of themselves would contradict its own definition (if it is not a member of itself, it is a member of itself, and if it is a member of itself, it is not). This contradiction is now known as Russell's paradox. One important method of resolving this paradox was proposed by Ernst Zermelo. Zermelo set theory was the first axiomatic set theory. It was developed into the now-canonical Zermelo–Fraenkel set theory (ZF). Russell's paradox symbolically is as follows:

$$\text{Let } R = \{x \mid x \notin x\}, \text{ then } R \in R \in R \notin R$$

The monumental Principia Mathematica, a three-volume work on the foundations of mathematics, written by Russell and Alfred North Whitehead and published 1910–13 also included an attempt to resolve the paradox, by means of an elaborate system of types: a set of elements is of a different type than is each of its elements (set is not the element; one element is not the set) and one cannot speak of the "set of all sets". The *Principia* was an attempt to derive all mathematical truths from a well-defined set of axioms and inference rules in symbolic logic.

Metamathematical Period

Kurt Gödel

The names of Gödel and Tarski dominate the 1930s, a crucial period in the development of metamathematics – the study of mathematics using mathematical methods to produce metatheories, or mathematical theories about other mathematical theories. Early investigations into metamathematics had been driven by Hilbert's program. Work on metamathematics culminated in the work of Gödel, who in 1929 showed that a given first-order sentence is deducible if and only if it is logically valid – i.e. it is true in every structure for its language. This is known as Gödel's completeness theorem. A year later, he proved two important theorems, which showed Hibert's program to be unattainable in its original form. The first is that no consistent system of axioms whose theorems can be listed by an effective procedure such as an algorithm or computer program is capable of proving all facts about the natural numbers. For any such system, there will always be statements about the natural numbers that are true, but that are unprovable within the system. The second is that if such a system is also capable of proving certain basic facts about the natural numbers, then the system cannot prove the consistency of the system itself. These two results are known as Gödel's incompleteness theorems, or simply *Gödel's Theorem*. Later in the decade, Gödel developed the concept of set-theoretic constructibility, as part of his proof that the axiom of choice and the continuum hypothesis are consistent with Zermelo–Fraenkel set theory. In proof theory, Gerhard Gentzen developed natural deduction and the sequent calculus. The former attempts to model logical reasoning as it 'naturally' occurs in practice and is most easily applied to intuitionistic logic, while the latter was devised to clarify the derivation of logical proofs in any formal system. Since Gentzen's work, natural deduction and sequent calculi have been widely applied in the fields of proof theory, mathematical logic and computer science. Gentzen also proved normalization and cut-elimination theorems for intuitionistic and classical logic which could be used to reduce logical proofs to a normal form.

Alfred Tarski

Alfred Tarski, a pupil of Łukasiewicz, is best known for his definition of truth and logical consequence, and the semantic concept of logical satisfaction. In 1933, he published (in Polish) *The concept of truth in formalized languages*, in which he proposed his semantic theory of truth: a sentence such as "snow is white" is true if and only if snow is white. Tarski's theory separated the metalanguage, which makes the statement about truth, from the object language, which contains the sentence whose truth is being asserted, and gave a correspondence (the T-schema) between phrases in the object language and elements of an interpretation. Tarski's approach to the difficult idea of explaining truth has been enduringly influential in logic and philosophy, especially in the development of model theory. Tarski also produced important work on the methodology of deductive systems, and on fundamental principles such as completeness, decidability, consistency and definability. According to Anita Feferman, Tarski "changed the face of logic in the twentieth century".

Alonzo Church and Alan Turing proposed formal models of computability, giving independent negative solutions to Hilbert's *Entscheidungsproblem* in 1936 and 1937, respectively. The *Entscheidungsproblem* asked for a procedure that, given any formal mathematical statement, would algorithmically determine whether the statement is true. Church and Turing proved there is no such procedure; Turing's paper introduced the halting problem as a key example of a mathematical problem without an algorithmic solution.

Church's system for computation developed into the modern λ-calculus, while the Turing machine became a standard model for a general-purpose computing device. It was soon shown that many other proposed models of computation were equivalent in power to those proposed by Church and Turing. These results led to the Church–Turing thesis that any deterministic algorithm that can be carried out by a human can be carried out by a Turing machine. Church proved additional undecidability results, showing that both Peano arithmetic and first-order logic are undecidable. Later work by Emil Post and Stephen Cole Kleene in the 1940s extended the scope of computability theory and introduced the concept of degrees of unsolvability.

The results of the first few decades of the twentieth century also had an impact upon analytic philosophy and philosophical logic, particularly from the 1950s onwards, in subjects such as modal logic, temporal logic, deontic logic, and relevance logic.

Logic After WWII

Saul Kripke

After World War II, mathematical logic branched into four inter-related but separate areas of research: model theory, proof theory, computability theory, and set theory.

In set theory, the method of forcing revolutionized the field by providing a robust method for constructing models and obtaining independence results. Paul Cohen introduced this method in 1963 to prove the independence of the continuum hypothesis and the axiom of choice from Zermelo–Fraenkel set theory. His technique, which was simplified and extended soon after its introduction, has since been applied to many other problems in all areas of mathematical logic.

Computability theory had its roots in the work of Turing, Church, Kleene, and Post in the 1930s and 40s. It developed into a study of abstract computability, which became known as recursion theory. The priority method, discovered independently by Albert Muchnik and Richard Friedberg in the 1950s, led to major advances in the understanding of the degrees of unsolvability and related structures. Research into higher-order computability theory demonstrated its connections to set theory. The fields of constructive analysis and computable analysis were developed to study the effective content of classical mathematical theorems; these in turn inspired the program of reverse mathematics. A separate branch of computability theory, computational complexity theory, was also characterized in logical terms as a result of investigations into descriptive complexity.

Model theory applies the methods of mathematical logic to study models of particular mathematical theories. Alfred Tarski published much pioneering work in the field, which is named after a series of papers he published under the title *Contributions to the theory of models*. In the 1960s, Abraham Robinson used model-theoretic techniques to develop calculus and analysis based on infinitesimals, a problem that first had been proposed by Leibniz.

In proof theory, the relationship between classical mathematics and intuitionistic mathematics was clarified via tools such as the realizability method invented by Georg Kreisel and Gödel's *Dialectica* interpretation. This work inspired the contemporary area of proof mining. The Curry-Howard correspondence emerged as a deep analogy between logic and computation, including a correspondence between systems of natural deduction and typed lambda calculi used in computer science. As a result, research into this class of formal systems began to address both logical and computational aspects; this area of research came to be known as modern type theory. Advances were also made in ordinal analysis and the study of independence results in arithmetic such as the Paris–Harrington theorem.

This was also a period, particularly in the 1950s and afterwards, when the ideas of mathematical logic begin to influence philosophical thinking. For example, tense logic is a formalised system for representing, and reasoning about, propositions qualified in terms of time. The philosopher Arthur Prior played a significant role in its development in the 1960s. Modal logics extend the scope of formal logic to include the elements of modality (for example, possibility and necessity). The ideas of Saul Kripke, particularly about possible worlds, and the formal system now called Kripke semantics have had a profound impact on analytic philosophy. His best known and most influential work is *Naming and Necessity* (1980). Deontic logics are closely related to modal logics: they attempt to capture the logical features of obligation, permission and related concepts. Although some basic novelties syncretizing mathematical and philosophical logic were shown by Bolzano in the early 1800s, it was Ernst Mally, a pupil of Alexius Meinong, who was to propose the first formal deontic system in his *Grundgesetze des Sollens*, based on the syntax of Whitehead's and Russell's propositional calculus.

Another logical system founded after World War II was fuzzy logic by Azerbaijani mathematician Lotfi Asker Zadeh in 1965.

Logic in the East

Logic in India

Logic began independently in ancient India and continued to develop to early modern times without any known influence from Greek logic. Medhatithi Gautama (c. 6th century BC) founded the *anviksiki* school of logic. The *Mahabharata* (12.173.45), around the 5th century BC, refers to the *anviksiki* and *tarka* schools of logic. Pāṇini (c. 5th century BC) developed a form of logic (to which Boolean logic has some similarities) for his formulation of Sanskrit grammar. Logic is described by Chanakya (c. 350-283 BC) in his *Arthashastra* as an independent field of inquiry.

Two of the six Indian schools of thought deal with logic: Nyaya and Vaisheshika. The

Nyaya Sutras of Aksapada Gautama (c. 2nd century AD) constitute the core texts of the Nyaya school, one of the six orthodox schools of Hindu philosophy. This realist school developed a rigid five-member schema of inference involving an initial premise, a reason, an example, an application, and a conclusion. The idealist Buddhist philosophy became the chief opponent to the Naiyayikas. Nagarjuna (c. 150-250 AD), the founder of the Madhyamika ("Middle Way") developed an analysis known as the catu⯑ko⯑i (Sanskrit), a "four-cornered" system of argumentation that involves the systematic examination and rejection of each of the 4 possibilities of a proposition, P:

1. P; that is, being.

2. not P; that is, not being.

3. P and not P; that is, being and not being.

4. not (P or not P); that is, neither being nor not being.

> It is interesting to note that under propositional logic, De Morgan's laws imply that this is equivalent to the third case (P and not P), and is therefore superfluous; there are actually only 3 cases to consider.

However, Dignaga (c 480-540 AD) is sometimes said to have developed a formal syllogism, and it was through him and his successor, Dharmakirti, that Buddhist logic reached its height; it is contested whether their analysis actually constitutes a formal syllogistic system. In particular, their analysis centered on the definition of an inference-warranting relation, "vyapti", also known as invariable concomitance or pervasion. To this end, a doctrine known as "apoha" or differentiation was developed. This involved what might be called inclusion and exclusion of defining properties.

The difficulties involved in this enterprise, in part, stimulated the neo-scholastic school of Navya-Nyaya, which developed a formal analysis of inference in the sixteenth century. This later school began around eastern India and Bengal, and developed theories resembling modern logic, such as Gottlob Frege's "distinction between sense and reference of proper names" and his "definition of number," as well as the Navya-Nyaya theory of "restrictive conditions for universals" anticipating some of the developments in modern set theory. Since 1824, Indian logic attracted the attention of many Western scholars, and has had an influence on important 19th-century logicians such as Charles Babbage, Augustus De Morgan, and particularly George Boole, as confirmed by his wife Mary Everest Boole, who wrote in 1901 an "open letter to Dr Bose", which was titled "Indian Thought and Western Science in the Nineteenth Century" and stated: "Think what must have been the effect of the intense Hinduizing of three such men as Babbage, De Morgan and George Boole on the mathematical atmosphere of 1830-1865".

Dignāga's famous "wheel of reason" (*Hetucakra*) is a method of indicating when one thing (such as smoke) can be taken as an invariable sign of another thing (like fire), but the inference is often inductive and based on past observation. Matilal remarks that Dignāga's analysis is much like John Stuart Mill's Joint Method of Agreement and Difference, which is inductive.

In addition, the traditional five-member Indian syllogism, though deductively valid, has repetitions that are unnecessary to its logical validity. As a result, some commentators see the traditional Indian syllogism as a rhetorical form that is entirely natural in many cultures of the world, and yet not as a logical form—not in the sense that all logically unnecessary elements have been omitted for the sake of analysis.

Logic in China

In China, a contemporary of Confucius, Mozi, "Master Mo", is credited with founding the Mohist school, whose canons dealt with issues relating to valid inference and the conditions of correct conclusions. In particular, one of the schools that grew out of Mohism, the Logicians, are credited by some scholars for their early investigation of formal logic. Due to the harsh rule of Legalism in the subsequent Qin Dynasty, this line of investigation disappeared in China until the introduction of Indian philosophy by Buddhists.

References

- Łukasiewicz, Jan (1957). Aristotle's syllogistic from the standpoint of modern formal logic (2nd ed.). Oxford University Press. p. 7. ISBN 978-0-19-824144-7.

- Whitehead, Alfred North; Russell, Bertrand (1967). Principia Mathematica to *56. Cambridge University Press. ISBN 0-521-62606-4.

- For a more modern treatment, see Hamilton, A. G. (1980). Logic for Mathematicians. Cambridge University Press. ISBN 0-521-29291-3.

- Bergmann, Merrie; Moor, James; Nelson, Jack (2009). The Logic Book (Fifth ed.). New York, NY: McGraw-Hill. ISBN 978-0-07-353563-0.

- Mendelson, Elliott (1964). "Quantification Theory: Completeness Theorems". Introduction to Mathematical Logic. Van Nostrand. ISBN 0-412-80830-7.

- Magnani, L. "Abduction, Reason, and Science: Processes of Discovery and Explanation". Kluwer Academic Plenum Publishers, New York, 2001. xvii. 205 pages. Hard cover, ISBN 0-306-46514-0.

- Simo Knuuttila (1981). "Reforging the great chain of being: studies of the history of modal theories". Springer Science & Business. p.71. ISBN 90-277-1125-9

- Michael Fisher, Dov M. Gabbay, Lluís Vila (2005). "Handbook of temporal reasoning in artificial intelligence". Elsevier. p.119. ISBN 0-444-51493-7

- Harold Joseph Berman (1983). "Law and revolution: the formation of the Western legal tradition". Harvard University Press. p.133. ISBN 0-674-51776-8

- Stolyar, Abram A. (1983). Introduction to Elementary Mathematical Logic. Dover Publications. p. 3. ISBN 0-486-64561-4.

- Barnes, Jonathan (1995). The Cambridge Companion to Aristotle. Cambridge University Press. p. 27. ISBN 0-521-42294-9.

- Brookshear, J. Glenn (1989). "Computability: Foundations of Recursive Function Theory". Theory of computation: formal languages, automata, and complexity. Redwood City, Calif.: Benjamin/ Cummings Pub. Co. ISBN 0-8053-0143-7.

- Goldman, Alvin I. (1986), Epistemology and Cognition, Harvard University Press, p. 293, ISBN 9780674258969, untrained subjects are prone to commit various sorts of fallacies and mistakes.

- Demetriou, A.; Efklides, A., eds. (1994), Intelligence, Mind, and Reasoning: Structure and Development, Advances in Psychology, 106, Elsevier, p. 194, ISBN 9780080867601.

- Hegel, G. W. F (1971) [1817]. Philosophy of Mind. Encyclopedia of the Philosophical Sciences. trans. William Wallace. Oxford: Clarendon Press. p. 174. ISBN 0-19-875014-5.

- Joseph E. Brenner (3 August 2008). Logic in Reality. Springer. pp. 28–30. ISBN 978-1-4020-8374-7. Retrieved 9 April 2012.

- Ernie Lepore, Kirk Ludwig (2002). "What is logical form?". In Gerhard Preyer, Georg Peter. Logical form and language. Clarendon Press. p. 54. ISBN 978-0-19-924555-0. preprint

- Hurley, Patrick J. (1988). A concise introduction to logic. Belmont, Calif.: Wadsworth Pub. Co. ISBN 0-534-08928-3.

- Robert C. Pinto (2001). Argument, inference and dialectic: collected papers on informal logic. Springer. p. 84. ISBN 978-0-7923-7005-5.

7

Types of Logic

Informal logic is a branch that helps in developing criteria for analyzing and interpreting the construction of an argument. Mathematical logic, term logic, BL and noise-based logic are some of the aspects elucidated in the section. The chapter provides a plethora of themes on the types of logic for a better comprehension.

Informal Logic

Informal logic, intuitively, refers to the principles of logic and logical thought outside of a formal setting. However, perhaps because of the "informal" in the title, the precise definition of "informal logic" is a matter of some dispute. Ralph H. Johnson and J. Anthony Blair define informal logic as "a branch of logic whose task is to develop non-formal standards, criteria, procedures for the analysis, interpretation, evaluation, criticism and construction of argumentation." This definition reflects what had been implicit in their practice and what others were doing in their informal logic texts.

Informal logic is associated with (informal) fallacies, critical thinking, the Thinking Skills Movement and the interdisciplinary inquiry known as argumentation theory. Frans H. van Eemeren writes that the label "informal logic" covers a "collection of normative approaches to the study of reasoning in ordinary language that remain closer to the practice of argumentation than formal logic."

History

Informal logic as a distinguished enterprise under this name emerged roughly in the late 1970s as a sub-field of philosophy. The naming of the field was preceded by the appearance of a number of textbooks that rejected the symbolic approach to logic on pedagogical grounds as inappropriate and unhelpful for introductory textbooks on logic for a general audience, for example Howard Kahane's *Logic and Contemporary Rhetoric*, subtitled "The Use of Reason in Everyday Life", first published in 1971. Kahane's textbook was described on the notice of his death in the *Proceedings And Addresses of the American Philosophical Association* (2002) as "a text in informal logic, [that] was intended to enable students to cope with the misleading rhetoric one frequently finds in the media and in political discourse. It was organized around a discussion of fallacies, and was meant to be a practical instrument for dealing with the problems of everyday life. [It has] ... gone through many editions; [it is] ... still in print; and the thousands upon thousands of students who have taken courses in which his text [was] ... used can thank Howard for contributing to their ability to dissect arguments and avoid the deceptions of deceitful rhetoric. He tried to put into practice the ideal of discourse that aims at truth rather than merely at persuasion. (Hausman et al. 2002)" Other textbooks from the era taking this approach were Michael Scriven's *Reasoning* (Edgepress, 1976) and *Logical Self-Defense* by Ralph Johnson and J. Anthony Blair, first published in 1977. Earlier precursors in this tradition can be considered Monroe Beardsley's *Practical Logic* (1950) and Stephen Toulmin's *The Uses of Argument* (1958).

The field perhaps became recognized under its current name with the *First International Symposium on Informal Logic* held in 1978. Although initially motivated by a new pedagogical approach to undergraduate logic textbooks, the scope of the field was basically defined by a list of 13 problems and issues which Blair and Johnson included as an appendix to their keynote address at this symposium:

- the theory of logical criticism

- the theory of argument

- the theory of fallacy

- the fallacy approach vs. the critical thinking approach

- the viability of the inductive/deductive dichotomy

- the ethics of argumentation and logical criticism

- the problem of assumptions and missing premises

- the problem of context

- methods of extracting arguments from context

- methods of displaying arguments

- the problem of pedagogy

- the nature, division and scope of informal logic

- the relationship of informal logic to other inquiries

David Hitchcock argues that the naming of the field was unfortunate, and that *philosophy of argument* would have been more appropriate. He argues that more undergraduate students in North America study informal logic than any other branch of philosophy, but that as of 2003 informal logic (or philosophy of argument) was not recognized as separate sub-field by the World Congress of Philosophy. Frans H. van Eemeren wrote that "informal logic" is mainly an approach to argumentation advanced by a group of US and Canadian philosophers and largely based on the previous works of Stephen Toulmin and to a lesser extent those of Chaïm Perelman.

Proposed Definitions

Johnson and Blair (2000) proposed the following definition: "Informal logic designates that branch of logic whose task is to develop non-formal[2] standards, criteria, procedures for the analysis, interpretation, evaluation, critique and construction of argumentation in everyday discourse." Their meaning of non-formal[2] is taken from Barth and Krabbe (1982), which is explained below.

To understand the definition above, one must understand "informal" which takes its meaning in contrast to its counterpart "formal." (This point was not made for a very long time, hence the nature of informal logic remained opaque, even to those involved in it, for a period of time.) Here it is helpful to have recourse to Barth and Krabbe (1982:14f) where they distinguish three senses of the term "form." By "form[1]," Barth and Krabbe mean the sense of the term which derives from the Platonic idea of form—the ultimate metaphysical unit. Barth and Krabbe claim that most traditional logic is formal in this sense. That is, syllogistic logic is a logic of terms where the terms could naturally be understood as place-holders for Platonic (or Aristotelian) forms. In this first sense of "form," almost all logic is informal (not-formal). Understanding informal logic this way would be much too broad to be useful.

By "form[2]," Barth and Krabbe mean the form of sentences and statements as these are understood in modern systems of logic. Here validity is the focus: if the premises are true, the conclusion must then also be true. Now validity has to do with the logical form of the statement that makes up the argument. In this sense of "formal," most modern and contemporary logic is "formal." That is, such logics canonize the notion of logical form, and the notion of validity plays the central normative role. In this second sense of form, informal logic is not-formal, because it abandons the notion of logical form as the key to understanding the structure of arguments, and likewise retires validity as normative for the purposes of the evaluation of argument. It seems to many that validity is too stringent a requirement, that there are good arguments in which the conclusion is supported by the premises even though it does not follow necessarily from them (as validity requires). An argument in which the conclusion is thought to be "beyond reasonable doubt, given the premises" is sufficient in law to cause a person to be sentenced to death, even though it does not meet the standard of logical validity. This type of argument, based on accumulation of evidence rather than pure deduction, is called a conductive argument.

By "form[3]," Barth and Krabbe mean to refer to "procedures which are somehow regulated or regimented, which take place according to some set of rules." Barth and Krabbe say that "we do not defend formality[3] of all kinds and under all circumstances." Rather "we defend the thesis that verbal dialectics must have a certain form (i.e., must proceed according to certain rules) in order that one can speak of the discussion as being won or lost" (19). In this third sense of "form", informal logic can be formal, for there is nothing in the informal logic enterprise that stands opposed to the idea that argumentative discourse should be subject to norms, i.e., subject to rules, criteria, standards or procedures. Informal logic does present standards for the evaluation of argument, procedures for detecting missing premises etc.

Johnson and Blair (2000) noticed a limitation of their own definition, particularly with respect to "everyday discourse", which could indicate that it does not seek to understand specialized, domain-specific arguments made in natural languages. Consequently, they have argued that the crucial divide is between arguments made in formal languages and those made in natural languages.

Fisher and Scriven (1997) proposed a more encompassing definition, seeing informal logic as "the discipline which studies the practice of critical thinking and provides its intellectual spine". By "critical thinking" they understand "skilled and active interpretation and evaluation of observations and communications, information and argumentation."

Criticisms

Some hold the view that informal logic is not a branch or subdiscipline of logic, or even the view that there cannot be such a thing as informal logic. Massey criticizes informal logic on the grounds that it has no theory underpinning it. Informal logic, he says, requires detailed classification schemes to organize it, which in other disciplines is provided by the underlying theory. He maintains that there is no method of establishing the invalidity of an argument aside from the formal method, and that the study of fallacies may be of more interest to other disciplines, like psychology, than to philosophy and logic.

Relation to Critical Thinking

Since the 1980s, informal logic has been partnered and even equated, in the minds of many, with critical thinking. The precise definition of "critical thinking" is a subject of much dispute. Critical thinking, as defined by Johnson, is the evaluation of an intellectual product (an argument, an explanation, a theory) in terms of its strengths and weaknesses. While critical thinking will include evaluation of arguments and hence require skills of argumentation including informal logic, critical thinking requires additional abilities not supplied by informal logic, such as the ability to obtain and assess information and to clarify meaning. Also, many believe that critical thinking requires certain dispositions. Understood in this way, "critical thinking" is a broad term for the attitudes and skills that are involved in analyzing and evaluating arguments. The critical thinking movement promotes critical thinking as an educational ideal. The movement emerged with great force in the '80s in North America as part of an ongoing critique of education as regards the thinking skills not being taught.

Relation to Argumentation Theory

The social, communicative practice of argumentation can and should be distinguished from implication (or entailment)—a relationship between propositions; and from inference—a mental activity typically thought of as the drawing of a conclusion from premises. Informal logic may thus be said to be a logic of argumentation, as distinguished from implication and inference.

Argumentation theory (or the theory of argumentation) has come to be the term that designates the theoretical study of argumentation. This study is interdisciplinary in the sense that no one discipline will be able to provide a complete account. A full appreciation of argumentation requires insights from logic (both formal and informal), rhetoric, communication theory, linguistics, psychology, and, increasingly, computer science. Since the 1970s, there has been significant agreement that there are three basic approaches to argumentation theory: the logical, the rhetorical and the dialectical. According to Wenzel, the logical approach deals with the product, the dialectical with the process, and the rhetorical with the procedure. Thus, informal logic is one contributor to this inquiry, being most especially concerned with the norms of argument.

Mathematical Logic

Mathematical logic is a subfield of mathematics exploring the applications of formal logic to mathematics. It bears close connections to metamathematics, the foundations of mathematics, and theoretical computer science. The unifying themes in mathematical logic include the study of the expressive power of formal systems and the deductive power of formal proof systems.

Mathematical logic is often divided into the fields of set theory, model theory, recursion theory, and proof theory. These areas share basic results on logic, particularly first-order logic, and definability. In computer science (particularly in the ACM Classification) mathematical logic encompasses additional topics not detailed in this article.

Since its inception, mathematical logic has both contributed to, and has been motivated by, the study of foundations of mathematics. This study began in the late 19th century with the development of axiomatic frameworks for geometry, arithmetic, and analysis. In the early 20th century it was shaped by David Hilbert's program to prove the consistency of foundational theories. Results of Kurt Gödel, Gerhard Gentzen, and others provided partial resolution to the program, and clarified the issues involved in proving consistency. Work in set theory showed that almost all ordinary mathematics can be formalized in terms of sets, although there are some theorems that cannot be proven in common axiom systems for set theory. Contemporary work in the foundations of mathematics often focuses on establishing which parts of mathematics can be formalized in particular formal systems (as in reverse mathematics) rather than trying to find theories in which all of mathematics can be developed.

Subfields and Scope

The *Handbook of Mathematical Logic* (Barwise 1989) makes a rough division of contemporary mathematical logic into four areas:

1. set theory

2. model theory

3. recursion theory, and

4. proof theory and constructive mathematics (considered as parts of a single area).

Each area has a distinct focus, although many techniques and results are shared among multiple areas. The borderlines amongst these fields, and the lines separating mathematical logic and other fields of mathematics, are not always sharp. Gödel's incompleteness theorem marks not only a milestone in recursion theory and proof theory, but has also led to Löb's theorem in modal logic. The method of forcing is employed in set theory, model theory, and recursion theory, as well as in the study of intuitionistic mathematics.

The mathematical field of category theory uses many formal axiomatic methods, and includes the study of categorical logic, but category theory is not ordinarily considered a subfield of mathematical logic. Because of its applicability in diverse fields of mathematics, mathematicians including Saunders Mac Lane have proposed category theory as a foundational system for mathematics, independent of set theory. These foundations use toposes, which resemble generalized models of set theory that may employ classical or nonclassical logic.

History

Mathematical logic emerged in the mid-19th century as a subfield of mathematics independent of the traditional study of logic (Ferreirós 2001, p. 443). Before this emergence, logic was studied with rhetoric, through the syllogism, and with philosophy. The first half of the 20th century saw an explosion of fundamental results, accompanied by vigorous debate over the foundations of mathematics.

Early History

Theories of logic were developed in many cultures in history, including China, India, Greece and the Islamic world. In 18th-century Europe, attempts to treat the operations of formal logic in a symbolic or algebraic way had been made by philosophical mathematicians including Leibniz and Lambert, but their labors remained isolated and little known.

19Th Century

In the middle of the nineteenth century, George Boole and then Augustus De Morgan presented systematic mathematical treatments of logic. Their work, building on work by algebraists such as George Peacock, extended the traditional Aristotelian doctrine of logic into a sufficient framework for the study of foundations of mathematics (Katz 1998, p. 686).

Charles Sanders Peirce built upon the work of Boole to develop a logical system for relations and quantifiers, which he published in several papers from 1870 to 1885. Gottlob Frege presented an independent development of logic with quantifiers in his *Begriffsschrift*, published in 1879, a work generally considered as marking a turning point in the history of logic. Frege's work remained obscure, however, until Bertrand Russell began to promote it near the turn of the century. The two-dimensional notation Frege developed was never widely adopted and is unused in contemporary texts.

From 1890 to 1905, Ernst Schröder published *Vorlesungen über die Algebra der Logik* in three volumes. This work summarized and extended the work of Boole, De Morgan, and Peirce, and was a comprehensive reference to symbolic logic as it was understood at the end of the 19th century.

Foundational Theories

Concerns that mathematics had not been built on a proper foundation led to the development of axiomatic systems for fundamental areas of mathematics such as arithmetic, analysis, and geometry.

In logic, the term *arithmetic* refers to the theory of the natural numbers. Giuseppe Peano (1889) published a set of axioms for arithmetic that came to bear his name (Peano axioms), using a variation of the logical system of Boole and Schröder but adding quantifiers. Peano was unaware of Frege's work at the time. Around the same time Richard Dedekind showed that the natural numbers are uniquely characterized by their induction properties. Dedekind (1888) proposed a different characterization, which lacked the formal logical character of Peano's axioms. Dedekind's work, however, proved theorems inaccessible in Peano's system, including the uniqueness of the set of natural numbers (up to isomorphism) and the recursive definitions of addition and multiplication from the successor function and mathematical induction.

In the mid-19th century, flaws in Euclid's axioms for geometry became known (Katz 1998, p. 774). In addition to the independence of the parallel postulate, established by Nikolai Lobachevsky in 1826 (Lobachevsky 1840), mathematicians discovered that certain theorems taken for granted by Euclid were not in fact provable from his axioms. Among these is the theorem that a line contains at least two points, or that circles of the same radius whose centers are separated by that radius must intersect. Hilbert (1899) developed a complete set of axioms for geometry, building on previous work by Pasch (1882). The success in axiomatizing geometry motivated Hilbert to seek complete axiomatizations of other areas of mathematics, such as the natural numbers and the real line. This would prove to be a major area of research in the first half of the 20th century.

The 19th century saw great advances in the theory of real analysis, including theories of convergence of functions and Fourier series. Mathematicians such as Karl Weierstrass began to construct functions that stretched intuition, such as nowhere-differentiable continuous functions. Previous conceptions of a function as a rule for computation, or a smooth graph, were no longer adequate. Weierstrass began to advocate the arithmetization of analysis, which sought to axiomatize analysis using properties of the natural numbers. The modern (ϵ, δ)-definition of limit and continuous functions was already developed by Bolzano in 1817 (Felscher 2000), but remained relatively unknown. Cauchy in 1821 defined continuity in terms of infinitesimals. In 1858, Dedekind proposed a definition of the real numbers in terms of Dedekind cuts of rational numbers (Dedekind 1872), a definition still employed in contemporary texts.

Georg Cantor developed the fundamental concepts of infinite set theory. His early results developed the theory of cardinality and proved that the reals and the natural numbers have different cardinalities (Cantor 1874). Over the next twenty years, Cantor developed a theory of transfinite numbers in a series of publications. In 1891, he published a new proof of the uncountability of the real numbers that introduced the diagonal argument, and used this method to prove Cantor's theorem that no set can have the same cardinality as its powerset. Cantor believed that every set could be well-ordered, but was unable to produce a proof for this result, leaving it as an open problem in 1895 (Katz 1998, p. 807).

20th Century

In the early decades of the 20th century, the main areas of study were set theory and formal logic. The discovery of paradoxes in informal set theory caused some to wonder whether mathematics itself is inconsistent, and to look for proofs of consistency.

In 1900, Hilbert posed a famous list of 23 problems for the next century. The first two of these were to resolve the continuum hypothesis and prove the consistency of elementary arithmetic, respectively; the tenth was to produce a method that could decide whether a multivariate polynomial equation over the integers has a solution. Subsequent work to resolve these problems shaped the direction of mathematical logic, as did the effort to resolve Hilbert's *Entscheidungsproblem*, posed in 1928. This problem asked for a procedure that would decide, given a formalized mathematical statement, whether the statement is true or false.

Set Theory and Paradoxes

Ernst Zermelo (1904) gave a proof that every set could be well-ordered, a result Georg Cantor had been unable to obtain. To achieve the proof, Zermelo introduced the axiom of choice, which drew heated debate and research among mathematicians and the pioneers of set theory. The immediate criticism of the method led Zermelo to publish a second exposition of his result, directly addressing criticisms of his proof (Zermelo 1908a). This paper led to the general acceptance of the axiom of choice in the mathematics community.

Skepticism about the axiom of choice was reinforced by recently discovered paradoxes in naive set theory. Cesare Burali-Forti (1897) was the first to state a paradox: the Burali-Forti paradox shows that the collection of all ordinal numbers cannot form a set. Very soon thereafter, Bertrand Russell discovered Russell's paradox in 1901, and Jules Richard (1905) discovered Richard's paradox.

Zermelo (1908b) provided the first set of axioms for set theory. These axioms, together with the additional axiom of replacement proposed by Abraham Fraenkel, are now called Zermelo–Fraenkel set theory (ZF). Zermelo's axioms incorporated the principle of limitation of size to avoid Russell's paradox.

In 1910, the first volume of *Principia Mathematica* by Russell and Alfred North Whitehead was published. This seminal work developed the theory of functions and cardinality in a completely formal framework of type theory, which Russell and Whitehead developed in an effort to avoid the paradoxes. *Principia Mathematica* is considered one of the most influential works of the 20th century, although the framework of type theory did not prove popular as a foundational theory for mathematics (Ferreirós 2001, p. 445).

Fraenkel (1922) proved that the axiom of choice cannot be proved from the remaining axioms of Zermelo's set theory with urelements. Later work by Paul Cohen (1966) showed that the addition of urelements is not needed, and the axiom of choice is unprovable in ZF. Cohen's proof developed the method of forcing, which is now an important tool for establishing independence results in set theory.

Symbolic Logic

Leopold Löwenheim (1915) and Thoralf Skolem (1920) obtained the Löwenheim–Skolem theorem, which says that first-order logic cannot control the cardinalities of infinite structures. Skolem realized that this theorem would apply to first-order formalizations of set theory, and that it implies any such formalization has a countable model. This counterintuitive fact became known as Skolem's paradox.

In his doctoral thesis, Kurt Gödel (1929) proved the completeness theorem, which establishes a correspondence between syntax and semantics in first-order logic. Gödel used the completeness theorem to prove the compactness theorem, demonstrating the finitary nature of first-order logical consequence. These results helped establish first-order logic as the dominant logic used by mathematicians.

In 1931, Gödel published *On Formally Undecidable Propositions of Principia Mathematica and Related Systems*, which proved the incompleteness (in a different meaning of the word) of all sufficiently strong, effective first-order theories. This result, known as Gödel's incompleteness theorem, establishes severe limitations on axiomatic foundations for mathematics, striking a strong blow to Hilbert's program. It showed the impossibility of providing a consistency proof of arithmetic within any formal theory of arithmetic. Hilbert, however, did not acknowledge the importance of the incompleteness theorem for some time.

Gödel's theorem shows that a consistency proof of any sufficiently strong, effective axiom system cannot be obtained in the system itself, if the system is consistent, nor in any weaker system. This leaves open the possibility of consistency proofs that cannot be formalized within the system they consider. Gentzen (1936) proved the consistency of arithmetic using a finitistic system together with a principle of transfinite induction. Gentzen's result introduced the ideas of cut elimination and proof-theoretic ordinals, which became key tools in proof theory. Gödel (1958) gave a different consistency proof, which reduces the consistency of classical arithmetic to that of intuitionistic arithmetic in higher types.

Beginnings of the Other Branches

Alfred Tarski developed the basics of model theory.

Beginning in 1935, a group of prominent mathematicians collaborated under the pseudonym Nicolas Bourbaki to publish a series of encyclopedic mathematics texts. These texts, written in an austere and axiomatic style, emphasized rigorous presentation and set-theoretic foundations. Terminology coined by these texts, such as the words *bijection*, *injection*, and *surjection*, and the set-theoretic foundations the texts employed, were widely adopted throughout mathematics.

The study of computability came to be known as recursion theory, because early formalizations by Gödel and Kleene relied on recursive definitions of functions. When these definitions were shown equivalent to Turing's formalization involving Turing machines, it became clear that a new concept – the computable function – had been discovered, and that this definition was robust enough to admit numerous independent characterizations. In his work on the incompleteness theorems in 1931, Gödel lacked a rigorous concept of an effective formal system; he immediately realized that the new definitions of computability could be used for this purpose, allowing him to state the incompleteness theorems in generality that could only be implied in the original paper.

Numerous results in recursion theory were obtained in the 1940s by Stephen Cole Kleene and Emil Leon Post. Kleene (1943) introduced the concepts of relative computability, foreshadowed by Turing (1939), and the arithmetical hierarchy. Kleene later generalized recursion theory to higher-order functionals. Kleene and Kreisel studied formal versions of intuitionistic mathematics, particularly in the context of proof theory.

Formal logical Systems

At its core, mathematical logic deals with mathematical concepts expressed using formal logical systems. These systems, though they differ in many details, share the common property of considering only expressions in a fixed formal language. The systems of propositional logic and first-order logic are the most widely studied today, because of their applicability to foundations of mathematics and because of their desirable proof-theoretic properties. Stronger classical logics such as second-order logic or infinitary logic are also studied, along with nonclassical logics such as intuitionistic logic.

First-order Logic

First-order logic is a particular formal system of logic. Its syntax involves only finite expressions as well-formed formulas, while its semantics are characterized by the limitation of all quantifiers to a fixed domain of discourse.

Early results from formal logic established limitations of first-order logic. The Löwenheim–Skolem theorem (1919) showed that if a set of sentences in a countable first-order language has an infinite model then it has at least one model of each infinite cardinality. This shows that it is impossible for a set of first-order axioms to characterize the natural numbers, the real numbers, or any other infinite structure up to isomorphism. As the goal of early foundational studies was to produce axiomatic theories for all parts of mathematics, this limitation was particularly stark.

Gödel's completeness theorem (Gödel 1929) established the equivalence between semantic and syntactic definitions of logical consequence in first-order logic. It shows that if a particular sentence is true in every model that satisfies a particular set of axioms, then there must be a finite deduction of the sentence from the axioms. The compactness theorem first appeared as a lemma in Gödel's proof of the completeness theorem, and it took many years before logicians grasped its significance and began to apply it routinely. It says that a set of sentences has a model if and only if every finite subset has a model, or in other words that an inconsistent set of formulas must have a finite inconsistent subset. The completeness and compactness theorems allow for sophisticated analysis of logical consequence in first-order logic and the development of model theory, and they are a key reason for the prominence of first-order logic in mathematics.

Gödel's incompleteness theorems (Gödel 1931) establish additional limits on first-order axiomatizations. The first incompleteness theorem states that for any consistent, effectively given (defined below) logical system that is capable of interpreting arithmetic (that is, of expressing the Peano axioms) there exists a statement (the Gödel sentence) which is true (in the sense that it holds for the natural numbers) but not provable within that logical system (and which indeed may fail in some non-standard models of arithmetic which may be consistent with the logical system.) Here a logical system is said to be effectively given if it is possible to decide, given any formula in the language of the system, whether the formula is an axiom, and one which can express the Peano axioms is called "sufficiently strong." When applied to first-order logic, the first incompleteness theorem implies that any sufficiently strong, consistent, effective first-order theory has models that are not elementarily equivalent, a stronger limitation than the one established by the Löwenheim–Skolem theorem. The second incompleteness theorem states that no sufficiently strong, consistent, effective axiom system for arithmetic can prove its own consistency, which has been interpreted to show that Hilbert's program cannot be completed.

Other Classical Logics

Many logics besides first-order logic are studied. These include infinitary logics, which allow for formulas to provide an infinite amount of information, and higher-order logics, which include a portion of set theory directly in their semantics.

The most well studied infinitary logic is $L_{\omega_1,\omega}$.. In this logic, quantifiers may only be nested to finite depths, as in first-order logic, but formulas may have finite or countably infinite conjunctions and disjunctions within them. Thus, for example, it is possible to say that an object is a whole number using a formula of $L_{\omega_1,\omega}$ such as

$$(x = 0) \vee (x = 1) \vee (x = 2) \vee \cdots.$$

Higher-order logics allow for quantification not only of elements of the domain of discourse, but subsets of the domain of discourse, sets of such subsets, and other objects of higher type. The semantics are defined so that, rather than having a separate domain for each higher-type quantifier to range over, the quantifiers instead range over all objects of the appropriate type. The logics studied before the development of first-order logic, for example Frege's logic, had similar set-theoretic aspects. Although higher-order logics are more expressive, allowing complete axiomatizations of structures such as the natural numbers, they do not satisfy analogues of the completeness and compactness theorems from first-order logic, and are thus less amenable to proof-theoretic analysis.

Another type of logics are fixed-point logics that allow inductive definitions, like one writes for primitive recursive functions.

One can formally define an extension of first-order logic — a notion which encompasses all logics in this section because they behave like first-order logic in certain fundamental ways, but does not encompass all logics in general, e.g. it does not encompass intuitionistic, modal or fuzzy logic. Lindström's theorem implies that the only extension of first-order logic satisfying both the compactness theorem and the Downward Löwenheim–Skolem theorem is first-order logic.

Nonclassical and Modal Logic

Modal logics include additional modal operators, such as an operator which states that a particular formula is not only true, but necessarily true. Although modal logic is not often used to axiomatize mathematics, it has been used to study the properties of first-order provability (Solovay 1976) and set-theoretic forcing (Hamkins and Löwe 2007).

Intuitionistic logic was developed by Heyting to study Brouwer's program of intuitionism, in which Brouwer himself avoided formalization. Intuitionistic logic specifically does not include the law of the excluded middle, which states that each sentence is either true or its negation is true. Kleene's work with the proof theory of intuitionistic logic showed that constructive information can be recovered from intuitionistic proofs. For example, any provably total function in intuitionistic arithmetic is computable; this is not true in classical theories of arithmetic such as Peano arithmetic.

Algebraic Logic

Algebraic logic uses the methods of abstract algebra to study the semantics of formal logics. A fundamental example is the use of Boolean algebras to represent truth values in classical propositional logic, and the use of Heyting algebras to represent truth values in intuitionistic propositional logic. Stronger logics, such as first-order logic and higher-order logic, are studied using more complicated algebraic structures such as cylindric algebras.

Set Theory

Set theory is the study of sets, which are abstract collections of objects. Many of the basic notions, such as ordinal and cardinal numbers, were developed informally by Cantor before formal axiomatizations of set theory were developed. The first such axiomatization, due to Zermelo (1908b), was extended slightly to become Zermelo–Fraenkel set theory (ZF), which is now the most widely used foundational theory for mathematics.

Other formalizations of set theory have been proposed, including von Neumann–Bernays–Gödel set theory (NBG), Morse–Kelley set theory (MK), and New Foundations (NF). Of these, ZF, NBG, and MK are similar in describing a cumulative hierarchy of sets. New Foundations takes a different approach; it allows objects such as the set of all sets at the cost of restrictions on its set-existence axioms. The system of Kripke–Platek set theory is closely related to generalized recursion theory.

Two famous statements in set theory are the axiom of choice and the continuum hypothesis. The axiom of choice, first stated by Zermelo (1904), was proved independent of ZF by Fraenkel (1922), but has come to be widely accepted by mathematicians. It states that given a collection of nonempty sets there is a single set C that contains exactly one element from each set in the collection. The set C is said to "choose" one element from each set in the collection. While the ability to make such a choice is considered obvious by some, since each set in the collection is nonempty, the lack of a general, concrete rule by which the choice can be made renders the axiom nonconstructive. Stefan Banach and Alfred Tarski (1924) showed that the axiom of choice can be used to decompose a solid ball into a finite number of pieces which can then be rearranged, with no scaling, to make two solid balls of the original size. This theorem, known as the Banach–Tarski paradox, is one of many counterintuitive results of the axiom of choice.

The continuum hypothesis, first proposed as a conjecture by Cantor, was listed by David Hilbert as one of his 23 problems in 1900. Gödel showed that the continuum hypothesis cannot be disproven from the axioms of Zermelo–Fraenkel set theory (with or without the axiom of choice), by developing the constructible universe of set theory in which the continuum hypothesis must hold. In 1963, Paul Cohen showed that the continuum hypothesis cannot be proven from the axioms of Zermelo–Fraenkel set theory (Cohen 1966). This independence result did not completely settle Hilbert's question, however, as it is possible that new axioms for set theory could resolve the hypothesis. Recent work along these lines has been conducted by W. Hugh Woodin, although its importance is not yet clear (Woodin 2001).

Contemporary research in set theory includes the study of large cardinals and determinacy. Large cardinals are cardinal numbers with particular properties so strong that the existence of such cardinals cannot be proved in ZFC. The existence of the smallest large cardinal typically studied, an inaccessible cardinal, already implies the consistency of ZFC. Despite the fact that large cardinals have extremely high cardinality, their existence has many ramifications for the structure of the real line. *Determinacy* refers to the possible existence of winning strategies for certain two-player games (the games are said to be *determined*). The existence of these strategies implies structural properties of the real line and other Polish spaces.

Model Theory

Model theory studies the models of various formal theories. Here a theory is a set of formulas in a particular formal logic and signature, while a model is a structure that gives a concrete interpretation of the theory. Model theory is closely related to universal algebra and algebraic geometry, although the methods of model theory focus more on logical considerations than those fields.

The set of all models of a particular theory is called an elementary class; classical model theory seeks to determine the properties of models in a particular elementary class, or determine whether certain classes of structures form elementary classes.

The method of quantifier elimination can be used to show that definable sets in particular theories cannot be too complicated. Tarski (1948) established quantifier elimination for real-closed fields, a result which also shows the theory of the field of real numbers is decidable. (He also noted that his methods were equally applicable to algebraically closed fields of arbitrary characteristic.) A modern subfield developing from this is concerned with o-minimal structures.

Morley's categoricity theorem, proved by Michael D. Morley (1965), states that if a first-order theory in a countable language is categorical in some uncountable cardinality, i.e. all models of this cardinality are isomorphic, then it is categorical in all uncountable cardinalities.

A trivial consequence of the continuum hypothesis is that a complete theory with less than continuum many nonisomorphic countable models can have only countably many. Vaught's conjecture, named after Robert Lawson Vaught, says that this is true even independently of the continuum hypothesis. Many special cases of this conjecture have been established.

Recursion Theory

Recursion theory, also called computability theory, studies the properties of computable functions and the Turing degrees, which divide the uncomputable functions into sets that have the same level of uncomputability. Recursion theory also includes the study of generalized computability and definability. Recursion theory grew from the work of Alonzo Church and Alan Turing in the 1930s, which was greatly extended by Kleene and Post in the 1940s.

Classical recursion theory focuses on the computability of functions from the natural numbers to the natural numbers. The fundamental results establish a robust, canonical class of computable functions with numerous independent, equivalent characterizations using Turing machines, λ calculus, and other systems. More advanced results concern the structure of the Turing degrees and the lattice of recursively enumerable sets.

Generalized recursion theory extends the ideas of recursion theory to computations that are no longer necessarily finite. It includes the study of computability in higher types as well as areas such as hyperarithmetical theory and α-recursion theory.

Contemporary research in recursion theory includes the study of applications such as algorithmic randomness, computable model theory, and reverse mathematics, as well as new results in pure recursion theory.

Algorithmically Unsolvable Problems

An important subfield of recursion theory studies algorithmic unsolvability; a decision problem or function problem is algorithmically unsolvable if there is no possible computable algorithm that returns the correct answer for all legal inputs to the problem. The first results about unsolvability, obtained independently by Church and Turing in 1936, showed that the Entscheidungsproblem is algorithmically unsolvable. Turing proved this by establishing the unsolvability of the halting problem, a result with far-ranging implications in both recursion theory and computer science.

There are many known examples of undecidable problems from ordinary mathematics. The word problem for groups was proved algorithmically unsolvable by Pyotr Novikov in 1955 and independently by W. Boone in 1959. The busy beaver problem, developed by Tibor Radó in 1962, is another well-known example.

Hilbert's tenth problem asked for an algorithm to determine whether a multivariate polynomial equation with integer coefficients has a solution in the integers. Partial progress was made by Julia Robinson, Martin Davis and Hilary Putnam. The algorithmic unsolvability of the problem was proved by Yuri Matiyasevich in 1970 (Davis 1973).

Proof theory and Constructive Mathematics

Proof theory is the study of formal proofs in various logical deduction systems. These proofs are represented as formal mathematical objects, facilitating their analysis by mathematical techniques. Several deduction systems are commonly considered, including Hilbert-style deduction systems, systems of natural deduction, and the sequent calculus developed by Gentzen.

The study of constructive mathematics, in the context of mathematical logic, includes the study of systems in non-classical logic such as intuitionistic logic, as well as the study of predicative systems. An early proponent of predicativism was Hermann Weyl, who showed it is possible to develop a large part of real analysis using only predicative methods (Weyl 1918).

Because proofs are entirely finitary, whereas truth in a structure is not, it is common for work in constructive mathematics to emphasize provability. The relationship between provability in classical (or nonconstructive) systems and provability in intuitionistic (or constructive, respectively) systems is of particular interest. Results such as the Gödel–Gentzen negative translation show that it is possible to embed (or *translate*) classical logic into intuitionistic logic, allowing some properties about intuitionistic proofs to be transferred back to classical proofs.

Recent developments in proof theory include the study of proof mining by Ulrich Kohlenbach and the study of proof-theoretic ordinals by Michael Rathjen.

Connections with Computer Science

The study of computability theory in computer science is closely related to the study of computability in mathematical logic. There is a difference of emphasis, however. Computer scientists often focus on concrete programming languages and feasible computability, while researchers in mathematical logic often focus on computability as a theoretical concept and on noncomputability.

The theory of semantics of programming languages is related to model theory, as is program verification (in particular, model checking). The Curry–Howard isomorphism between proofs and programs relates to proof theory, especially intuitionistic logic. Formal calculi such as the lambda calculus and combinatory logic are now studied as idealized programming languages.

Computer science also contributes to mathematics by developing techniques for the automatic checking or even finding of proofs, such as automated theorem proving and logic programming.

Descriptive complexity theory relates logics to computational complexity. The first significant result in this area, Fagin's theorem (1974) established that NP is precisely the set of languages expressible by sentences of existential second-order logic.

Foundations of Mathematics

In the 19th century, mathematicians became aware of logical gaps and inconsistencies in their field. It was shown that Euclid's axioms for geometry, which had been taught for centuries as an example of the axiomatic method, were incomplete. The use of infinitesimals, and the very definition of function, came into question in analysis, as pathological examples such as Weierstrass' nowhere-differentiable continuous function were discovered.

Cantor's study of arbitrary infinite sets also drew criticism. Leopold Kronecker famously stated "God made the integers; all else is the work of man," endorsing a return to the study of finite, concrete objects in mathematics. Although Kronecker's argument was carried forward by constructivists in the 20th century, the mathematical community as a whole rejected them. David Hilbert argued in favor of the study of the infinite, saying "No one shall expel us from the Paradise that Cantor has created."

Mathematicians began to search for axiom systems that could be used to formalize large parts of mathematics. In addition to removing ambiguity from previously naive terms such as function, it was hoped that this axiomatization would allow for consistency proofs. In the 19th century, the main method of proving the consistency of a set of axioms was to provide a model for it. Thus, for example, non-Euclidean geometry can be proved consistent by defining *point* to mean a point on a fixed sphere and *line* to mean a great circle on the sphere. The resulting structure, a model of elliptic geometry, satisfies the axioms of plane geometry except the parallel postulate.

With the development of formal logic, Hilbert asked whether it would be possible to prove that an axiom system is consistent by analyzing the structure of possible proofs in the system, and showing through this analysis that it is impossible to prove a contradiction. This idea led to the study of proof theory. Moreover, Hilbert proposed that the analysis should be entirely concrete, using the term *finitary* to refer to the methods he would allow but not precisely defining them. This project, known as Hilbert's program, was seriously affected by Gödel's incompleteness theorems, which show that the consistency of formal theories of arithmetic cannot be established using methods formalizable in those theories. Gentzen showed that it is possible to produce a proof of the consistency of arithmetic in a finitary system augmented with axioms of transfinite induction, and the techniques he developed to do so were seminal in proof theory.

A second thread in the history of foundations of mathematics involves nonclassical logics and constructive mathematics. The study of constructive mathematics includes many different programs with various definitions of *constructive*. At the most accommodating end, proofs in ZF set theory that do not use the axiom of choice are called constructive by many mathematicians. More limited versions of constructivism limit themselves to natural numbers, number-theoretic functions, and sets of natural numbers (which can be used to represent real numbers, facilitating the study of mathematical analysis). A common idea is that a concrete means of computing the values of the function must be known before the function itself can be said to exist.

In the early 20th century, Luitzen Egbertus Jan Brouwer founded intuitionism as a philosophy of mathematics. This philosophy, poorly understood at first, stated that in order for a mathematical statement to be true to a mathematician, that person must be able to *intuit* the statement, to not only believe its truth but understand the reason for its truth. A consequence of this definition of truth was the rejection of the law of the excluded middle, for there are statements that, according to Brouwer, could not be claimed to be true while their negations also could not be claimed true. Brouwer's philosophy was influential, and the cause of bitter disputes among prominent mathematicians. Later, Kleene and Kreisel would study formalized versions of intuitionistic logic (Brouwer rejected formalization, and presented his work in unformalized natural language). With the advent of the BHK interpretation and Kripke models, intuitionism became easier to reconcile with classical mathematics.

Modal Logic

Modal logic is a type of formal logic primarily developed in the 1960s that extends classical propositional and predicate logic to include operators expressing modality. A modal—a word that expresses a modality—qualifies a statement. For example, the statement "John is happy" might be qualified by saying that John is usually happy, in which case the term "usually" is functioning as a modal. The traditional alethic modalities, or modalities of truth, include possibility ("Possibly, p", "It is possible that p"), necessity ("Necessarily, p", "It is necessary that p"), and impossibility ("Impossibly, p", "It is impossible that p"). Other modalities that have been formalized in modal logic include temporal modalities, or modalities of time (notably, "It was the case that p", "It has always been that p", "It will be that p", "It will always be that p"), deontic modalities (notably, "It is obligatory that p", and "It is permissible that p"), epistemic modalities, or modalities of knowledge ("It is known that p") and doxastic modalities, or modalities of belief ("It is believed that p").

A formal modal logic represents modalities using modal operators. For example, "It might rain today" and "It is possible that rain will fall today" both contain the notion of possibility. In a modal logic this is represented as an operator, "Possibly", attached to the sentence "It will rain today".

The basic unary (1-place) modal operators are usually written "□" for "Necessarily" and "◇" for "Possibly". In a classical modal logic, each can be expressed by the other with negation:

$$P \leftrightarrow P;$$

$$P \leftrightarrow P.$$

Thus it is *possible* that it will rain today if and only if it is *not necessary* that it will *not* rain today; and it is *necessary* that it will rain today if and only if it is *not possible* that it will *not* rain today. Alternative symbols used for the modal operators are "L" for "Necessarily" and "M" for "Possibly".

Development of Modal Logic

In addition to his non-modal syllogistic, Aristotle also developed a modal syllogistic in Book I of his *Prior Analytics* (chs 8–22), which Theophrastus attempted to improve. There are also passages in Aristotle's work, such as the famous sea-battle argument in *De Interpretatione* §9, that are now seen as anticipations of the connection of modal logic with potentiality and time. In the Hellenistic period, the logicians Diodorus Cronus, Philo the Dialectician and the Stoic Chrysippus each developed a modal system that accounted for the interdefinability of possibility and necessity, accepted axiom T, and combined elements of modal logic and temporal logic in attempts to solve the notorious Master Argument. The earliest formal system of modal logic was developed by Avicenna, who ultimately developed a theory of "temporally modal" syllogistic. Modal logic as a self-aware subject owes much to the writings of the Scholastics, in particular William of Ockham and John Duns Scotus, who reasoned informally in a modal manner, mainly to analyze statements about essence and accident.

C. I. Lewis founded modern modal logic in his 1910 Harvard thesis and in a series of scholarly articles beginning in 1912. This work culminated in his 1932 book *Symbolic Logic* (with C. H. Langford), which introduced the five systems *S1* through *S5*.

Ruth C. Barcan (later Ruth Barcan Marcus) developed the first axiomatic systems of quantified modal logic — first and second order extensions of Lewis' *S2*, *S4*, and *S5*.

The contemporary era in modal semantics began in 1959, when Saul Kripke (then only a 19-year-old Harvard University undergraduate) introduced the now-standard Kripke semantics for modal logics. These are commonly referred to as "possible worlds" semantics. Kripke and A. N. Prior had previously corresponded at some length. Kripke semantics is basically simple, but proofs are eased using semantic-tableaux or analytic tableaux, as explained by E. W. Beth.

A. N. Prior created modern temporal logic, closely related to modal logic, in 1957 by adding modal operators [F] and [P] meaning "eventually" and "previously". Vaughan Pratt introduced dynamic logic in 1976. In 1977, Amir Pnueli proposed using temporal logic to formalise the behaviour of continually operating concurrent programs. Flavors of temporal logic include propositional dynamic logic (PDL), propositional linear temporal logic (PLTL), linear temporal logic (LTL), computational tree logic (CTL), Hennessy–Milner logic, and *T*.

The mathematical structure of modal logic, namely Boolean algebras augmented with unary operations (often called modal algebras), began to emerge with J. C. C. McKinsey's 1941 proof that *S2* and *S4* are decidable, and reached full flower in the work of Alfred Tarski and his student Bjarni Jónsson (Jónsson and Tarski 1951–52). This work revealed that *S4* and *S5* are models of interior algebra, a proper extension of Boolean algebra originally designed to capture the properties of the interior and closure operators of topology. Texts on modal logic typically do little more than mention its connections with the study of Boolean algebras and topology.

Semantics

Model Theory

The semantics for modal logic are usually given as follows: First we define a *frame*, which consists of a non-empty set, *G*, whose members are generally called possible worlds, and a binary relation, *R*, that holds (or not) between the possible worlds of *G*. This binary relation is called the *accessibility relation*. For example, *w R u* means that the world *u* is accessible from world *w*. That is to say, the state of affairs known as *u* is a live possibility for *w*. This gives a pair, $\langle G, R \rangle$. Some formulations of modal logic also include a constant term in *G*, conventionally called "the actual world", which is often symbolized as $w*$.

Next, the *frame* is extended to a *model* by specifying the truth-values of all propositions at each of the worlds in G. We do so by defining a relation v between possible worlds and positive literals. If there is a world w such that $v(w, P)$, , then P is true at w. A model is thus an ordered triple, $\langle G, R, v \rangle$.

According to these semantics, a truth is *necessary* with respect to a possible world w if it is true at every world that is accessible to w, and *possible* if it is true at some world that is accessible to w. Possibility thereby depends upon the accessibility relation R, which allows us to express the relative nature of possibility. For example, we might say that given our laws of physics it is not possible for humans to travel faster than the speed of light, but that given other circumstances it could have been possible to do so. Using the accessibility relation we can translate this scenario as follows: At all of the worlds accessible to our own world, it is not the case that humans can travel faster than the speed of light, but at one of these accessible worlds there is *another* world accessible from *those* worlds but not accessible from our own at which humans can travel faster than the speed of light.

It should also be noted that the definition of □ makes vacuously true certain sentences, since when it speaks of "every world that is accessible to w" it takes for granted the usual mathematical interpretation of the word "every". Hence, if a world w doesn't have any accessible worlds, any sentence beginning with □ is true.

The different systems of modal logic are distinguished by the properties of their corresponding accessibility relations. There are several systems that have been espoused (often called *frame conditions*). An accessibility relation is:

- reflexive iff $w R w$, for every w in G

- symmetric iff $w R u$ implies $u R w$, for all w and u in G

- transitive iff $w R u$ and $u R q$ together imply $w R q$, for all w, u, q in G.

- serial iff, for each w in G there is some u in G such that $w R u$.

- Euclidean iff, for every u, t, and w, $w R u$ and $w R t$ implies $u R t$ (note that it also implies: $t R u$)

The logics that stem from these frame conditions are:

- K := no conditions

- D := serial

- T := reflexive

- S4 := reflexive and transitive

- S5 := reflexive and Euclidean

The Euclidean property along with reflexivity yields symmetry and transitivity. (The Euclidean property can be obtained, as well, from symmetry and transitivity.) Hence if the accessibility relation R is reflexive and Euclidean, R is provably symmetric and transitive as well. Hence for models of S5, R is an equivalence relation, because R is reflexive, symmetric and transitive.

We can prove that these frames produce the same set of valid sentences as do the frames where all worlds can see all other worlds of W (*i.e.*, where R is a "total" relation). This gives the corresponding *modal graph* which is total complete (*i.e.*, no more edges (relations) can be added). For example, in any modal logic based on frame conditions:

$$w \vDash P \text{ if and only if for some element } u \text{ of } G, \text{ it holds that } u \vDash P \text{ and } w \, R \, u.$$

If we consider frames based on the total relation we can just say that

$$w \vDash P \text{ if and only if for some element } u \text{ of } G, \text{ it holds that } u \vDash P.$$

We can drop the accessibility clause from the latter stipulation because in such total frames it is trivially true of all w and u that $w \, R \, u$. But note that this does not have to be the case in all S5 frames, which can still consist of multiple parts that are fully connected among themselves but still disconnected from each other.

All of these logical systems can also be defined axiomatically, as is shown in the next section. For example, in S5, the axioms $P \Rightarrow \Box P$, $\Box P \Rightarrow \Box \Box P$ and $\Box P \Rightarrow P$ (corresponding to *symmetry*, *transitivity* and *reflexivity*, respectively) hold, whereas at least one of these axioms does not hold in each of the other, weaker logics.

Axiomatic Systems

The first formalizations of modal logic were axiomatic. Numerous variations with very different properties have been proposed since C. I. Lewis began working in the area in 1910. Hughes and Cresswell (1996), for example, describe 42 normal and 25 non-normal modal logics. Zeman (1973) describes some systems Hughes and Cresswell omit.

Modern treatments of modal logic begin by augmenting the propositional calculus with two unary operations, one denoting "necessity" and the other "possibility". The notation of C. I. Lewis, much employed since, denotes "necessarily p" by a prefixed "box" ($\Box p$) whose scope is established by parentheses. Likewise, a prefixed "diamond" ($\Diamond p$) denotes "possibly p". Regardless of notation, each of these operators is definable in terms of the other in classical modal logic:

- $\Box p$ (necessarily p) is equivalent to $\neg \Diamond \neg p$ ("not possible that not-p")

- $\Diamond p$ (possibly p) is equivalent to $\neg \Box \neg p$ ("not necessarily not-p")

Hence $\Box$ and $\Diamond$ form a dual pair of operators.

In many modal logics, the necessity and possibility operators satisfy the following analogues of de Morgan's laws from Boolean algebra:

"It is not necessary that X" is logically equivalent to "It is possible that not X".

"It is not possible that X" is logically equivalent to "It is necessary that not X".

Precisely what axioms and rules must be added to the propositional calculus to create a usable system of modal logic is a matter of philosophical opinion, often driven by the theorems one wishes to prove; or, in computer science, it is a matter of what sort of computational or deductive system one wishes to model. Many modal logics, known collectively as normal modal logics, include the following rule and axiom:

- N, Necessitation Rule: If p is a theorem (of any system invoking N), then $\Box p$ is likewise a theorem.

- K, Distribution Axiom: $\Box(p \to q) \to (\Box p \to \Box q)$.

The weakest normal modal logic, named K in honor of Saul Kripke, is simply the propositional calculus augmented by $\Box$, the rule N, and the axiom K. K is weak in that it fails to determine whether a proposition can be necessary but only contingently necessary. That is, it is not a theorem of K that if $\Box p$ is true then $\Box\Box p$ is true, i.e., that necessary truths are "necessarily necessary". If such perplexities are deemed forced and artificial, this defect of K is not a great one. In any case, different answers to such questions yield different systems of modal logic.

Adding axioms to K gives rise to other well-known modal systems. One cannot prove in K that if "p is necessary" then p is true. The axiom T remedies this defect:

- T, Reflexivity Axiom: $\Box p \to p$ (If p is necessary, then p is the case.)

T holds in most but not all modal logics. Zeman (1973) describes a few exceptions, such as $S1^o$.

Other well-known elementary axioms are:

- 4: $p \to p$
- B: $p \to p$
- D: $p \to p$
- 5: $p \to p$

These yield the systems (axioms in bold, systems in italics):

- $K := K + N$
- $T := K + T$

- $S4 := T + 4$

- $S5 := S4 + 5$

- $D := K + \text{D}.$

K through $S5$ form a nested hierarchy of systems, making up the core of normal modal logic. But specific rules or sets of rules may be appropriate for specific systems. For example, in deontic logic, $\Box p \to p$ (If it ought to be that p, then it is permitted that p) seems appropriate, but we should probably not include that $p \to \Box p$.. In fact, to do so is to commit the naturalistic fallacy (i.e. to state that what is natural is also good, by saying that if p is the case, p ought to be permitted).

The commonly employed system $S5$ simply makes all modal truths necessary. For example, if p is possible, then it is "necessary" that p is possible. Also, if p is necessary, then it is necessary that p is necessary. Other systems of modal logic have been formulated, in part because $S5$ does not describe every kind of modality of interest.

Structural Proof Theory

Sequent calculi and systems of natural deduction have been developed for several modal logics, but it has proven hard to combine generality with other features expected of good structural proof theories, such as purity (the proof theory does not introduce extra-logical notions such as labels) and analyticity (the logical rules support a clean notion of analytic proof). More complex calculi have been applied to modal logic to achieve generality.

Decision Methods

Analytic tableaux provide the most popular decision method for modal logics.

Alethic Logic

Modalities of necessity and possibility are called *alethic* modalities. They are also sometimes called *special* modalities, from the Latin *species*. Modal logic was first developed to deal with these concepts, and only afterward was extended to others. For this reason, or perhaps for their familiarity and simplicity, necessity and possibility are often casually treated as *the* subject matter of modal logic. Moreover, it is easier to make sense of relativizing necessity, e.g. to legal, physical, nomological, epistemic, and so on, than it is to make sense of relativizing other notions.

In classical modal logic, a proposition is said to be

- possible if and only if it is *not necessarily false* (regardless of whether it is actually true or actually false);

- necessary if and only if it is *not possibly false*; and

- contingent if and only if it is *not necessarily false* and *not necessarily true* (i.e. possible but not necessarily true);

- impossible if and only if it is *not possibly true* (i.e. false and necessarily false).

In classical modal logic, therefore, either the notion of possibility or necessity may be taken to be basic, where these other notions are defined in terms of it in the manner of De Morgan duality. Intuitionistic modal logic treats possibility and necessity as not perfectly symmetric.

For those with difficulty with the concept of something being possible but not true, the meaning of these terms may be made more comprehensible by thinking of multiple "possible worlds" (in the sense of Leibniz) or "alternate universes"; something "necessary" is true in all possible worlds, something "possible" is true in at least one possible world. These "possible world semantics" are formalized with Kripke semantics.

Alternatively think of configurations of objects and materials that could have been differently arranged. E.g. the tea is on the top shelf and the coffee is on a lower shelf, but it might have been the case that both were on the lower shelf, or both on the top shelf, or the tea on the lower shelf and the coffee on the top shelf. Moreover, there might have been nothing on the top shelf (at that time), or there might have been nothing but a mouse there. Anyone who considers rearranging contents of a cupboard, or furniture of a room, or possible items to serve for a meal is using this concept of alethic possibility. If you try to draw a planar convex closed figure with four sides and exactly three corners you will find that is impossible: an example of alethic impossibility. Such a figure with four sides necessarily has four corners: that's an example of alethic necessity. Ordinary uses of these concepts do not seem to require the ability to think about collections of complete possible universes, merely rearrangements of portions of this universe.

Physical Possibility

Something is physically, or nomically, possible if it is permitted by the laws of physics. For example, current theory is thought to allow for there to be an atom with an atomic number of 126, even if there are no such atoms in existence. In contrast, while it is logically possible (i.e. probably via Alcubierre drive or worm holes) to accelerate beyond the speed of light, modern science stipulates that it is not physically possible for material particles or information.

Metaphysical Possibility

Philosophers ponder the properties that objects have independently of those dictated by scientific laws. For example, it might be metaphysically necessary, as some who advocate physicalism have thought, that all thinking beings have bodies and can experience the passage of time. Saul Kripke has argued that every person necessarily has the parents they do have: anyone with different parents would not be the same person.

Metaphysical possibility has been thought to be more restricting than bare logical possibility (i.e., fewer things are metaphysically possible than are logically possible). However, its exact relation (if any) to logical possibility or to physical possibility is a matter of dispute. Philosophers also disagree over whether metaphysical truths are necessary merely "by definition", or whether they reflect some underlying deep facts about the world, or something else entirely.

Confusion with Epistemic Modalities

Alethic modalities and epistemic modalities are often expressed in English using the same words. "It is possible that bigfoot exists" can mean either "Bigfoot *could* exist, whether or not bigfoot does in fact exist" (alethic), or more likely, "For all I know, bigfoot exists" (epistemic).

It has been questioned whether these modalities should be considered distinct from each other. The criticism states that there is no real difference between "the truth in the world" (alethic) and "the truth in an individual's mind" (epistemic). An investigation has not found a single language in which alethic and epistemic modalities are formally distinguished, as by the means of a grammatical mood.

Epistemic Logic

Epistemic modalities (from the Greek *episteme*, knowledge), deal with the *certainty* of sentences. The □ operator is translated as "x knows that...", and the ◇ operator is translated as "For all x knows, it may be true that..." In ordinary speech both metaphysical and epistemic modalities are often expressed in similar words; the following contrasts may help:

A person, Jones, might reasonably say *both*: (1) "No, it is *not* possible that Bigfoot exists; I am quite certain of that"; *and*, (2) "Sure, Bigfoot possibly *could* exist". What Jones means by (1) is that given all the available information, there is no question remaining as to whether Bigfoot exists. This is an epistemic claim. By (2) he makes the *metaphysical* claim that it is *possible for* Bigfoot to exist, *even though he does not* (which is not equivalent to "it is *possible that* Bigfoot exists – for all I know", which contradicts (1)).

From the other direction, Jones might say, (3) "It is *possible* that Goldbach's conjecture is true; but also *possible* that it is false", and *also* (4) "if it *is* true, then it is necessarily true, and not possibly false". Here Jones means that it is *epistemically possible* that it is true or false, for all he knows (Goldbach's conjecture has not been proven either true or false), but if there *is* a proof (heretofore undiscovered), then it would show that it is not *logically* possible for Goldbach's conjecture to be false—there could be no set of numbers that violated it. Logical possibility is a form of *alethic* possibility; (4) makes a claim about whether it is possible (i.e., logically speaking) that a mathematical truth to have been false, but (3) only makes a claim about whether it is possible, for all Jones knows, (i.e., speaking of certitude) that the mathematical claim is specifically either true or false, and so again Jones does not contradict himself. It is worthwhile to observe that Jones is not necessarily correct: It is possible (epistemically) that Goldbach's conjecture is both true and unprovable.

Epistemic possibilities also bear on the actual world in a way that metaphysical possibilities do not. Metaphysical possibilities bear on ways the world *might have been,* but epistemic possibilities bear on the way the world *may be* (for all we know). Suppose, for example, that I want to know whether or not to take an umbrella before I leave. If you tell me "it is *possible that* it is raining outside" – in the sense of epistemic possibility – then that would weigh on whether or not I take the umbrella. But if you just tell me that "it is *possible for* it to rain outside" – in the sense of *metaphysical possibility* – then I am no better off for this bit of modal enlightenment.

Some features of epistemic modal logic are in debate. For example, if x knows that p, does x know that it knows that p? That is to say, should $\Box P \to \Box\Box P$ be an axiom in these systems? While the answer to this question is unclear, there is at least one axiom that is generally included in epistemic modal logic, because it is minimally true of all normal modal logics:

- K, *Distribution Axiom*: $\Box(p \to q) \to (\Box p \to \Box q)$..

Temporal Logic

Temporal logic is an approach to the semantics of expressions with tense, that is, expressions with qualifications of when. Some expressions, such as '2 + 2 = 4', are true at all times, while tensed expressions such as 'John is happy' are only true sometimes.

In temporal logic, tense constructions are treated in terms of modalities, where a standard method for formalizing talk of time is to use *two* pairs of operators, one for the past and one for the future (P will just mean 'it is presently the case that P'). For example:

> FP : It will sometimes be the case that P

> GP : It will always be the case that P

> PP : It was sometime the case that P

> HP : It has always been the case that P

There are then at least three modal logics that we can develop. For example, we can stipulate that,

> $P = P$ is the case at some time t

$\Box P = P$ is the case at every time t

Or we can trade these operators to deal only with the future (or past). For example,

$$_1 P = \text{F}P$$

$$\Box_1 P = \text{G}P$$

or,

$$_2 P = P \text{ and/or } \text{F}P$$

$$\Box_2 P = P \text{ and } \text{G}P$$

The operators F and G may seem initially foreign, but they create normal modal systems. Note that FP is the same as $\neg$G$\neg P$. We can combine the above operators to form complex statements. For example, P$P \rightarrow \Box$PP says (effectively), *Everything that is past and true is necessary.*

It seems reasonable to say that possibly it will rain tomorrow, and possibly it won't; on the other hand, since we can't change the past, if it is true that it rained yesterday, it probably isn't true that it may not have rained yesterday. It seems the past is "fixed", or necessary, in a way the future is not. This is sometimes referred to as accidental necessity. But if the past is "fixed", and everything that is in the future will eventually be in the past, then it seems plausible to say that future events are necessary too.

Similarly, the problem of future contingents considers the semantics of assertions about the future: is either of the propositions 'There will be a sea battle tomorrow', or 'There will not be a sea battle tomorrow' now true? Considering this thesis led Aristotle to reject the principle of bivalence for assertions concerning the future.

Additional binary operators are also relevant to temporal logics, *q.v.* Linear Temporal Logic.

Versions of temporal logic can be used in computer science to model computer operations and prove theorems about them. In one version, $\Diamond P$ means "at a future time in the computation it is possible that the computer state will be such that P is true"; $\Box P$ means "at all future times in the computation P will be true". In another version, $\Diamond P$ means "at the immediate next state of the computation, P might be true"; $\Box P$ means "at the immediate next state of the computation, P will be true". These differ in the choice of Accessibility relation. (P always means "P is true at the current computer state".) These two examples involve nondeterministic or not-fully-understood computations; there are many other modal logics specialized to different types of program analysis. Each one naturally leads to slightly different axioms.

Deontic Logic

Likewise talk of morality, or of obligation and norms generally, seems to have a modal structure. The difference between "You must do this" and "You may do this" looks a lot like the difference between "This is necessary" and "This is possible". Such logics are called *deontic*, from the Greek for "duty".

Deontic logics commonly lack the axiom T semantically corresponding to the reflexivity of the accessibility relation in Kripke semantics: in symbols, $\Box\phi \to \phi$. . Interpreting $\Box$ as "it is obligatory that", T informally says that every obligation is true. For example, if it is obligatory not to kill others (i.e. killing is morally forbidden), then T implies that people actually do not kill others. The consequent is obviously false.

Instead, using Kripke semantics, we say that though our own world does not realize all obligations, the worlds accessible to it do (i.e., T holds at these worlds). These worlds are called idealized worlds. P is obligatory with respect to our own world if at all idealized worlds accessible to our world, P holds. Though this was one of the first interpretations of the formal semantics, it has recently come under criticism.

One other principle that is often (at least traditionally) accepted as a deontic principle is D, $\Box\phi \to \phi$, , which corresponds to the seriality (or extendability or unboundedness) of the accessibility relation. It is an embodiment of the Kantian idea that "ought implies can". (Clearly the "can" can be interpreted in various senses, e.g. in a moral or alethic sense.)

Intuitive Problems with Deontic Logic

When we try and formalize ethics with standard modal logic, we run into some problems. Suppose that we have a proposition K: you have stolen some money, and another, Q: you have stolen a small amount of money. Now suppose we want to express the thought that "if you have stolen some money, it ought to be a small amount of money". There are two likely candidates,

(1) $(K \to Q)$

(2) $(K \to Q)$

But (1) and K together entail $\Box Q$, which says that it ought to be the case that you have stolen a small amount of money. This surely isn't right, because you ought not to have stolen anything at all. And (2) doesn't work either: If the right representation of "if you have stolen some money it ought to be a small amount" is (2), then the right representation of (3) "if you have stolen some money then it ought to be a large amount" is $\Box(K \to (K \wedge \neg Q))$. Now suppose (as seems reasonable) that you ought not to steal anything, or $\Box\neg K$. But then we can deduce $\Box(K \to (K \wedge \neg Q))$ via $\Box(\neg K) \to \Box(K \to K \wedge \neg K)$ and $\Box(K \wedge \neg K \to (K \wedge \neg Q))$ (the contrapositive of $Q \to K$); so sentence (3) follows from our hypothesis (of course the same logic shows sentence (2)). But that can't be right, and is not right when we use natural language. Telling someone they should not steal certainly does not imply that they should steal large amounts of money if they do engage in theft.

Doxastic Logic

Doxastic logic concerns the logic of belief (of some set of agents). The term doxastic is derived from the ancient Greek *doxa* which means "belief". Typically, a doxastic logic uses □, often written "B", to mean "It is believed that", or when relativized to a particular agent s, "It is believed by s that".

Other Modal Logics

Significantly, modal logics can be developed to accommodate most of these idioms; it is the fact of their common logical structure (the use of "intensional" sentential operators) that make them all varieties of the same thing.

The Ontology of Possibility

In the most common interpretation of modal logic, one considers "logically possible worlds". If a statement is true in all possible worlds, then it is a necessary truth. If a statement happens to be true in our world, but is not true in all possible worlds, then it is a contingent truth. A statement that is true in some possible world (not necessarily our own) is called a possible truth.

Under this "possible worlds idiom," to maintain that Bigfoot's existence is possible but not actual, one says, "There is some possible world in which Bigfoot exists; but in the actual world, Bigfoot does not exist". However, it is unclear what this claim commits us to. Are we really alleging the existence of possible worlds, every bit as real as our actual world, just not actual? Saul Kripke believes that 'possible world' is something of a misnomer – that the term 'possible world' is just a useful way of visualizing the concept of possibility. For him, the sentences "you could have rolled a 4 instead of a 6" and "there is a possible world where you rolled a 4, but you rolled a 6 in the actual world" are not significantly different statements, and neither commit us to the existence of a possible world. David Lewis, on the other hand, made himself notorious by biting the bullet, asserting that all merely possible worlds are as real as our own, and that what distinguishes our world as *actual* is simply that it is indeed our world – *this* world. That position is a major tenet of "modal realism". Some philosophers decline to endorse any version of modal realism, considering it ontologically extravagant, and prefer to seek various ways to paraphrase away these ontological commitments. Robert Adams holds that 'possible worlds' are better thought of as 'world-stories', or consistent sets of propositions. Thus, it is possible that you rolled a 4 if such a state of affairs can be described coherently.

Computer scientists will generally pick a highly specific interpretation of the modal operators specialized to the particular sort of computation being analysed. In place of "all worlds", you may have "all possible next states of the computer", or "all possible future states of the computer".

Further Applications

Modal logics have begun to be used in areas of the humanities such as literature, poetry, art and history.

Controversies

Nicholas Rescher has argued that Bertrand Russell rejected modal logic, and that this rejection led to the theory of modal logic languishing for decades. However, Jan Dejnozka has argued against this view, stating that a modal system which Dejnozka calls *MDL* is described in Russell's works, although Russell did believe the concept of modality to "come from confusing propositions with propositional functions," as he wrote in *The Analysis of Matter*.

Arthur Norman Prior warned his protégé Ruth Barcan to prepare well in the debates concerning quantified modal logic with Willard Van Orman Quine, due to the biases against modal logic.

Term Logic

In philosophy, term logic, also known as traditional logic, syllogistic logic or Aristotelian logic, is a loose name for an approach to logic that began with Aristotle and that was dominant until the advent of modern predicate logic in the late nineteenth century. This entry is an introduction to the term logic needed to understand philosophy texts written before predicate logic came to be seen as the only formal logic of interest. Readers lacking a grasp of the basic terminology and ideas of term logic can have difficulty understanding such texts, because their authors typically assumed an acquaintance with term logic.

Aristotle's System

Aristotle's logical work is collected in the six texts that are collectively known as the *Organon*. Two of these texts in particular, namely the *Prior Analytics* and *De Interpretatione*, contain the heart of Aristotle's treatment of judgements and formal inference, and it is principally this part of Aristotle's works that is about term logic. Modern work on Aristotle's logic builds on the tradition started in 1951 with the establishment by Jan Lukasiewicz of a revolutionary paradigm. The Jan Lukasiewicz approach was reinvigorated in the early 1970s by John Corcoran and Timothy Smiley - which informs modern translations of *Prior Analytics* by Robin Smith in 1989 and Gisela Striker in 2009.

Basics

The fundamental assumption behind the theory is that propositions are composed of two terms - hence the name "two-term theory" or "term logic" - and that the reasoning process is in turn built from propositions:

- The term is a part of speech representing something, but which is not true or false in its own right, such as "man" or "mortal".

- The proposition consists of two terms, in which one term (the "predicate") is "affirmed" or "denied" of the other (the "subject"), and which is capable of truth or falsity.

- The syllogism is an inference in which one proposition (the "conclusion") follows of necessity from two others (the "premises").

A proposition may be universal or particular, and it may be affirmative or negative. Traditionally, the four kinds of propositions are:

- A-type: Universal and affirmative ("Every philosopher is mortal")

- I-type: Particular and affirmative ("Some philosopher is mortal")

- E-type: Universal and negative ("Every philosopher is not immortal")

- O-type: Particular and negative ("Some philosopher is not immortal")

This was called the *fourfold scheme* of propositions. Aristotle's *original* square of opposition, however, does not lack existential import:

- A-type: Universal and affirmative ("Every philosopher is mortal")

- I-type: Particular and affirmative ("Some philosopher is mortal")

- E-type: Universal and negative ("No philosopher is mortal")

- O-type: Particular and negative ("Not every philosopher is mortal")

In the Stanford Encyclopedia of Philosophy article, "The Traditional Square of Opposition", Terence Parsons explains:

One central concern of the Aristotelian tradition in logic is the theory of the categorical syllogism. This is the theory of two-premised arguments in which the premises and conclusion share three terms among them, with each proposition containing two of them. It is distinctive of this enterprise that everybody agrees on which syllogisms are valid. The theory of the syllogism partly constrains the interpretation of the forms. For example, it determines that the *A* form has existential import, at least if the *I* form does. For one of the valid patterns (Darapti) is:

> Every C is B
>
> Every C is A
>
> So, some A is B

This is invalid if the A form lacks existential import, and valid if it has existential import. It is held to be valid, and so we know how the A form is to be interpreted. One then naturally asks about the O form; what do the syllogisms tell us about it? The answer is that they tell us nothing. This is because Aristotle did not discuss weakened forms of syllogisms, in which one concludes a particular proposition when one could already conclude the corresponding universal. For example, he does not mention the form:

> No C is B
>
> Every A is C
>
> So, some A is not B

If people had thoughtfully taken sides for or against the validity of this form, that would clearly be relevant to the understanding of the O form. But the weakened forms were typically ignored...

One other piece of subject-matter bears on the interpretation of the O form. People were interested in Aristotle's discussion of "infinite" negation, which is the use of negation to form a term from a term instead of a proposition from a proposition. In modern English we use "non" for this; we make "non-horse," which is true for exactly those things that are not horses. In medieval Latin "non" and "not" are the same word, and so the distinction required special discussion. It became common to use infinite negation, and logicians pondered its logic. Some writers in the twelfth century and thirteenth centuries adopted a principle called "conversion by contraposition." It states that

- 'Every S is P' is equivalent to 'Every non-P is non-S'
- 'Some S is not P' is equivalent to 'Some non-P is not non-S'

Unfortunately, this principle (which is not endorsed by Aristotle) conflicts with the idea that there may be empty or universal terms. For in the universal case it leads directly from the truth:

> Every man is a being

to the falsehood:

> Every non-being is a non-man

(which is false because the universal affirmative has existential import, and there are no non-beings). And in the particular case it leads from the truth (remember that the **O** form has no existential import):

> A chimera is not a man

To the falsehood:

> A non-man is not a non-chimera

These are [Jean] Buridan's examples, used in the fourteenth century to show the invalidity of contraposition. Unfortunately, by Buridan's time the principle of contraposition had been advocated by a number of authors. The doctrine is already present in several twelfth century tracts, and it is endorsed in the thirteenth century by Peter of Spain, whose work was republished for centuries, by William Sherwood, and by Roger Bacon. By the fourteenth century, problems associated with contraposition seem to be well-known, and authors generally cite the principle and note that it is not valid, but that it becomes valid with an additional assumption of existence of things falling under the subject term. For example, Paul of Venice in his eclectic and widely published *Logica Parva* from the end of the fourteenth century gives the traditional square with simple conversion but rejects conversion by contraposition, essentially for Buridan's reason.

> — *Terence Parsons, The Stanford Encyclopedia of Philosophy*

Term

A term (Greek *horos*) is the basic component of the proposition. The original meaning of the *horos* (and also of the Latin *terminus*) is "extreme" or "boundary". The two terms lie on the outside of the proposition, joined by the act of affirmation or denial. For early modern logicians like Arnauld (whose *Port-Royal Logic* was the best-known text of his day), it is a psychological entity like an "idea" or "concept". Mill considers it a word. To assert "all Greeks are men" is not to say that the concept of Greeks is the concept of men, or that word "Greeks" is the word "men". A proposition cannot be built from real things or ideas, but it is not just meaningless words either.

Proposition

In term logic, a "proposition" is simply a *form of language*: a particular kind of sentence, in which the subject and predicate are combined, so as to assert something true or false. It is not a thought, or an abstract entity. The word *"propositio"* is from the Latin, meaning the first premise of a syllogism. Aristotle uses the word premise (*protasis*) as a sentence affirming or denying one thing or another (*Posterior Analytics* 1. 1 24a 16), so a premise is also a form of words. However, as in modern philosophical logic, it means that which is asserted by the sentence. Writers before Frege and Russell, such as Bradley, sometimes spoke of the "judgment" as something distinct from a sentence, but this is not quite the same. As a further confusion the word "sentence" derives from the Latin, meaning an opinion or judgment, and so is equivalent to "proposition".

The *logical quality* of a proposition is whether it is affirmative (the predicate is affirmed of the subject) or negative (the predicate is denied of the subject). Thus *every philosopher is mortal* is affirmative, since the mortality of philosophers is affirmed universally, whereas *no philosopher is mortal* is negative by denying such mortality in particular.

The *quantity* of a proposition is whether it is universal (the predicate is affirmed or denied of all subjects or of "the whole") or particular (the predicate is affirmed or denied of some subject or a "part" thereof). In case where existential import is assumed, quantification implies the existence of at least one subject, unless disclaimed.

Singular Terms

For Aristotle, the distinction between singular and universal is a fundamental metaphysical one, and not merely grammatical. A singular term for Aristotle is primary substance, which can only be predicated of itself: (this) "Callias" or (this) "Socrates" are not predicable of any other thing, thus one does not say *every Socrates* one says *every human* (*De Int.* 7; *Meta.* D9, 1018a4). It may feature as a grammatical predicate, as in the sentence "the person coming this way is Callias". But it is still a *logical* subject.

He contrasts universal (*katholou*) secondary substance, genera, with primary substance, particular (*kath' hekaston*) specimens. The formal nature of universals, in so far as they can be generalized "always, or for the most part", is the subject matter of both scientific study and formal logic.

The essential feature of the syllogistic is that, of the four terms in the two premises, one must occur twice. Thus

> All Greeks are men

> All men are mortal.

The subject of one premise, must be the predicate of the other, and so it is necessary to eliminate from the logic any terms which cannot function both as subject and predicate, namely singular terms.

However, in a popular 17th century version of the syllogistic, Port-Royal Logic, singular terms were treated as universals:

> All men are mortals

> All Socrates are men

> All Socrates are mortals

This is clearly awkward, a weakness exploited by Frege in his devastating attack on the system.

The famous syllogism "Socrates is a man ...", is frequently quoted as though from Aristotle, but fact, it is nowhere in the *Organon*. Sextus Empiricus in his *Hyp. Pyrrh* (Outlines of Pyrronism) ii. 164 first mentions the related syllogism "Socrates is a human being,Every human being is an animal, Therefore, Socrates is an animal."

Influence on Philosophy

The Aristotelian logical system had a formidable influence on the late-philosophy of the French psychoanalyst Jacques Lacan. In the early 1970s, Lacan reworked Aristotle's term logic by way of Frege and Jacques Brunschwig to produce his four formulae of sexuation. While these formulae retain the formal arrangement of the square of opposition, they seek to undermine the universals of both qualities by the 'existence without essence' of Lacan's particular negative proposition.

Decline of Term Logic

Term logic began to decline in Europe during the Renaissance, when logicians like Rodolphus Agricola Phrisius (1444-1485) and Ramus (1515-1572) began to promote place logics. The logical tradition called Port-Royal Logic, or sometimes "traditional logic", saw propositions as combinations of ideas rather than of terms, but otherwise followed many of the conventions of term logic. It remained influential, especially in England, until the 19th century. Leibniz created a distinctive logical calculus, but nearly all of his work on logic remained unpublished and unremarked until Louis Couturat went through the Leibniz *Nachlass* around 1900, publishing his pioneering studies in logic.

19th-century attempts to algebraize logic, such as the work of Boole (1815-1864) and Venn (1834-1923), typically yielded systems highly influenced by the term-logic tradition. The first predicate logic was that of Frege's landmark *Begriffsschrift* (1879), little read before 1950, in part because of its eccentric notation. Modern predicate logic as we know it began in the 1880s with the writings of Charles Sanders Peirce, who influenced Peano (1858-1932) and even more, Ernst Schröder (1841-1902). It reached fruition in the hands of Bertrand Russell and A. N. Whitehead, whose *Principia Mathematica* (1910–13) made use of a variant of Peano's predicate logic.

Term logic also survived to some extent in traditional Roman Catholic education, especially in seminaries. Medieval Catholic theology, especially the writings of Thomas Aquinas, had a powerfully Aristotelean cast, and thus term logic became a part of Catholic theological reasoning. For example, Joyce's *Principles of Logic* (1908; 3rd edition 1949), written for use in Catholic seminaries, made no mention of Frege or of Bertrand Russell.

Revival

Some philosophers have complained that predicate logic:

- Is unnatural in a sense, in that its syntax does not follow the syntax of the sentences that figure in our everyday reasoning. It is, as Quine acknowledged, "Procrustean," employing an artificial language of function and argument, quantifier, and bound variable.

- Suffers from theoretical problems, probably the most serious being empty names and identity statements.

Even academic philosophers entirely in the mainstream, such as Gareth Evans, have written as follows:

> "I come to semantic investigations with a preference for *homophonic* theories; theories which try to take serious account of the syntactic and semantic devices which actually exist in the language ...I would prefer [such] a theory ... over a theory which is only able to deal with [sentences of the form "all A's are B's"] by "discovering" hidden logical constants ... The objection would not be that such [Fregean] truth conditions are not correct, but that, in a sense which we would all dearly love to have more exactly explained, the syntactic shape of the sentence is treated as so much misleading surface structure" (Evans 1977)

BL (Logic)

Basic fuzzy Logic (or shortly BL), the logic of continuous t-norms, is one of t-norm fuzzy logics. It belongs to the broader class of substructural logics, or logics of residuated lattices; it extends the logic of all left-continuous t-norms MTL.

Syntax

Language

The language of the propositional logic BL consists of countably many propositional variables and the following primitive logical connectives:

- Implication $\rightarrow$ (binary)

- Strong conjunction $\otimes$ (binary). The sign & is a more traditional notation for strong conjunction in the literature on fuzzy logic, while the notation $\otimes$ follows the tradition of substructural logics.

- Bottom $\bot$ (nullary — a propositional constant); 0 or $\overline{0}$ are common alternative signs and zero a common alternative name for the propositional constant (as the constants bottom and zero of substructural logics coincide in MTL).

The following are the most common defined logical connectives:

- Weak conjunction $\wedge$ (binary), also called lattice conjunction (as it is always realized by the lattice operation of meet in algebraic semantics). Unlike MTL and weaker substructural logics, weak conjunction is definable in BL as

$$A \wedge B \equiv A \otimes (A \to B)$$

- Negation $\neg$ (unary), defined as

$$\neg A \equiv A \to \bot$$

- Equivalence $\leftrightarrow$ (binary), defined as

$$A \leftrightarrow B \equiv (A \to B) \wedge (B \to A)$$

  As in MTL, the definition is equivalent to $(A \to B) \otimes (B \to A)$.

- (Weak) disjunction $\vee$ (binary), also called lattice disjunction (as it is always realized by the lattice operation of join in algebraic semantics), defined as

$$A \vee B \equiv ((A \to B) \to B) \wedge ((B \to A) \to A)$$

- Top $\top$ (nullary), also called one and denoted by or $\overline{1}$ (as the constants top and zero of substructural logics coincide in MTL), defined as

$$\top \equiv \bot \to \bot$$

Well-formed formulae of BL are defined as usual in propositional logics. In order to save parentheses, it is common to use the following order of precedence:

- Unary connectives (bind most closely)

- Binary connectives other than implication and equivalence

- Implication and equivalence (bind most loosely)

Axioms

A Hilbert-style deduction system for BL has been introduced by Petr Hájek (1998). Its single derivation rule is modus ponens:

from A and $A \to B$ derive B.

The following are its axiom schemata:

(BL1): $(A \to B) \to ((B \to C) \to (A \to C))$

(BL2): $A \otimes B \to A$

(BL3): $A \otimes B \to B \otimes A$

(BL4): $A \otimes (A \to B) \to B \otimes (B \to A)$

(BL5a): $(A \to (B \to C)) \to (A \otimes B \to C)$

(BL5b): $(A \otimes B \to C) \to (A \to (B \to C))$

(BL6): $((A \to B) \to C) \to (((B \to A) \to C) \to C)$

(BL7): $\bot \to A$

The axioms (BL2) and (BL3) of the original axiomatic system were shown to be redundant (Chvalovský, 2012) and (Cintula, 2005). All the other axioms were shown to be independent (Chvalovský, 2012).

Semantics

Like in other propositional t-norm fuzzy logics, algebraic semantics is predominantly used for BL, with three main classes of algebras with respect to which the logic is complete:

- General semantics, formed of all *BL-algebras* — that is, all algebras for which the logic is sound

- Linear semantics, formed of all *linear* BL-algebras — that is, all BL-algebras whose lattice order is linear

- Standard semantics, formed of all *standard* BL-algebras — that is, all BL-algebras whose lattice reduct is the real unit interval [0, 1] with the usual order; they are uniquely determined by the function that interprets strong conjunction, which can be any continuous t-norm

Noise-based Logic

Noise-based logic (NBL) is a new class of multivalued deterministic logic schemes where the logic values and bits are represented by different realizations of a stochastic process. The concept of noise-based logic and its name was created by Laszlo B. Kish. In its foundation paper it is noted that the idea was inspired by the stochasticity of brain signals and by the unconventional noise-based communication schemes, such as the Kish cypher.

The Noise-based logic Space and Hyperspace

The logic values are represented by multi-dimensional "vectors" (orthogonal functions) and their superposition, where the orthogonal basis vectors are independent noises. By the proper combination (products or set-theoretical products) of basis-noises, which are called *noise-bit*, a logic hyperspace can be constructed with $D(N) = 2^N$ number of dimensions, where N is the number of noise-bits. Thus N noise-bits in a single wire correspond to a system of 2^N classical bits that can express 2^{2^N} different logic values. Independent realizations of a stochastic process of zero mean have zero cross-correlation with each other and with other stochastic processes of zero mean. Thus the basis noise vectors are orthogonal not only to each other but they and all the noise-based logic states (superpositions) are orthogonal also to any background noises in the hardware. Therefore, the noise-based logic concept is robust against background noises, which is a property that can potentially offer a high energy-efficiency.

The Types of Signals Used in Noise-based Logic

In the paper, where noise-based logic was first introduced, generic stochastic-processes with zero mean were proposed and a system of orthogonal sinusoidal signals were also proposed as a deterministic-signal version of the logic system. The mathematical analysis about statistical errors and signal energy was limited to the cases of Gaussian noises and superpositions as logic signals in the basic logic space and their products and superpositions of their products in the logic hyperspace (In the subsequent brain logic scheme, the logic signals were (similarly to neural signals) unipolar spike sequences generated by a Poisson process, and set-theoretical unifications (superpositions) and intersections (products) of different spike sequences. Later, in the instantaneous noise-based logic schemes and computation works, random telegraph waves (periodic time, bipolar, with fixed absolute value of amplitude) were also utilized as one of the simplest stochastic processes available for NBL. With choosing unit amplitude and symmetric probabilities, the resulting random-telegraph wave has 0.5 probability to be in the +1 or in the -1 state which is held over the whole clock period.

The Noise-based Logic Gates

Noise-based logic gates can be classified according to the method the input identifies the logic value at the input. The first gates analyzed the statistical correlations between the input signal and the reference noises. The advantage of these is the robustness against background noise. The disadvantage is the slow speed and higher hardware complexity. The instantaneous logic gates are fast, they have low complexity but they are not robust against background noises. With either neural spike type signals or with bipolar random-telegraph waves of unity absolute amplitude, and randomness only in the sign of the amplitude offer very simple instantaneous logic gates. Then linear or analog devices unnecessary and the scheme can operate in the digital domain. However, whenever instantaneous logic must be interfaced with classical logic schemes, the interface must use correlator-based logic gates for an error-free signal.

Universality of Noise-based Logic

All the noise-based logic schemes listed above have been proven universal. The papers typically produce the NOT and the AND gates to prove universality, because having both of them is a satisfactory condition for the universality of a Boolean logic.

Computation by Noise-based Logic

The string verification work over a slow communication channel shows a powerful computing application where the methods is inherently based on calculating the hash function. The scheme is based on random telegraph waves and it is mentioned in the paper that the authors intuitively conclude that the *intelligence* of the brain is using similar operations to make a reasonably good decision based on a limited amount of information. It is also interesting to note that the superposition of the first $D(N) = 2^N$ integer numbers can be produced with only $2N$ operations, which the authors call "Achilles ankle operation" in the paper.

Computer Chip Realization of Noise-based Logic

Preliminary schemes have already been published to utilize noise-based logic in practical computers. However, it is obvious from these papers that this young field has yet a long way to go before it will be seen in everyday applications.

References

- Eschenroeder, Erin; Sarah Mills; Thao Nguyen (2006-09-30). William Frawley, ed. The Expression of Modality. The Expression of Cognitive Categories. Mouton de Gruyter. pp. 8–9. ISBN 3-11-018436-2. Retrieved 2010-01-03.

- Nuyts, Jan (November 2000). Epimestic Modality, Language, and Conceptualization: A Cognitive-pragmatic Perspective. Human Cognitive Processing. John Benjamins Publishing Co. p. 28. ISBN 90-272-2357-2.

- Blackburn, Patrick; de Rijke, Maarten; and Venema, Yde (2001) Modal Logic. Cambridge University Press. ISBN 0-521-80200-8

- Chagrov, Aleksandr; and Zakharyaschev, Michael (1997) Modal Logic. Oxford University Press. ISBN 0-19-853779-4

- Cresswell, M. J. (2001) "Modal Logic" in Goble, Lou; Ed., The Blackwell Guide to Philosophical Logic. Basil Blackwell: 136–58. ISBN 0-631-20693-0

- Thomson, Judith and Alex Byrne (2006). Content and Modality : Themes from the Philosophy of Robert Stalnaker. Oxford: Oxford University Press. p. 107. Retrieved 16 December 2014.

8

Applications of Fuzzy Logic

The applications of fuzzy logic explained in the chapter are neuro-fuzzy, fuzzy electronics, fuzzy mathematics, fuzzy extractor, distribution management system etc. Neuro-fuzzy is referred to as the combination of artificial neural networks and fuzzy logic. The aspects elucidated in this chapter are of vital importance, and provides a better understanding of fuzzy logic.

Neuro-fuzzy

In the field of artificial intelligence, neuro-fuzzy refers to combinations of artificial neural networks and fuzzy logic. Neuro-fuzzy was proposed by J. S. R. Jang. Neuro-fuzzy hybridization results in a hybrid intelligent system that synergizes these two techniques by combining the human-like reasoning style of fuzzy systems with the learning and connectionist structure of neural networks. Neuro-fuzzy hybridization is widely termed as Fuzzy Neural Network (FNN) or Neuro-Fuzzy System (NFS) in the literature. Neuro-fuzzy system (the more popular term is used henceforth) incorporates the human-like reasoning style of fuzzy systems through the use of fuzzy sets and a linguistic model consisting of a set of IF-THEN fuzzy rules. The main strength of neuro-fuzzy systems is that they are universal approximators with the ability to solicit interpretable IF-THEN rules.

The strength of neuro-fuzzy systems involves two contradictory requirements in fuzzy modeling: interpretability versus accuracy. In practice, one of the two properties prevails. The neuro-fuzzy in fuzzy modeling research field is divided into two areas: linguistic fuzzy modeling that is focused on interpretability, mainly the Mamdani model; and precise fuzzy modeling that is focused on accuracy, mainly the Takagi-Sugeno-Kang (TSK) model.

Although generally assumed to be the realization of a fuzzy system through connectionist networks, this term is also used to describe some other configurations including:

- Deriving fuzzy rules from trained RBF networks.

- Fuzzy logic based tuning of neural network training parameters.

- Fuzzy logic criteria for increasing a network size.

- Realising fuzzy membership function through clustering algorithms in unsupervised learning in SOMs and neural networks.

- Representing fuzzification, fuzzy inference and defuzzification through multi-layers feed-forward connectionist networks.

It must be pointed out that interpretability of the Mamdani-type neuro-fuzzy systems can be lost. To improve the interpretability of neuro-fuzzy systems, certain measures must be taken, wherein important aspects of interpretability of neuro-fuzzy systems are also discussed.

A recent research line addresses the data stream mining case, where neuro-fuzzy systems are sequentially updated with new incoming samples on demand and on-the-fly. Thereby, system updates do not only include a recursive adaptation of model parameters, but also a dynamic evolution and pruning of model components (neurons, rules), in order to handle concept drift and dynamically changing system behavior adequately and to keep the systems/models "up-to-date" anytime. Comprehensive surveys of various evolving neuro-fuzzy systems approaches can be found in and.

Pseudo Outer-product-based Fuzzy Neural Networks

Pseudo outer-product-based fuzzy neural networks ("POPFNN") are a family of neuro-fuzzy systems that are based on the linguistic fuzzy model.

Three members of POPFNN exist in the literature:

- POPFNN-AARS(S), which is based on the Approximate Analogical Reasoning Scheme

- POPFNN-CRI(S), which is based on commonly accepted fuzzy Compositional Rule of Inference

- POPFNN-TVR, which is based on Truth Value Restriction

The "POPFNN" architecture is a five-layer neural network where the layers from 1 to 5 are called: input linguistic layer, condition layer, rule layer, consequent layer, output linguistic layer. The fuzzification of the inputs and the defuzzification of the outputs are respectively performed by the input linguistic and output linguistic layers while the fuzzy inference is collectively performed by the rule, condition and consequence layers.

The learning process of POPFNN consists of three phases:

1. Fuzzy membership generation

2. Fuzzy rule identification

3. Supervised fine-tuning

Various fuzzy membership generation algorithms can be used: Learning Vector Quantization (LVQ), Fuzzy Kohonen Partitioning (FKP) or Discrete Incremental Clustering (DIC). Generally, the POP algorithm and its variant LazyPOP are used to identify the fuzzy rules.

Fuzzy Electronics

Fuzzy electronics is an electronic technology that uses fuzzy logic, instead of the two-state Boolean logic more commonly used in digital electronics. Fuzzy electronics is fuzzy logic implemented on dedicated hardware. This is to be compared with fuzzy logic implemented in software running on a conventional processor. Fuzzy electronics has a wide range of applications, including control systems and artificial intelligence.

History

The first fuzzy electronic circuit was built by Takeshi Yamakawa *et al.* in 1980 using discrete bipolar transistors. The first industrial fuzzy application was in a cement kiln in Denmark in 1982. The first VLSI fuzzy electronics was by Masaki Togai and Hiroyuki Watanabe in 1984. In 1987, Yamakawa built the first analog fuzzy controller. The first digital fuzzy processors came in 1988 by Togai (Russo, pp. 2-6).

Fuzzy Mathematics

Fuzzy mathematics forms a branch of mathematics related to fuzzy set theory and fuzzy logic. It started in 1965 after the publication of Lotfi Asker Zadeh's seminal work *Fuzzy sets*. A fuzzy subset A of a set X is a function $A:X{\rightarrow}L$, where L is the interval [0,1]. This function is also called a membership function. A membership function is a generalization of a characteristic function or an indicator function of a subset defined for $L = \{0,1\}$. More generally, one can use a complete lattice L in a definition of a fuzzy subset A .

The evolution of the fuzzification of mathematical concepts can be broken down into three stages:

1. straightforward fuzzification during the sixties and seventies,

2. the explosion of the possible choices in the generalization process during the eighties,

3. the standardization, axiomatization and L-fuzzification in the nineties.

Usually, a fuzzification of mathematical concepts is based on a generalization of these concepts from characteristic functions to membership functions. Let A and B be two fuzzy subsets of X. Intersection $A \cap B$ and union $A \cup B$ are defined as follows: $(A \cap B)(x) = \min(A(x),B(x))$, $(A \cup B)(x) = \max(A(x),B(x))$ for all $x \in X$. Instead of *min* and *max* one can use t-norm and t-conorm, respectively , for example, $min(a,b)$ can be replaced by multiplication ab. A straightforward fuzzification is usually based on *min* and *max* operations because in this case more properties of traditional mathematics can be extended to the fuzzy case.

A very important generalization principle used in fuzzification of algebraic operations is a closure property. Let $*$ be a binary operation on X. The closure property for a fuzzy subset A of X is that for all $x,y \in X$, $A(x*y) \geq \min(A(x),A(y))$. Let $(G,*)$ be a group and A a fuzzy subset of G. Then A is a fuzzy subgroup of G if for all x,y in G, $A(x*y^{-1}) \geq \min(A(x),A(y^{-1}))$.

A similar generalization principle is used, for example, for fuzzification of the transitivity property. Let R be a fuzzy relation in X, i.e. R is a fuzzy subset of $X \times X$. Then R is transitive if for all x,y,z in X, $R(x,z) \geq \min(R(x,y),R(y,z))$.

Some Fields of Mathematics Using Fuzzy set Theory

Fuzzy subgroupoids and fuzzy subgroups were introduced in 1971 by A. Rosenfeld . Hundreds of papers on related topics have been published.

Main results in fuzzy fields and fuzzy Galois theory are published in a 1998 paper.

Fuzzy topology was introduced by C.L. Chang in 1968 and further was studied in many papers.

Main concepts of fuzzy geometry were introduced by Tim Poston in 1971, A. Rosenfeld in 1974, by J.J. Buckley and E. Eslami in 1997 and by D. Ghosh and D. Chakraborty in 2012-14

Basic types of fuzzy relations were introduced by Zadeh in 1971.

The properties of fuzzy graphs have been studied by A. Kaufman, A. Rosenfel, and by R.T. Yeh and S.Y. Bang. Recent results can be found in a 2000 article.

Possibility theory, nonadditive measures, fuzzy measure theory and fuzzy integrals are studied in the cited articles and treatises.

Main results and references on formal fuzzy logic can be found in these citations.

Fuzzy Cognitive Map

Rod Tabers FCM depicting eleven factors of the American drug market

A fuzzy cognitive map is a cognitive map within which the relations between the elements (e.g. concepts, events, project resources) of a "mental landscape" can be used to compute the "strength of impact" of these elements. Fuzzy cognitive maps were introduced by Bart Kosko. Ron Axelord introduced Cognitive Maps as a formal way of representing social scientific knowledge and modeling decision making in social and political systems. Then brought in the computation fuzzy logic.

Details

Fuzzy cognitive maps are signed fuzzy digraphs. They may look at first blush like Hasse diagrams but they are not. Spreadsheets or tables are used to map FCMs into matrices for further computation. FCM is a technique used for causal knowledge acquisition and representation, it supports causal knowledge reasoning process and belong to the neuro-fuzzy system that aim at solving decision making problems, modeling and simulate complex systems. Learning algorithms have been proposed for training and updating FCMs weights mostly based on ideas coming from the field of Artificial Neural Networks. Adaptation and learning methodologies used to adapt the FCM model and adjust its weights. Kosko and Dickerson (Dickerson & Kosko, 1994) suggested the Differential Hebbian Learning (DHL) to train FCM. There have been proposed algorithms based on the initial Hebbian algorithm; others algorithms come from the field of genetic algorithms, swarm intelligence and evolutionary computation. Learning algorithms are used to overcome the shortcomings that the traditional FCM present i.e. decreasing the human intervention by suggested automated FCM candidates; or by activating only the most relevant concepts every execution time; or by making models more transparent and dynamic. .

Fuzzy cognitive maps (FCMs) have gained considerable research interest due to their ability in representing structured knowledge and model complex systems in various fields. This growing interest led to the need for enhancement and making more reliable models that can better represent real situations. A first simple application of FCMs is described in a book of William R. Taylor, where the war in Afghanistan and Iraq is analyzed. And in Bart Kosko's book *Fuzzy Thinking*, several Hasse diagrams illustrate the use of FCMs. As an example, one FCM quoted from Rod Taber describes 11 factors of the American cocaine market and the relations between these factors. For computations, Taylor uses pentavalent logic (scalar values out of {-1,-0.5,0,+0.5,+1}). That particular map of Taber uses trivalent logic (scalar values out of {-1,0,+1}). Taber et al. also illustrate the dynamics of map fusion and give a theorem on the convergence of combination in a related article

While applications in social sciences introduced FCMs to the public, they are used in a much wider range of applications, which all have to deal with creating and using models of uncertainty and complex processes and systems. Examples:

- In business FCMs can be used for product planning.

- In economics, FCMs support the use of game theory in more complex settings.

- In Medical applications to model systems, provide diagnosis , develop decision support systems and medical assessment.

- In Engineering for modeling and control mainly of complex systems

- In project planning FCMs help to analyze the mutual dependencies between project resources.

- In robotics FCMs support machines to develop fuzzy models of their environments and to use these models to make crisp decisions.

- In computer assisted learning FCMs enable computers to check whether students understand their lessons.

- In expert systems a few or many FCMs can be aggregated into one FCM in order to process estimates of knowledgeable persons.

- In IT project management, a FCM-based methodology helps to success modelling.

FCMappers - an international online community for the analysis and the visualization of fuzzy cognitive maps offer support for starting with FCM and also provide an MS-Excel-based tool that is able to check and analyse FCMs. The output is saved as Pajek file and can be visualized within 3rd party software like Pajek, Visone,... . They also offer to adapt the software to specific research needs. On their webpage you also will find a linklist for interesting scientific articles, related software, institutes, people and projects. The FCMappers have about one thousand registered members worldwide.

Additional FCM software tools, such as Mental Modeler, have recently been developed as a decision-support tool for use in social science research, collaborative decision-making, and natural resource planning.

Bipolar Fuzzy Cognitive Maps

Fuzzy cognitive maps have been further extended to bipolar fuzzy cognitive maps based on bipolar fuzzy sets and bipolar cognitive mapping.   Bipolar fuzzy set theory as an equilibrium-based extension to fuzzy sets is recognized by L. A. Zadeh.

Fuzzy Extractor

Fuzzy extractors convert biometric data into random strings, which makes it possible to apply cryptographic techniques for biometric security. They are used to encrypt and authenticate users records, with biometric inputs as a key. Historically, the first biometric system of this kind was designed by Juels and Wattenberg and was called "Fuzzy commitment", where the cryptographic key is decommitted using biometric data. "Fuzzy", in that context, implies that the value close to the original one can extract the committed value. Later, Juels and Sudan came up with Fuzzy vault schemes which are order invariant for the fuzzy commitment scheme but uses a Reed–Solomon code. Codeword is evaluated by polynomial and the secret message is inserted as the coefficients of the polynomial. The polynomial is evaluated for different values of a set of features of the biometric data. So Fuzzy commitment and Fuzzy Vault were per-cursor to Fuzzy extractors. Fuzzy extractor is a biometric tool to authenticate a user using its own biometric template as a key. They extract uniform and random string R from its input R that has tolerance for noise. If the input changes to R but is still close to R, the string R can still be re-constructed. When R is used first time to re-construct, it outputs a helper string R which can be made public without compromising the security of R (used for encryption and authentication key) and R (helper string) is stored to recover R. They remain secure even when the adversary modifies R (key agreement between a user and a server based only on a biometric input). This article is based on the papers "Fuzzy Extractors: A Brief Survey of Results from 2004 to 2006" and "Fuzzy Extractors: How to Generate Strong Keys from Biometrics and Other Noisy Data" by Yevgeniy Dodis, Rafail Ostrovsky, Leonid Reyzin and Adam Smith

Motivation

As fuzzy extractors deal with how to generate strong keys from Biometrics and other Noisy Data, it applies cryptography paradigms to biometric data and that means (1) Make little assumptions about the biometric data (these data comes from variety of sources and don't want adversary to exploit that so it is best to assume the input is unpredictable) (2) Apply cryptographic application techniques to the input. (for that fuzzy extractor converts biometric data into secret, uniformly random and reliably reproducible random string). According to "Fuzzy Extractors: How to Generate Strong Keys from Biometrics and Other Noisy Data" paper by Yevgeniy Dodis, Rafail Ostrovsky, Leonid Reyzin and Adam Smith – these techniques also have other broader applications (when noisy inputs are used) such as human memory, images used as passwords, keys from quantum channel. Based on the Differential Privacy paper by Cynthia Dwork (ICALP 2006) – fuzzy extractors have application in the proof of impossibility of strong notions of privacy for statistical databases.

Basic Definitions

Predictability

Predictability indicates probability that adversary can guess a secret key. Mathematically speaking, the predictability of a random variable A is $\max_a P[A = a]$. For example, if pair of random variable A and B, if the adversary knows b of B, then predictability of A will be $\max_a P[A = a \mid B = b]$. . So, Adversary can predict A

with $E_{b \leftarrow B}[\max_a P[A = a \mid B = b]]$.. Taking average over B as it is not under adversary control, but since knowing b makes A prediction adversarial, taking the worst case over A..

Min-entropy

Min-entropy indicates worst-case entropy. Mathematically speaking, it is defined as $H_{\infty}(A) = -\log(\max_a P[A = a])$. Random variables with min-entropy at least m is called m – source.

Statistical Distance

Statistical distance is measure of distinguishability. Mathematically speaking, it is be_ tween two probability distributions A and B is, $SD[A, B] = \dfrac{1}{2} \sum |P[A = v] - P[B = v]|$.

In any system if A is replaced by B,, it will behave as original system with probability at least $1 - SD[A, B]$.

Definition 1 (Strong Extractor)

Set M is strong randomness extractor. Randomized function Ext: $M \rightarrow \{0,1\}^l$ with randomness of length r is an (m, l, ϵ)-strong extractor if for all m – sources (Random variables with min-entropy at least m is called m – -source) W on $M(Ext(W; I), I) \approx_{\epsilon} (U_l, U_r)$, where $I = U_r$ is independent of W. Output of the extractor is a key generated from $w \leftarrow W$ with the seed $i \leftarrow I$. It behaves independent of other parts of the system with the probability of $1 - \epsilon$.. Strong extractors can extract at most $l = m - 2\log \dfrac{1}{\epsilon} + O(1)$ bits from arbitrary m -source.

Secure Sketch

Secure sketch makes it possible to reconstruct noisy input, so if the input is w and sketch is s,, given and value close to , it is possible to recover . But sketch doesn't give much information about , so it is secure. If is a metric space with distance function dis. Secure sketch recovers string from any close string without disclosing .

Definition 2 (Secure Sketch)

An $(m, \tilde{m}, t)$ secure sketch is a pair of efficient randomized procedures (SS – Sketch, Rec – Recover) such that – (1) The sketching procedure SS on input $w \in \mathbb{M}$ returns a string $s \in \{0,1\}^*$. The recovery procedure Rec takes an element $w' \in \mathbb{M}$ and $s \in \{0,1\}^*$.. (2) Correctness: If $s \in \{0,1\}^*$.then $dis(w, w') \leq t$. (3) Security: For any $Rec(w', SS(w)) = w$. -source over m –, the min-entropy of M given W is high: for any (W, E),, if $\tilde{H}_{\infty}(W \mid E) \geq m$,, then $\tilde{H}_{\infty}(W \mid SS(W), E) \geq \tilde{m}$.

Fuzzy Extractor

Fuzzy extractors do not recover the original input but generate string (which is close to uniform) from and its subsequent reproduction (using helper string) given any close to . Strong extractors are a special case of fuzzy extractors when = 0 and .

Definition 3 (Fuzzy Extractor)

An (m,l,t,ϵ) fuzzy extractor is a pair of efficient randomized procedures (Gen – Generate and Rep – Reproduce) such that: (1) Gen, given $w \in \mathbb{M}$, outputs an extracted string $R \in \{0,1\}^l$ and a helper string $P \in \{0,1\}^*$. (2) Correctness: If $dis(w,w') \leq t$ and $(R,P) \leftarrow Gen(w)$, then $Rep(w',P) = R..$ (3) Security: For all m-sources W over M, the string R is nearly uniform even given P, So $\tilde{H}_\infty(W \mid E) \geq m$, then $(R,P,E) \approx (U_1,P,E)..$

So Fuzzy extractors output almost uniform random bits which is prerequisite for using cryptographic applications (in terms of secret keys). Since output bits are slightly non-uniform, it can decrease security, but not more than the distance from the uniform and as long as that distance is sufficiently small – security still remains robust.

Secure Sketches and Fuzzy Extractors

Secure sketches can be used to construct fuzzy extractors. Like applying SS to to obtain and strong extractor Ext with randomness to to get . can be stored as helper string . can be reproduced by and can recover and can reproduce . Following Lemma formalize this.

Lemma 1 (Fuzzy Extractors from Sketches)

Assume (SS,Rec) is an $(M,m,\tilde{m},t)$ secure sketch and let Ext be an average-case $(n,\tilde{m},l,\epsilon)$ strong extractor. Then the following (Gen, Rep) is an (M,m,l,t,ϵ) fuzzy extractor: (1) Gen $(w,r,x): setP = (SS(w;r),x), R = Ext(w;x)$, and output (R,P). (2) Rep $(w',(s,x))$: recover $w = Rec(w',s)$ and output $R = Ext(w;x)$.

Proof: From the definition of secure sketch (Definition 2), $H_\infty(W \mid SS(W)) \geq \tilde{m}..$ And since Ext is an average-case (n,m,l,ϵ)-strong extractor.

Corollary 1

If (SS,Rec) is an $(M,m,\tilde{m},t) -$ secure sketch and Ext is an $(n,\tilde{m} - log(\frac{1}{\delta}),l,\epsilon) -$ strong

extractor, then the above construction (Gen,Rep) is a $(M,m,l,t,\epsilon + \delta)$ fuzzy extractor. Reference paper "Fuzzy Extractors: How to Generate Strong Keys from Biometrics and Other Noisy Data" by Yevgeniy Dodis, Rafail Ostrovsky, Leonid Reyzin and Adam Smith (2008) includes many generic combinatorial bounds on secure sketches and fuzzy extractors

Basic Constructions

Due to their error tolerant properties, a secure sketches can be treated, analyzed, and constructed like a general error correcting code or for linear codes, where is the length of codewords, is the length of the message to be codded, is the distance between codewords, and is the alphabet. If is the universe of possible words then it may be possible to find an error correcting code that has a unique codeword for every and have a Hamming distance of . The first step for constructing a secure sketch is determining the type of errors that will likely occur and then choosing a distance to measure.

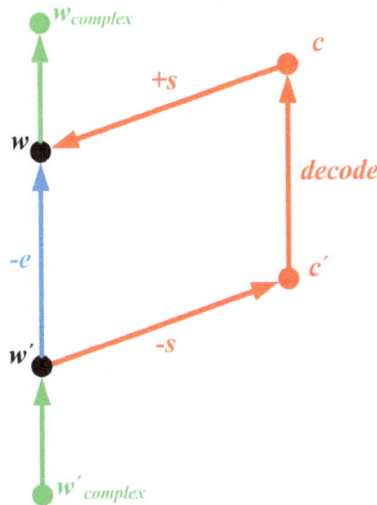

Red is the code-offset construction, blue is the syndrome construction, green represents edit distance and other complex constructions.

Hamming Distance Constructions

When there is no chance of data being deleted and only being corrupted then the best measurement to use for error correction is Hamming distance. There are two common constructions for correcting Hamming errors depending on whether the code is linear or not. Both constructions start with an error correcting code that has a distance of where is the number of tolerated errors.

Code-offset Construction

When using a general code, assign a uniformly random codeword to each , then let which is the shift needed to change into . To fix errors in subtract from then correct the errors in the resulting incorrect codeword to get and finally add to to get . This means . This construction can achieve the best possible tradeoff between error tolerance and entropy loss when and a Reed–Solomon code is used resulting in an entropy loss of , and the only way to improve upon this is to find a code better than Reed–Solomon.

Syndrome Construction

When using a linear code let the be the syndrome of . To correct find a vector such that , then .

Set Difference Constructions

When working with a very large alphabet or very long strings resulting in a very large universe , it may be more efficient to treat and as sets and look at set differences to correct errors. To work with a large set it is useful to look at its characteristic vector , which is a binary vector of length that has a value of 1 when an element and , or 0 when . The best way to decrease the size of a secure sketch when is large is make large since the size is determined by . A good code to base this construction on is a BCH code where and so , it is also useful that BCH codes can be decode in sub-linear time.

Pin Sketch Construction

Let $SS(w) = s = syn(x_w)$. To correct w' first find , then find a set v where $SS(w') = s' = syn(x_{w'})$, finally compute the symmetric difference to get $syn(x_v) = s' - s$, . While this is not the only construction to use set difference it is the easiest one to use.

Edit Distance Constructions

When data can be corrupted or deleted the best measurement to use is edit distance. To make a construction based on edit distance it is easiest to start with a construction for set difference or hamming distance as an intermediate correction step and then build the edit distance construction around that.

Other Distance Measure Constructions

There are many other types of errors and distances that can be measured which can be used to model other situations. Most of these other possible constructions are like edit distance constructions where they build upon simpler constructions.

Improving Error-tolerance Via Relaxed Notions of Correctness

It is possible to show that the error-tolerance of a secure sketch can be improved by applying a probabilistic method to error correction and only needing errors to be correctable with a high probability. This will show that it is possible to exceed the Plotkin bound which is limited to correcting errors, and approach Shannon's bound allowing for nearly corrections. To achieve this better error correction a less restrictive error distribution model must be used.

Random Errors

For this most restrictive model use a BSCto create a that a probability at each position in that the bit received is wrong. This model can show that entropy loss is limited to , where is the binary entropy function, and if min-entropy then errors can be tolerated, for some constant .

Input-dependent Errors

For this model errors do not have a known distribution and can be from an adversary, the only constraints are and that a corrupted word depends only on the input and not on the secure sketch. It can be shown for this error model that there will never be more than errors since this model can account for all complex noise processes, meaning that Shannon's bound can be reached, to do this a random permutation is prepended to the secure sketch that will reduce entropy loss.

Computationally Bounded Errors

This differs from the input dependent model by having errors that depend on both the input and the secure sketch, and an adversary is limited to polynomial time algorithms for introducing errors. Since algorithms that can run in better than polynomial time are not currently feasible in the real world, then a positive result using this error model would guarantee that any errors can be fixed. This is the least restrictive model the only known way to approach Shannon's bound is to use list-decodable codes although this may not always be useful in practice since returning a list instead of a single codeword may not always be acceptable.

Privacy Guarantees

In general a secure system attempts to leak as little information as possible to an adversary. In the case of biometrics if information about the biometric reading is leaked the adversary may be able to learn personal information about a user. For example an adversary notices that there is a certain pattern in the helper strings that implies the ethnicity of the user. We can consider this additional information a function . If an adversary were to learn a helper string, it must be ensured that, from this data he can not infer any data about the person from which the biometric reading was taken.

Correlation between Helper String and Biometric Input

Ideally the helper string would reveal no information about the biometric input . This is only possible when every subsequent biometric reading is identical to the original . In this case there is actually no need for the helper string, so it is easy to generate a string that is in no way correlated to .

Since it is desirable to accept biometric input similar to the helper string must be somehow correlated. The more different and are allowed to be, the more correlation there will be between and , the more correlated they are the more information reveals about . We can consider this information to be a function . The best possible solution is to make sure the adversary can't learn anything useful from the helper string.

Gen(W) as a Probabilistic Map

A probabilistic map hides the results of functions with a small amount of leakage . The leakage is the difference in probability two adversaries have of guessing some function when one knows the probabilistic map and one does not. Formally:

If the function is a probabilistic map, then even if an adversary knows both the helper string and the secret string they are only negligibly more likely figure something out about the subject as if they knew nothing. The string is supposed to kept secret, so even if it is leaked (which should be very unlikely) the adversary can still figure out nothing useful about the subject, as long as is small. We can considerto be any correlation between the biometric input and some physical characteristic of the person. Setting in the above equation changes it to:

$$| Pr[A_1(Y(W)) = f(W)] - Pr[A_2() = f(W)] | \le \epsilon$$

This means that if one adversary has and a second adversary knows nothing, their best guesses at are only apart.

Uniform Fuzzy Extractors

Uniform fuzzy extractors are a special case of fuzzy extractors, where the output of are negligibly different from strings picked from the uniform distribution, i.e.

Uniform Secure Sketches

Since secure sketches imply fuzzy extractors, constructing a uniform secure sketch allows for the easy construction of a uniform fuzzy extractor. In a uniform secure sketch the sketch procedure is a randomness extractor . Where is the biometric input and is the random seed. Since randomness extractors output a string that appears to be from a uniform distribution they hide all the information about their input.

Applications

Extractor sketches can be used to construct -fuzzy perfectly one-way hash functions. When used as a hash function the input is the object you want to hash. The that outputs is the hash value. If one wanted to verify that a within from the original , they would verify that . -fuzzy perfectly one-way hash functions are special hash functions where they accept any input with at most errors, compared to traditional hash functions which only accept when the input matches the original exactly. Traditional cryptographic hash functions attempt to guarantee that is it is computationally infeasible to find two different inputs that hash to the same value. Fuzzy perfectly one-way hash functions make an analogous claim. They make it computationally infeasible two find two inputs, that are more than Hamming distance apart and hash to the same value.

Protection Against Active Attacks

An active attack could be one where the adversary can modify the helper string . If the adversary is able to change to another string that is also acceptable to the reproduce function, it cause to output an incorrect secret string . Robust fuzzy extractors solve this problem by allowing the reproduce function to fail, if a modified helper string is provided as input.

Robust Fuzzy Extractors

One method of constructing robust fuzzy extractors is to use hash functions. This construction requires two hash functions and . The functions produces the helper string by appending the output of a secure sketch to the hash of both the reading and secure sketch . It generates the secret string by applying the second hash function to and . Formally: $Gen(w) : s = SS(w), return : P = (s, H_1(w, s)), R = H_2(w, s)$

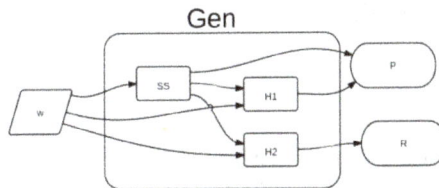

Gen

The reproduce function also makes use of the hash functions and . In addition to verifying the biometric input is similar enough to the one recovered using the function, it also verifies that hash in the second part of was actually derived from and . If both of those conditions are met it returns which is itself the second hash function applied to and . Formally:

Get and $\tilde{P}; \tilde{w} = Rec(w', \tilde{s}).$ from $\Delta(\tilde{w}, w') \leq t$ If $\tilde{h} = H_1(\tilde{w}, \tilde{s})$ and $return : H_2(\tilde{w}, \tilde{s})$ then $return : H_2(\tilde{w}, \tilde{s})$ else

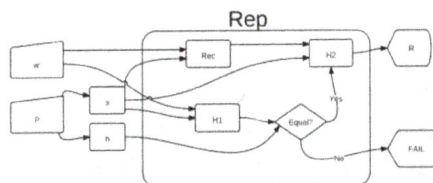

Rep

If has been tampered with it will be obvious because, will output fail with very high probability. To cause the algorithm accept a different an adversary would have to find a such that . Since hash function are believed to be one way functions, it is

computationally infeasible to find such a . Seeing would provide the adversary with
no useful informations. Since, again, hash function are one way functions, it is computationally infeasible for the adversary to reverse the hash function and figure out
. Part of is the secure sketch, but by definition the sketch reveals negligible information about its input. Similarly seeing would provide the adversary with no useful
information as the adversary wouldn't be able to reverse the hash function and see
the biometric input.

Fuzzy Locating System

Fuzzy locating is a rough but reliable method based on appropriate measuring technology for estimating a location of an object. The concept of precise or *crisp locating*
is replaced with respect to the operational requirements and the economic viability.
In most cases the knowledge of exact coordinates does not contribute to operations,
but the spatial or planar relation between entities is relevant. Hence fuzzy locating determines the radial distances between entities involved in an operational process and
reduces the required accuracy of measurement to basic qualities of *close*, *near* or *far*
and to relations simple as *in* or *out*. However such segregation shall be achieved with
high reliability and sound repetition.

Basics

The term *fuzzy* relates to rough spatial coincidence or contiguity assessment compared
to the alternative *crisp locating*, which derives precise coordinates of a location of an
object. The fuzzy part of the locating process balances the physical and the mathematical portions of processing measurement data of the objects involved and a priori knowledge with the operational ambience.

The result of fuzzy locating shall suffice for operational support and not for metric con-
firmation of measures taken at earlier occasions. However, available information is
exploited as *a priori* knowledge. Fuzzy locating compares with the distinction respectively segregation of mathematical logic with the terms *crisp sets* and *crisp relations* or
fuzzy sets and *fuzzy relations*.

Systematic and stochastic errors occurring under operational requirements and conditions turn virtually precise measures in a friendly ambience to fuzzy metrics. In
even worse ambient conditions, which mostly applies to wireless propagation in ISM
bands, this leads to erratic results and various misinterpretation. In consequence, the
trade-off between technical effort and achieved operational support adjusts inevitably
to physical limitations as well as to weaknesses in mathematical modeling. Better resolved balancing deliberately neglects classical terms of precision in favour of a strong
commitment for operational unambiguity.

The Techno-economic Challenge

Generally, precision is obtained at expense. The balancing of capital expenditure and of operational cost shall take into account not what is possible, but what is necessary. A better designed balancing leads to the less precise fuzzy locating at much lesser expense: As a more general approach, fuzzy locating is a method for best estimating a roughly determined location of an object as a distinction of operational contiguity. Contiguity means more or less a handy distance between an individual and an object. Three basic situations generally apply:

- Basic task is, for example, current presence in a room, well discriminated from other adjacent rooms.

- The more challenging requirement is to segregate the presence of an object or an individual at just one of several work positions in the very same room or with any other known and well referred place in this room.

- The more generalized approach hence is the spatial relationship between an object and an individual in any ambience, defining the actual location of either the object or alternatively the individual as the point of reference.

In all three cases the absolute coordinates are not of interest, as long as the discrimination of rooms or work positions is reliably achieved.

Wireless, Optical and Acoustical Approaches

For the locating task, an object to be located must be at least equipped with a wireless tag in a wireless communication environment. Each operating wireless target in a wireless communication environment may contribute. Prerequisite for radio frequency based wireless cooperation is some cohesion of the wireless nodes in a networking concept. Each wireless target has at least one physical propagation parameter that varies with location. Better qualified approaches make use of more than one physical parameter.

Alternatively, optical and acoustical solutions are known. Variation of parameters is partly deterministic with varying distances between wireless nodes. A location estimate approximating the real location of the transmitter, preferably under real time constraints, is determined on the basis of a stochastical model of propagation and a model for the process of observation in a noisy ambience and on a chosen set of observed deterministic parameters of transmission and propagation.

Implemented Examples

Several suppliers offer the so-called electronic leash solution. This serves for wirelessly tethering mobile appliances with each other. The RSSI estimate serves for a radial metrics but without any certified calibration. Setting an alarm on unintentional loss is the key service offered with this concept. An advanced aspect has been launched with

Bluetooth low energy for better economized battery life cycle. Special trimming serves for two years operation from a button cell.

Comparison to Metric Locating

Metric or crisp locating determines a spatial or planar relationship between independently moving or residing entities (usually addressed as targets) by means of qualified methods for measuring distances. This is the topic for example of

- satellite positioning systems, as GPS or Galileo

- real time locating systems (RTLS) as defined with ISO/IEC 19762-5 or

- inertial navigation systems (INS)

These technologies generally make use of a travel time measurement as the approach with best resolution and precision. Further enhancement is achieved with time differences discriminated for several paths. Such basic or enhanced travel time measurement requires a multiplicity of measures for unambiguous locating.

All of these sophisticated physical methods of measuring are hampered with a challenge caused by motion and caused by transmitter population. This makes restrictions effective in time both for observation and capture and for communication of measurement data. In consequence, the pecuniary and the technical effort adjusts to physical limitations and the limited metric precision with a special aspect to operational clarity. Such balancing neglects classical terms of metric precision, prevents from over-interpreting erratic measures and provides sufficient escape.

Additionally the model of propagation contributes to the achieved results. In satellite based systems, direct line of sight is generally required, without escape. That determines the restriction with applying such approach between buildings or, even worse, inside buildings: The highest precision does not compensate for bad visibility. Whenever the path of propagation gets cranked, the result of time measurement gets biased.

In comparison to locating on the move, exactly determining a location with highest possible precision is the topic of geodesy and surveying. These disciplines traditionally do not deal with motion and may integrate over long time. The terms of 'locating', 'positioning' and 'navigating' or 'surveying' are commonly used in almost equivalence, hence neglecting that the sense of these terms is different concerning sensor and actor functions and motion conditions.

Radio Signal Strength Indication (RSSI) as a Coarse Metric

For many purposes, a distinct and reliable determination of a location relative to somewhat rastered position on a floor or just a room in a building will be sufficient for sound fuzzy locating by single measurements.

Typically fuzzy locating coincides with simple power level measurement, usually configured as unilateration. A combination of multiple distance measurement, as a multilateration, based on power level measurement, appears unbalanced. The effort for multiple measures aiming at an unambiguous multilateration process will not be justified with the achievable precision of the results of power level measures.

Though leaving some ambiguity with sparse measurements, the contiguity may be assessed applying a priori knowledge. Such includes primarily tracking motion over time. As a generalized approach the fuzzy locating based on power levels measured with wireless nodes will roughly suffice for coarse guessing a location, where an object or a person with a wireless node resides in contiguity to other wireless nodes in a wireless environment.

Therefore each wireless node recognises the received power level with the distance to the transmitting wireless node and this parameter can be measured. Some additional calibration parameters serve as a basis for the statistical model of the parameters of propagation in a known neighborhood of distributed wireless nodes and of other passive objects, which influence propagation. A location estimate, which approximates the approach of the transmitting node, will be described with the stochastic model of propagation and the statistics of the monitored parameters.

Offered Discrimination of Positions

For applications where no need for absolute coordinates determination is assessed, the implementing of a more simple solution is advantageous. Compared to multilateration as the concept of *crisp locating*, the other option is *fuzzy locating*, where just one distance delivers the relation between detector and detected object. This most simple approach is unilateration. However, such unilateration approach never delivers the angular position with reference to the detector. Many solutions are available today.

Offered Qualities with Auxiliary Mapping

Increasing accuracy means increasing cost. The most indirect approach is the increase of distributed anchor nodes. The first direct approach simply is a fixed excitation through wall-mounted wireless nodes or optical excitors. That will provide a sound room discrimination in any case. The second direct approach is the position discrimination using apparently available infrastructure objects, as with networked work stations yet equipped with Bluetooth transponders.

The easy escape beyond increase of accuracy in measuring is the accuracy gain with mapping. Such mapping seldom suffices when performed statically. The more advantageous approach is a combination of initial mapping based on floor plans or area maps with an intelligent update based on updates obtained from actual measuring. That is offered with the concept of *simultaneous Locating and Mapping*.

Physical Restrictions

Measurement of propagation parameters is generally heavily loaded with various noise components. Such noise may be stochastic ([white noise]) or deterministic or a mixture of noises with limited bandwidth ([pink noise]).

As with all wireless systems, qualities of measurement in any one aspect contradict to qualities in some other aspects:

- Especially extension of reach collides with reducing the separation limits.

- Also increase of probing duration collides with motion speed.

- Generally the requirement for unambiguous locating requires a set of measured distances.

Such contracdictions result in challenges for systems performance and in hard restrictions concerning timely parallel availability of certain quality levels. Facing limitations as inevitable, the implementer of a locating systems has to determine the operational requirements first and then has to make a choice under a set of alternatives and must scale the adjoint limitations. The outcome is always a compromise with trade offs usually in budgetary effort and in technical precision. Any chosen alternative generally will exclude certain other technical options from operational availability.

Noisy Ambience as a General Condition

In technical terms, operational environments are generally noisy. For measuring, that is not a friendly ambience. Systematic and stochastic errors under operational requirements and conditions turn virtually precise measures to fuzzy metrics. The measuring results get in worse in a densely populated ambience, especially in the vicinity of other electrically active objects. Even physically passive surfaces contribute to the measurement problem. All physical effects tend to contribute unsteady and non linear behavior. Hence the physical measurement errors lead to biased or erratic results under such noisy conditions.

Frequency Spread as an Option

One escape from collision problems with wireless networks is the separation of the measuring and the communications processes with allocation to different frequencies, however requiring respective dual transponder capabilities. Another escape from collision problems with wireless networks is the spread of signals with stochastic coding.

Time as a Restriction for Measuring

Even if targets do not move, time is a restriction with the performing of a locating function. The first impact is that of allowed minimal distinct time differences that define the theoretically best resolution. This varies conceptually between phase discrimination in fractions of a cycle to be measured and full cycles to be counted. In competition for non-colliding transmission, time may appear as the main aspect with systems that use the very same frequency band. Other stochastic frequency allocation may ease the thrive for results, but normally coincide with lower allowance for power according to the set conditions of unlicensed usage.

Allowable time differences mostly vary with motion of the observed target. As for locating with absolute coordinates in a noisy environment several measurements are required and for disambiguation in space generally four measurements from independent reference points determine a target location, time appears as the sparse parameter.

Sequencing as a Restriction

In general, the strict sequencing of tasks appears with single tasking in one processor. Similarly the factual sequencing on only one frequency results from anti collision procedures. Both types of sequencing produce some dilation of time (with anti collision) or some dilution of location (with moving targets), while the respective wireless processes are performed by each target.

Bandwidth as a Restriction

The measuring of signals with steady modulation is bound to the bandwidth of the modulated carrier. The measuring of chirped signals is equivalently bound to the bandwidth of the transmitted pulses. In both methods, the available bandwidth will limit the precision of measurements.

Resolution as a Restriction

Technical means for measuring offer limited resolution and respective digitizing errors. This limits again the quality of results. Any way to overcome such limits raises system cost. Hence the escape again is not in improving the technical effort, but directs to the mathematical yield.

Battery Life Cycle as a Limiting Factor

The use of primary or secondary cells in wireless nodes limits both the time of operation as well as the life cycle without change for fresh batteries or just recharging. The mode of operation will be designed accordingly to widen the span of battery supply. That may be achieved by sleep up mode with respective wake up circuitry, operating without receiver in connectionless beacon mode, low repetition cycles and optimally low transmission power. An integrated loading circuit raises cost but saves the cost for external contacts. An unchangeable primary battery improves by lower self discharge compared to secondary cells, but causes the need for complete replacement or at least of the casing facultatively.

Motion as a Dynamic Challenge

When a target moves at a certain speed, the sequential measuring of distances from such transmitter target to a set of responder targets may deliver distance data for the subsequent locations at each measuring directly back to the transmitter target. This effect is independent from architecture of the network.

However, a measuring triggered from the transmitter target but performed almost in parallel by a set of receiver targets delivers a much better result under motion conditions, but requires either a server function for collecting the resulting data or requires additional response back to the triggering transceiver target.

The other escape is to apply a procedure to bundle the required measurements for each target in direct sequence thus reducing one effect of motion challenge by saving the preparation times for a reporting communications link. If not, then the competition for non-colliding transmission will lengthen the time span for each set of transmissions.

Population as Restriction

When several targets move independently in the same area or space and same wireless reach and also request locating independently and potentially in timely conflict using on the very same frequency for communications and for measurement, then the required measurements in one single ambience may collide. One escape again is the separation of the measuring and the communications processes with allocation to different frequencies, however requiring respective dual transponder capabilities.

Line of Sight as a Problem

In any case, line of sight is required for correct distance measurements. This may be eased by using auxiliary targets, but then increases the count of measurements. And the usage of auxiliary targets burdens the results with an increase of numeric inaccuracy.

Multipath Propagation as a Problem

Multipath propagation is inevitable with wireless systems. The reception from any transmitter and the response to any transmission are both challenged by the option of multiple propagation paths. If there is none but a single cranked path, there is no desired result at all, and the option to discriminate false measures from proper measures fails completely.

Typical issues with multipath propagation are fading, dither, diffraction, combining as non-linearity effects for the distance model. Additionally with power level measure the transmission through walls delivers rough errors, even with travel time measurement such error occurs.

Mathematical Requirements and Options

Mathematics serves for everything that cannot be covered by physics approaches. The assumption that a most qualified electro technical approach solves all problems arising from measurement is naïve and does not lead to sufficient results. At least thriving for best performance only at the expense of electronics is not an economized approach.

An operationally sufficient locating system will balance benefit and effort. Measurement and estimate shall take motion into account. This must not include the measuring of motion itself, but proper assessment of current and past motion to estimates. All estimate approximating the real location of the target is determined on the basis of a statistical model for the observed stochastic processes. Such model and estimation will use the set of observed propagation parameters. Some calibration data may serve as a basis for a statistical model of the propagation parameters. Such calibration is performed versus a spatial distribution of radio energy and with aspect to a known spatial distribution of corresponding targets. Other passive objects affecting propagation interfere with wireless operation and measurement.

Filtering as a Basic requirement

Any measurement is always biased with disturbances from ambient radiation out of electrical units with switches, from other wireless units and from stationary equipment as computers. To eliminate erratic results, some estimation based on past behavior, current dynamic properties and with reference to coupling mechanisms is recommended. State of the art for such tracking is e.g. extended Kalman filtering. Approaches that do not apply filtering produce no reasonable results. However, scalar filtering uses the model of residence in a fixed location or stationary motion. If abrupt turning the direction of motion, the filter algorithm may totally fail until filtering has recovered from tracking a sufficient walk in the new direction has been performed.

Statistics as a Means for Estimation

In case of biased signals the prerequisite for filtering is some statistic estimation, which serves for eliminating the large errors and smoothes a sequence of measurement results. This may be integrated with filtering, as far as the eliminating of large errors does not bias the filtering process under any conditions.

Quadratic Equations as a Problem

The determining equations are quadratic ones, thus requiring at least one more equation (n+1) than defined by dimension (n). This leads to a minimum requirement of three equations for planar problems and four equations for spatial problems.

Over Determination as a Support

The common approach to locating calculations may be the inversion of the Euclidic distance equations. However, such deterministic approach does not serve for the balancing in over determined equation systems. The easy approach is the exploitation of Gauss' least squares principle with the multi dimensional scaling according to Torgerson.

Wireless Coexistence

Many offered systems architectures and product offerings use license free ISM frequency bands and reside in similar channel patterns. The operating of fuzzy locating shall not compromise the communications options. Some restrictions apply not to infringe this requirement.

Technology Approaches and Options

The second step after scalar calculation is the involvement of model data according to the dimensionality of the motion. If reference is made to targets in other planes but the plane on which the moving targets may operate, such model must be a three-dimensional model. For model based operations, there are several options.

Coincident Locating as the Initial Option

Imagine a worker operating with a handheld reader of any type. The person is skilled to capture the identity of an object and used reader will report the capturing with time stamp. Such report discloses the location of capture as far as this information is reported in contents. The mandatory condition could be some automatic means to capture the location at the moment of identifying. In all other cases the quality and reliability of the location report limits the validity of the e.g. vocally reported data.

In all implementations of automatic data acquisition and locating systems the option of locating a handheld reader in the moment of manual triggering shall be foreseen as the fall back option. Otherwise the robustness of automatic data acquisition systems operation is bound to availability of automatic operation only.

Choke Point Locating as the Poorest Option

A choke point is a static bottleneck in process flow designs. There the passage of individuals and/or objects may indicate the identity of such entity to a steadily installed identifier unit. This approach under all conditions is restricted to just one location.

Politics and Sales force may describe that as locating, but it is definitively still just identifying.

Power Mapping as a First Poor Option

Propagation of radio signals happens according to Maxwell's equations and includes attenuation in atmosphere proportional to distance, Such concept is the basis for power mapping. The irregularities from local ambient conditions may be taken into account by power measurement in the operational area to correct the theoretically linear attenuation with distance. However, this approach does not work with an accuracy of better than 10% of the calculated distances in the range of propagation, thus leading to accuracies in the range of some meters.

Time Distance Equivalence for Radiation

Propagation of radio signals happens according to Maxwell's equations and includes travel time in atmosphere proportional to distance, Such concept is the basis for precise distance measurement. The irregularities from local ambient conditions are not dominant, thus the approach is more precise than power measurement. So this approach works with an accuracy of better than 1% of the calculated distances in the range of propagation, thus leading to accuracies in the range of some centimeters. However, this approach serves as well for line of sight propagation as for indirect reflected propagation.

Space Model as a Strong Option

To escape the biasing with secondary paths, there must be some reasoning that excludes the physically impossible locations from sets of results from locating. Simply, all calculated locations in material will be assessed as erratic, all calculated locations at distances not possible with inherent speed limits will be assessed as erratic and all locations above ground will be assessed impossible for floor operation. The requirement for space modeling leads to depicting the operational planes different from the limits to such planes, as walls, racks, and other installations.

Statistical Model to Exploit the Measurements

There is no chance to base stable results for location on single measurements. Statistics allow for

- combining subsequent measurement results to form a track
- smoothing a single location from subsequent measurement results
- iterating stable results from coarse first estimates

The methods for computing a set of results are described in context of various applications not just with locating technologies.

Fuzzy Reasoning with Discrete Spatial Compartments

As far as locating just has to support discrimination of rooms where a target may reside, the continuous model approaches may be combined with reasoning procedures to eliminate improbable results and to exclude operationally invalid locations from potential depicting a scenery. The known methods of inference apply to such processing.

Geometric Mapping Contributions to Reasoning

As far as the ambient operational conditions are stable, a geometric mapping of the neighborhood may support the reasoning. Then all massive obstacles describe the residual space of operation. As well such mapping will support the systematic and well determined consideration of multipath propagation effects. Hence geometric mapping derives the major gain compared to Bayes' estimators.

Adaptive Approaches

A common approach as preparatory mapping requires steady conditions and a constant ambience. This crucial condition is not fulfilled in dynamic operational theatres. However, a robust solution will always detect and investigate the actual conditions and reconnoiter the present ambience. For robot navigation, the methods of adaptive systems design, hence application of learning functions, is state of the art. However, adaptation requires time. A fully adaptive solution not applying a priori knowledge will be rather slow and will show limited dynamics. A balanced combination of adaptive functions will allow for best performance in a generally known ambience and cope for all changes that occurred after last encounter.

Operational Requirements

Locating arises from operational challenges. Traditional understanding of well kept enterprises with well educated staff is undermined with a thriving for reduced skills to achieve lesser cost. In result and in addition with continuing socio-economical disparities the processes and objects under control are threatened by negligence, fraud and theft.

Evidence

A simple indication of presence is given with the signal used for locating an object or a person. However, as presence may be temporary, a time stamp is required to adduce evidence in retrospective.

Cooperation Requirement

Persons carrying transponders or tagged objects with transponders might not be willing to be observed though having agreed earlier to this process. Then cooperation may be technically required, but individually denied. The robustness of the detection hence shall not be dependent to such cooperation. Especially covering the transponder or tearing off the transponder or otherwise tampering must be sensed automatically.

Proof of Presence

The presence of an object or a person in an operational vicinity is a strong demand. Absence of required resources generally affects planned processes. Therefore the proof of presence may be performed as far as possible before binding of additional resources happens.

Co-locating of Staff

Specially team work is bound to availability of required staff. The persons involved in a scheduled operation are well skilled to determine who is missing, but locating the missing parties is not that easy and may be strongly improved by system support.

Discrimination of Rooms

To allow for operation the respective room shall coincide with the scheduled action. Any request to operate under restrictions outside the planned confined area is suspect and may challenge security of processes and of secured knowledge. Locating the acting entities in the named confinement contributes to fulfilling security requirements.

Coincidence of Presence and Challenge

A person may try to access secured data, material or other resources outside the well secured rooms or areas. However, control may not always secure the subordination of the user to given orders. Locating simply in close distance or just in contiguity to allowed work positions may confirm the request for access as a basic feature.

Other application is coincidence of service provider, e.g.physician, with service requestor, e.g. patient, in a hospital. After identifying both persons as estimating their radial distance then access to the patient's file may be granted without any error. Such function would not be viable with *precise* or *crisp* systems as precision and allowable cost are in contradiction with absolute coordinate estimates.

Quality Requirements

The above listed terms will show that the definition of desired precision and accuracy, of repeatability and delays alone does not comply with a proper definition of requirements under aspects of cost. However, other terms of quality apply without restrictions.

Tamper Proof Identity

Basic requirement for any means to support locating is a tamper proof inherent identity of the carrying target with secured access. The secrecy of the identity prevents from plump copying threat and the tamper protection prevents from manipulating the target.

Self Identifying Authenticity

Persons who pursue to access data and applications normally authenticate themselves. Such authentication is generally bound to known locations, where the persons are authorized to perform work. Locating persons when they challenge authorization procedures is and advantage to prevent from fraud and theft.

Object identifying Security

Numerous means are known to identify objects. Normally the location where hand held units are operated are just roughly determined by the access point where connection to network is made. However such locating is still an improvement in many operations to secure knowledge about whereabouts of objects upon identifying.

Tracking Capability

In larger context of spatially distributed services and especially in logistics, numerous objects are in use in parallel, in different locations or on the move. Especially with transportation whereabouts of objects are understood as an essential to achieve high quality of service. As far as trust is with the forwarder, no problem exists on the journey and locating may happen just on leave and upon arrival. But third party infringements may collide with this assumption and generate a demand for permanent tracking on the journey as well. Then fuzzy locating is an economized and sufficient approach, which shall not provide location data with high metric accuracies, but status information with checkable and justifiable evidence.

Tracing Capability

In case single objects are lost, the capability to trace the whereabouts is another option to get access to the missing target again. However, this tracing is performed on yet available data and no means will deliver the data from the past without respective precautions. Especially in transportation whereabouts of lost objects are understood as an essential to retrieve the missing belongings.

Alert on Deviation

Easily any deviation from planned course, set route and scheduled arrival may lead to an alert. This requires timely locating and comparison of captured data with planning.

Expert System

A Symbolics Lisp Machine: an early platform for expert systems. Note the unusual *space-cadet keyboard*.

In artificial intelligence, an expert system is a computer system that emulates the decision-making ability of a human expert. Expert systems are designed to solve complex problems by reasoning about knowledge, represented mainly as if–then rules rather than through conventional procedural code. The first expert systems were created in the 1970s and then proliferated in the 1980s. Expert systems were among the first truly successful forms of artificial intelligence (AI) software.

An expert system is divided into two subsystems: the inference engine and the knowledge base. The knowledge base represents facts and rules. The inference engine applies the rules to the known facts to deduce new facts. Inference engines can also include explanation and debugging abilities.

History

Expert systems were introduced by the Stanford Heuristic Programming Project led by Edward Feigenbaum, who is sometimes termed the "father of expert systems"; other key early contributors were Edward Shortliffe, Bruce Buchanan, and Randall Davis. The Stanford researchers tried to identify domains where expertise was highly valued and complex, such as diagnosing infectious diseases (Mycin) and identifying unknown organic molecules (Dendral). Although that "intelligent systems derive their power from the knowledge they possess rather than from the specific formalisms and inference schemes they use" – as Feigenbaum said – seems in retrospect a rather straight-forward insight, it was a significant step forward then, since until then, research had been focused on attempts to develop very general-purpose problem solvers, such as those described by Allen Newell and Herb Simon. Expert systems became some of the first truly successful forms of artificial intelligence (AI) software.

Research on expert systems was also active in France. While in the US the focus tended to be on rule-based systems, first on systems hard coded on top of LISP programming environments and then on expert system shells developed by vendors such as Intel-licorp, in France research focused more on systems developed in Prolog. The advantage of expert system shells was that they were somewhat easier for nonprogrammers to use. The advantage of Prolog environments was that they weren't focused only on *if-then* rules; Prolog environments provided a much fuller realization of a complete First Order Logic environment.

In the 1980s, expert systems proliferated. Universities offered expert system courses and two thirds of the Fortune 500 companies applied the technology in daily business activities. Interest was international with the Fifth Generation Computer Systems proj-ect in Japan and increased research funding in Europe.

In 1981, the first IBM PC, with the PC DOS operating system, was introduced. The imbalance between the high affordability of the relatively powerful chips in the PC, compared to the much more expensive cost of processing power in the mainframes that dominated the corporate IT world at the time, created a new type of architecture for corporate computing, termed the client-server model. Calculations and reasoning could be performed at a fraction of the price of a mainframe using a PC. This model also enabled business units to bypass corporate IT departments and directly build their own applications. As a result, client server had a tremendous impact on the ex-pert systems market. Expert systems were already outliers in much of the business world, requiring new skills that many IT departments did not have and were not eager to develop. They were a natural fit for new PC-based shells that promised to put ap-plication development into the hands of end users and experts. Until then, the main development environment for expert systems had been high end Lisp machines from Xerox, Symbolics, and Texas Instruments. With the rise of the PC and client server computing, vendors such as Intellicorp and Inference Corporation shifted their priori-ties to developing PC based tools. Also, new vendors, often financed by venture capital (such as Aion Corporation, Neuron Data, Exsys, and many others), started appearing regularly.

In the 1990s and beyond, the term *expert system* and the idea of a standalone AI system mostly dropped from the IT lexicon. There are two interpretations of this. One is that "expert systems failed": the IT world moved on because expert systems didn't deliver on their over hyped promise. The other is the mirror opposite, that expert systems were simply victims of their success: as IT professionals grasped concepts such as rule engines, such tools migrated from being standalone tools for developing special purpose *expert* systems, to being one of many standard tools. Many of the leading major business application suite vendors (such as SAP, Siebel, and Oracle) integrated expert system abilities into their suite of products as a way of specifying business logic – rule engines are no longer simply for defining the rules an expert would use but for any type of complex, volatile, and critical business logic; they often go hand in hand with business process automation and integration environments.

Software Architecture

An expert system is an example of a knowledge-based system. Expert systems were the first commercial systems to use a knowledge-based architecture. A knowledge-based system is essentially composed of two sub-systems: the knowledge base and the inference engine.

The knowledge base represents facts about the world. In early expert systems such as Mycin and Dendral, these facts were represented mainly as flat assertions about variables. In later expert systems developed with commercial shells, the knowledge base took on more structure and used concepts from object-oriented programming. The world was represented as classes, subclasses, and instances and assertions were replaced by values of object instances. The rules worked by querying and asserting values of the objects.

The inference engine is an automated reasoning system that evaluates the current state of the knowledge-base, applies relevant rules, and then asserts new knowledge into the knowledge base. The inference engine may also include abilities for explanation, so that it can explain to a user the chain of reasoning used to arrive at a particular conclusion by tracing back over the firing of rules that resulted in the assertion.

There are mainly two modes for an inference engine: forward chaining and backward chaining. The different approaches are dictated by whether the inference engine is being driven by the antecedent (left hand side) or the consequent (right hand side) of the rule. In forward chaining an antecedent fires and asserts the consequent. For example, consider the following rule:

$$R1 : Man(x) => Mortal(x)$$

A simple example of forward chaining would be to assert Man(Socrates) to the system and then trigger the inference engine. It would match R1 and assert Mortal(Socrates) into the knowledge base.

Backward chaining is a bit less straight forward. In backward chaining the system looks at possible conclusions and works backward to see if they might be true. So if the system was trying to determine if Mortal(Socrates) is true it would find R1 and query the knowledge base to see if Man(Socrates) is true. One of the early innovations of expert systems shells was to integrate inference engines with a user interface. This could be especially powerful with backward chaining. If the system needs to know a particular fact but doesn't it can simply generate an input screen and ask the user if the information is known. So in this example, it could use R1 to ask the user if Socrates was a Man and then use that new information accordingly.

The use of rules to explicitly represent knowledge also enabled explanation abilities. In the simple example above if the system had used R1 to assert that Socrates was Mortal and a user wished to understand why Socrates was mortal they could query the system and the system would look back at the rules which fired to cause the assertion and present those rules to the user as an explanation. In English if the user asked "Why is Socrates Mortal?" the system would reply "Because all men are mortal and Socrates is a man". A significant area for research was the generation of explanations from the knowledge base in natural English rather than simply by showing the more formal but less intuitive rules.

As expert systems evolved, many new techniques were incorporated into various types of inference engines. Some of the most important of these were:

- Truth maintenance. These systems record the dependencies in a knowledge-base so that when facts are altered, dependent knowledge can be altered accordingly. For example, if the system learns that Socrates is no longer known to be a man it will revoke the assertion that Socrates is mortal.

- Hypothetical reasoning. In this, the knowledge base can be divided up into many possible views, a.k.a. worlds. This allows the inference engine to explore multiple possibilities in parallel. For example, the system may want to explore the consequences of both assertions, what will be true if Socrates is a Man and what will be true if he is not?

- Fuzzy logic. One of the first extensions of simply using rules to represent knowledge was also to associate a probability with each rule. So, not to assert that Socrates is mortal, but to assert Socrates *may* be mortal with some probability value. Simple probabilities were extended in some systems with sophisticated mechanisms for uncertain reasoning and combination of probabilities.

- Ontology classification. With the addition of object classes to the knowledge base, a new type of reasoning was possible. Along with reasoning simply about object values, the system could also reason about object structures. In this simple example, Man can represent an object class and R1 can be redefined as a rule that defines the class of all men. These types of special purpose inference engines are termed classifiers. Although they were not highly used in expert systems, classifiers are very powerful for unstructured volatile domains, and are a key technology for the Internet and the emerging Semantic Web.

Advantages

The goal of knowledge-based systems is to make the critical information required for the system to work explicit rather than implicit. In a traditional computer program the logic is embedded in code that can typically only be reviewed by an IT specialist. With an expert system the goal was to specify the rules in a format that was intuitive and easily understood, reviewed, and even edited by domain experts rather than IT experts. The benefits of this explicit knowledge representation were rapid development and ease of maintenance.

Ease of maintenance is the most obvious benefit. This was achieved in two ways. First, by removing the need to write conventional code, many of the normal problems that can be caused by even small changes to a system could be avoided with expert systems. Essentially, the logical flow of the program (at least at the highest level) was simply a given for the system, simply invoke the inference engine. This also was a reason for the second benefit: rapid prototyping. With an expert system shell it was possible to enter a few rules and have a prototype developed in days rather than the months or year typically associated with complex IT projects.

A claim for expert system shells that was often made was that they removed the need for trained programmers and that experts could develop systems themselves. In reality, this was seldom if ever true. While the rules for an expert system were more comprehensible than typical computer code, they still had a formal syntax where a misplaced comma or other character could cause havoc as with any other computer language. Also, as expert systems moved from prototypes in the lab to deployment in the business world, issues of integration and maintenance became far more critical. Inevitably demands to integrate with, and take advantage of, large legacy databases and systems arose. To accomplish this, integration required the same skills as any other type of system.

Disadvantages

The most common disadvantage cited for expert systems in the academic literature is the knowledge acquisition problem. Obtaining the time of domain experts for any software application is always difficult, but for expert systems it was especially difficult because the experts were by definition highly valued and in constant demand by the organization. As a result of this problem, a great deal of research in the later years of expert systems was focused on tools for knowledge acquisition, to help automate the process of designing, debugging, and maintaining rules defined by experts. However, when looking at the life-cycle of expert systems in actual use, other problems – essentially the same problems as those of any other large system – seem at least as critical as knowledge acquisition: integration, access to large databases, and performance.

Performance was especially problematic because early expert systems were built using tools such as Lisp, which executed interpreted (rather than compiled) code. Interpreting provided an extremely powerful development environment but with the drawback that it was virtually impossible to match the efficiency of the fastest compiled languages, such as C. System and database integration were difficult for early expert systems because the tools were mostly in languages and platforms that were neither familiar to nor welcome in most corporate IT environments – programming languages such as Lisp and Prolog, and hardware platforms such as Lisp machines and personal computers. As a result, much effort in the later stages of expert system tool development was focused on integrating with legacy environments such as COBOL and large database systems, and on porting to more standard platforms. These issues were resolved mainly by the client-server paradigm shift, as PCs were gradually accepted in the IT environment as a legitimate platform for serious business system development and as affordable minicomputer servers provided the processing power needed for AI applications.

Applications

Hayes-Roth divides expert systems applications into 10 categories illustrated in the following table. The example applications were not in the original Hayes-Roth table, and some of them arose well afterward. Any application that is not footnoted is described in the Hayes-Roth book. Also, while these categories provide an intuitive framework to describe the space of expert systems applications, they are not rigid categories, and in some cases an application may show traits of more than one category.

Category	Problem addressed	Examples
Interpretation	Inferring situation descriptions from sensor data	Hearsay (speech recognition), PROSPECTOR
Prediction	Inferring likely consequences of given situations	Preterm Birth Risk Assessment
Diagnosis	Inferring system malfunctions from observables	CADUCEUS, MYCIN, PUFF, Mistral, Eydenet, Kaleidos
Design	Configuring objects under constraints	Dendral, Mortgage Loan Advisor, R1 (DEC VAX Configuration)
Planning	Designing actions	Mission Planning for Autonomous Underwater Vehicle
Monitoring	Comparing observations to plan vulnerabilities	REACTOR
Debugging	Providing incremental solutions for complex problems	SAINT, MATHLAB, MACSYMA
Repair	Executing a plan to administer a prescribed remedy	Toxic Spill Crisis Management
Instruction	Diagnosing, assessing, and repairing student behavior	SMH.PAL, Intelligent Clinical Training, STEAMER

Control	Interpreting, predicting, repairing, and monitoring system behaviors	Real Time Process Control, Space Shuttle Mission Control

Hearsay was an early attempt at solving voice recognition through an expert systems approach. For the most part this category or expert systems was not all that successful. Hearsay and all interpretation systems are essentially pattern recognition systems—looking for patterns in noisy data. In the case of Hearsay recognizing phonemes in an audio stream. Other early examples were analyzing sonar data to detect Russian submarines. These kinds of systems proved much more amenable to a neural network AI solution than a rule-based approach.

CADUCEUS and MYCIN were medical diagnosis systems. The user describes their symptoms to the computer as they would to a doctor and the computer returns a medical diagnosis.

Dendral was a tool to study hypothesis formation in the identification of organic molecules. The general problem it solved—designing a solution given a set of constraints—was one of the most successful areas for early expert systems applied to business domains such as salespeople configuring Digital Equipment Corporation (DEC) VAX computers and mortgage loan application development.

SMH.PAL is an expert system for the assessment of students with multiple disabilities.

Mistral is an expert system to monitor dam safety, developed in the 90's by Ismes (Italy). It gets data from an automatic monitoring system and performs a diagnosis of the state of the dam. Its first copy, installed in 1992 on the Ridracoli Dam (Italy), is still operational 24/7/365. It has been installed on several dams in Italy and abroad (e.g., Itaipu Dam in Brazil), and on landslide sites under the name of Eydenet, and on monuments under the name of Kaleidos. Mistral is a registered trade mark of CESI.

Fuzzy Markup Language

Fuzzy Markup Language (FML) is a specific purpose markup language based on XML, used for describing the structure and behavior of a fuzzy system independently of the hardware architecture devoted to host and run it.

Overview

FML was designed and developed by Giovanni Acampora during his Ph.D. course in Computer Science, at University of Salerno, Italy, in 2004. The original idea inspired Giovanni Acampora to create FML was the necessity of creating a cooperative fuzzy-based framework aimed at automatically controlling a living environment characterized by a plethora of heterogeneous devices whose interactions were devoted to maximize the human comfort under energy saving constraints. This framework represented one of the first concrete examples of Ambient Intelligence. Beyond this pioneering application, the major advantage of using XML to describe a fuzzy system is hardware/software interoperability. Indeed, all that is needed to read an FML file is the appropriate schema for that file, and an FML parser. This markup approach makes it much easier to exchange fuzzy systems between software: for example, a machine learning application could extract fuzzy rules which could then be read directly into a fuzzy inference engine or uploaded into a fuzzy controller. Also, with technologies like XSLT, it is possible to compile the FML into the programming language of your choice, ready for embedding into whatever application you please. As stated by Mike Watts on his popular Computational Intelligence blog:

"Although Acampora's motivation for developing FML seems to be to develop embedded fuzzy controllers for ambient intelligence applications, FML could be a real boon for developers of fuzzy rule extraction algorithms: from my own experience during my PhD, I know that having to design a file format and implement the appropriate parsers for rule extraction and fuzzy inference engines can be a real pain, taking as much time as implementing the rule extraction algorithm itself. I would much rather have used something like FML for my work."

A complete overview of FML and related applications can be found in the book titled *On the power of Fuzzy Markup Language* edited by Giovanni Acampora, Chang-Shing Lee, Vincenzo Loia and Mei-Hui Wang, and published by Springer in the series *Studies on Fuzziness and Soft Computing*.

FML at Work: Syntax, Grammar and Hardware Synthesis

FML allows fuzzy systems to be coded through a collection of correlated semantic tags capable of modeling the different components of a classical fuzzy controller such as knowledge base, rule base, fuzzy variables and fuzzy rules. Therefore, the FML tags used to build a fuzzy controller represent the set of lexemes used to create fuzzy expressions. In order to design a well-formed XML-based language, an FML context-free grammar is defined by means of a XML schema which defines name, type and attributes characterized each XML element. However, since an FML program represents only a static view of a fuzzy logic controller, the so-called eXtensible Stylesheet Language Translator (XSLT) is provided to change this static view to a computable version. Indeed, XSLTs modules are able to convert the FML-based fuzzy controller in a general purpose computer language using an XSL file containing the translation description. At this level, the control is executable for the hardware. In short, FML is essentially composed by three layers:

- XML in order to create a new markup language for fuzzy logic control;

- a XML Schema in order to define the legal building blocks;

- eXtensible Stylesheet Language Transformations (XSLT) in order to convert a fuzzy controller description into a specific programming language.

FML Syntax

FML syntax is composed of XML tags and attributes which describe the different components of a fuzzy logic controller listed below:

- fuzzy knowledge base;

- fuzzy rule base;

- inference engine

- fuzzification subsystem;

- defuzzification subsystem.

In detail, the opening tag of each FML program is `<FuzzyController>` which represents the fuzzy controller under modeling. This tag has two attributes: *name* and *ip*. The first attribute permits to specify the name of fuzzy controller and *ip* is used to define the location of controller in a computer network. The fuzzy knowledge base is defined by means of the tag `<KnowledgeBase>` which maintains the set of fuzzy concepts used to model the fuzzy rule base. In order to define the fuzzy concept related controlled system, `<KnowledgeBase>` tag uses a set of nested tags:

- `<FuzzyVariable>` defines the fuzzy concept;

- `<FuzzyTerm>` defines a linguistic term describing the fuzzy concept;

- a set of tags defining a shape of fuzzy sets are related to fuzzy terms.

The attributes of `<FuzzyVariable>` tag are: *name, scale, domainLeft, domainRight, type* and, for only an output, *accumulation, defuzzifier* and *defaultValue*. The *name* attribute defines the name of fuzzy concept, for instance, *temperature*; *scale* is used to define the scale used to measure the fuzzy concept, for instance, *Celsius degree*; *domainLeft* and *domainRight* are used to model the universe of discourse of fuzzy concept, that is, the set of real values related to fuzzy concept, for instance [0°,40°] in the case of Celsius degree; the position of fuzzy concept into rule (consequent part or antecedent part) is defined by *type* attribute (input/output); *accumulation* attribute defines the method of accumulation that is a method that permits the combination of results of a variable of each rule in a final result; *defuzzifier* attribute defines the method used to execute the conversion from a fuzzy set, obtained after aggregation process, into a numerical value to give it in output to system; *defaultValue* attribute defines a real value used only when no rule has fired for the variable at issue. As for tag `<FuzzyTerm>`, it uses two attributes: *name* used to identify the linguistic value associate with fuzzy concept and *complement*, a boolean attribute that defines, if it is true, it is necessary to consider the complement of membership function defined by given parameters. Fuzzy shape tags, used to complete the definition of fuzzy concept, are:

- `<TRIANGULARSHAPE>`

- `<RIGHTLINEARSHAPE>`

- `<LEFTLINEARSHAPE>`

- `<PISHAPE>`

- `<GAUSSIANSHAPE>`

- `<RIGHTGAUSSIANSHAPE>`

- `<LEFTGAUSSIANSHAPE>`

- `<TRAPEZOIDSHAPE>`

- `<SSHAPE>`

- `<ZSHAPE>`

- `<RECTANGULARSHAPE>`

- `<SINGLETONSHAPE>`

Every shaping tag uses a set of attributes which defines the real outline of corresponding fuzzy set. The number of these attributes depends on the chosen fuzzy set shape.

In order to make an example, let us consider the *Tipper Inference System* described in Mathwork Matlab Fuzzy Logic Toolbox Tutorial. This Mamdani system is used to regulate the tipping in, for example, a restaurant. It has got two variables in input (*food* and *service*) and one in output (*tip*). FML code for modeling part of knowledge base of this fuzzy system containing variables *food* and *tip* is shown below.

```xml
<?xml version="1.0" encoding="UTF-8"?>

<FuzzyController name="newSystem" ip="127.0.0.1">

    <KnowledgeBase>

        <FuzzyVariable  name="food"  domainleft="0.0"  domain-
right="10.0" scale="" type="input">

            <FuzzyTerm name="delicious" complement="false">

                <LeftLinearShape Param1="5.5" Param2="10.0"/>

            </FuzzyTerm>

            <FuzzyTerm name="rancid" complement="false">

                <TriangularShape Param1="0.0" Param2="2.0"
Param3="5.5"/>
```

```
                        </FuzzyTerm>

                </FuzzyVariable>

                . . . . . . . . . .

        <FuzzyVariable    name="tip"    domainleft="0.0"    domain-
right="20.0" scale="Euro" defaultValue="0.0" defuzzifier="COG"

                        accumulation="MAX"   type="output">

                <FuzzyTerm name="average" complement="false">

                        <TriangularShape Param1="5.0" Param2="10.0"
Param3="15.0"/>

                </FuzzyTerm>

                <FuzzyTerm name="cheap" complement="false">

                        <TriangularShape Param1="0.0" Param2="5.0"
Param3="10.0"/>

                </FuzzyTerm>

                <FuzzyTerm name="generous" complement="false">

                        <TriangularShape Param1="10.0" Param2="15.0"
Param3="20.0"/>

                </FuzzyTerm>

        </FuzzyVariable>

    <KnowledgeBase>

        . . . . . . . . . . .

</FuzzyController>
```

A special tag that can furthermore be used to define a fuzzy shape is <UserShape>. This tag is used to customize fuzzy shape (custom shape). The custom shape modeling is performed via a set of <Point> tags that lists the extreme points of geometric area defining the custom fuzzy shape. Obviously, the attributes used in <Point> tag are x and y coordinates. As for rule base component, FML allows to define a set of rule bases, each one of them describes a different behavior of system. The root of each rule base is modeled by <RuleBase> tag which defines a fuzzy rule set. The <RuleBase> tag uses five attributes: *name, type, activationMethod, andMethod* and *orMethod*. Obviously, the *name* attribute uniquely identifies the rule base. The *type* attribute permits to specify the kind of fuzzy controller (Mamdani or TSK) respect to the rule base at issue. The *activationMethod* attribute defines the method used to implication process; the *and-*

Method and *orMethod* attribute define, respectively, the *and* and *or* algorithm to use by default. In order to define the single rule the <Rule> tag is used. The attributes used by the <Rule> tag are: *name, connector, operator* and *weight*. The *name* attribute permits to identify the rule; *connector* is used to define the logical operator used to connect the different clauses in antecedent part (and/or); *operator* defines the algorithm to use for chosen connector; *weight* defines the importance of rule during inference engine step. The definition of antecedent and consequent rule part is obtained by using <Antecedent> and <Consequent> tags. <Clause> tag is used to model the fuzzy clauses in antecedent and consequent part. This tag use the attribute *modifier* to describe a modification to term used in the clause. The possible values for this attribute are: *above, below, extremely, intensify, more or less, norm, not, plus, slightly, somewhat, very, none*. To complete the definition of fuzzy clause the nested <Variable> and <Term> tags have to be used. A sequence of <Rule> tags realizes a fuzzy rule base.

As example, let us consider a Mamdani rule composed by *(food is rancid) OR (service is very poor)* as antecedent and *tip is cheap* as consequent. The antecedent part is formed by two clauses: *(food is rancid)* and *(service is poor)*. The first antecedent clause uses *food* as variable and *rancid* as fuzzy term, whereas, the second antecedent clause uses *service* as a variable, *poor* as fuzzy term and *very* as modifier; the consequent clause uses *tip* as a fuzzy variable and *cheap* as a fuzzy term. The complete rule is:

IF *(food is rancid)* **OR** *(service is very poor)* **THEN** *(tip is cheap)*.

Let us see how FML defines a rule base with this rule.

```
<RuleBase name="Rulebase1" activationMethod="MIN" andMethod="MIN"
orMethod="MAX" type="mamdani">

     <Rule name="reg1" connector="or" operator="MAX" weight="1.0">

          <Antecedent>

               <Clause>

                    <Variable>food</Variable>

                    <Term>rancid</Term>

               </Clause>

               <Clause modifier="very">

                    <Variable>service</Variable>

                    <Term>poor</Term>

               </Clause>

          </Antecedent>
```

```
                <Consequent>

                    <Clause>

                        <Variable>tip</Variable>

                        <Term>cheap</Term>

                    </Clause>

                </Consequent>

            </Rule>

            . . . . . . . . . . .

</RuleBase>
```

Now, let us see a Takagi-Sugeno-Kang system that regulates the same issue. The most important difference with Mamdani system is the definition of a different output variable *tip*. The <TSKVariable> tag is used to define an output variable that can be used in a rule of a Tsk system. This tag has the same attributes of a Mamdani output variable except for the *domainleft* and *domainright* attribute because a variable of this kind (called tsk-variable) hasn't a universe of discourse. The nested <TSKTerm> tag represents a linear function and so it is completely different from <FuzzyTerm>. The <TSKValue> tag is used to define the coefficients of linear function. The following crunch of FML code shows the definition of output variable *tip* in a Tsk system.

```
<?xml version="1.0" encoding="UTF-8"?>

<FuzzyController name="newSystem" ip="127.0.0.1">

    <KnowledgeBase>

        . . . . . . .

        <TSKVariable name="tip" scale="null" accumulation="MAX"
defuzzifier="WA" type="output">

            <TSKTerm name="average" order="0">

                <TSKValue>1.6</TSKValue>

            </TSKTerm>

            <TSKTerm name="cheap" order="1">

                <TSKValue>1.9</TSKValue>

                <TSKValue>5.6</TSKValue>

                <TSKValue>6.0</TSKValue>
```

```
        </TSKTerm>

        <TSKTerm name="generous" order="1">

            <TSKValue>0.6</TSKValue>

            <TSKValue>1.3</TSKValue>

            <TSKValue>1.0</TSKValue>

        </TSKTerm>

      </TSKVariable>

    <KnowledgeBase>

    . . . . . . . . . .

</FuzzyController>
```

The FML definition of rule base component in a Tsk system doesn't change a lot. The only different thing is that the <Clause> tag doesn't have the modifier attribute.

As example, let us consider a tsk rule composed by *(food is rancid) OR (service is very poor)* as antecedent and, as consequent, *tip=1.9+5.6*food+6.0*service* that can be written as *tip is cheap* in an implicitly way. So the rule can be written in this way:

IF *(food is rancid)* OR *(service is very poor)* THEN *(tip is cheap)*.

Let us see how FML defines a rule base with this rule.

```
<RuleBase name="Rulebase1" activationMethod="MIN" andMethod="MIN"
orMethod="MAX" type="tsk">

<Rule name="reg1" connector="or" operator="MAX" weight="1.0">

            <Antecedent>

            <Clause>

                <Variable>food</Variable>

                <Term>rancid</Term>

            </Clause>

            <Clause>

                <Variable>service</Variable>

                <Term>poor</Term>

            </Clause>
```

```
        </Antecedent>

        <Consequent>

            <Clause>

                <Variable>tip</Variable>

                <Term>cheap</Term>

            </Clause>

        </Consequent>

    </Rule>

    . . . . . . . . . . .

</RuleBase>
```

FML Grammar

The FML tags used to build a fuzzy controller represent the set of lexemes used to create fuzzy expressions. However, in order to realize a well-formed XML-based language, an FML context-free grammar is necessary and described in the following. The FML context-free grammar is modeled by XML file in the form of a XML Schema Document (XSD) which expresses the set of rules to which a document must conform in order to be considered a *valid* FML document. Based on the previous definition, a portion of the FML XSD regarding the knowledge base definition is given below.

```
<?xml version="1.0" encoding="UTF-8"?>

<xs:schema xmlns:xs="http://www.w3.org/2001/XMLSchema">

    . . . . . . . .

        <xs:complexType name="KnowledgeBaseType">

            <xs:sequence>

                <xs:choice minOccurs="0" maxOccurs="unbounded">

                    <xs:element name="FuzzyVariable" type="-
FuzzyVariableType"/>

                    <xs:element name="TSKVariable" type="TSK-
VariableType"/>

                </xs:choice>

            </xs:sequence>

        </xs:complexType>
```

```xml
<xs:complexType name="FuzzyVariableType">
    <xs:sequence>
        <xs:element name="FuzzyTerm" type="-
FuzzyTermType" maxOccurs="unbounded"/>
    </xs:sequence>
    <xs:attribute name="name" type="xs:string"
use="required"/>
    <xs:attribute name="defuzzifier" default="COG">
        <xs:simpleType>
            <xs:restriction base="xs:string">
                <xs:pattern value="MM|COG|-
COA|WA|Custom"/>
            </xs:restriction>
        </xs:simpleType>
    </xs:attribute>
    <xs:attribute name="accumulation" default="MAX">
        <xs:simpleType>
            <xs:restriction base="xs:string">
                <xs:pattern value="MAX|-
SUM"/>
            </xs:restriction>
        </xs:simpleType>
    </xs:attribute>
    <xs:attribute name="scale" type="xs:string" />
    <xs:attribute name="domainleft" type="xs:float"
use="required"/>
    <xs:attribute name="domainright" type="xs:float"
use="required"/>
    <xs:attribute name="defaultValue" type="xs:float"
default="0"/>
    <xs:attribute name="type" default="input">
```

```
                        <xs:simpleType>
                                <xs:restriction base="xs:string">
                                        <xs:pattern value="input|out-
put"/>
                                </xs:restriction>
                        </xs:simpleType>
                </xs:attribute>
        </xs:complexType>
        <xs:complexType name="FuzzyTermType">
                <xs:choice>
                        <xs:element          name="RightLinearShape"
type="TwoParamType"/>
                        <xs:element           name="LeftLinearShape"
type="TwoParamType"/>
                        <xs:element name="PIShape" type="TwoParam-
Type"/>
                        <xs:element name="TriangularShape" type="-
ThreeParamType"/>
                        <xs:element name="GaussianShape" type="Two-
ParamType"/>
                        <xs:element         name="RightGaussianShape"
type="TwoParamType"/>
                        <xs:element          name="LeftGaussianShape"
type="TwoParamType"/>
                        <xs:element name="TrapezoidShape" type="-
FourParamType"/>
                        <xs:element name="SingletonShape" type="O-
neParamType"/>
                        <xs:element           name="RectangularShape"
type="TwoParamType"/>
                        <xs:element name="ZShape" type="TwoParam-
Type"/>
```

```
                    <xs:element  name="SShape"  type="TwoParam-
Type"/>

                    <xs:element  name="UserShape"  type="User-
ShapeType"/>

            </xs:choice>

        <xs:complexType name="TwoParamType">

            <xs:attribute    name="Param1"    type="xs:float"
use="required"/>

            <xs:attribute    name="Param2"    type="xs:float"
use="required"/>

        </xs:complexType>

        <xs:complexType name="ThreeParamType">

            <xs:attribute    name="Param1"    type="xs:float"
use="required"/>

            <xs:attribute    name="Param2"    type="xs:float"
use="required"/>

            <xs:attribute    name="Param3"    type="xs:float"
use="required"/>

        </xs:complexType>

        <xs:complexType name="FourParamType">

            <xs:attribute    name="Param1"    type="xs:float"
use="required"/>

            <xs:attribute    name="Param2"    type="xs:float"
use="required"/>

            <xs:attribute    name="Param3"    type="xs:float"
use="required"/>

            <xs:attribute    name="Param4"    type="xs:float"
use="required"/>

        </xs:complexType>

        <xs:complexType name="UserShapeType">

            <xs:sequence>

                <xs:element  name="Point"  type="PointType"
```

```
minOccurs="2" maxOccurs="unbounded"/>
                </xs:sequence>
        </xs:complexType>
        <xs:complexType name="PointType">
                <xs:attribute   name="x"   type="xs:float"   use="re-
quired"/>
                <xs:attribute   name="y"   type="xs:float"   use="re-
quired"/>
        </xs:complexType>
        <xs:complexType name="RuleBaseType">
                <xs:attribute        name="name"        type="xs:string"
use="required"/>
                <xs:attribute        name="activationMethod"        de-
fault="MIN">
                        <xs:simpleType>
                                <xs:restriction base="xs:string">
                                        <xs:pattern                val-
ue="PROD|MIN"/>
                                </xs:restriction>
                        </xs:simpleType>
                </xs:attribute>
                <xs:attribute name="andMethod" default="MIN">
                        <xs:simpleType>
                                <xs:restriction base="xs:string">
                                        <xs:pattern                val-
ue="PROD|MIN"/>
                                </xs:restriction>
                        </xs:simpleType>
                </xs:attribute>
                <xs:attribute name="orMethod" default="MAX">
```

```xml
                    <xs:simpleType>
                        <xs:restriction base="xs:string">
                            <xs:pattern    value="PROBOR|-
MAX"/>
                        </xs:restriction>
                    </xs:simpleType>
                </xs:attribute>
                <xs:attribute name="type" use="required">
                    <xs:simpleType>
                        <xs:restriction        base="x-
s:string">
                            <xs:pattern       val-
ue="TSK|Tsk|tsk|Mamdani|mamdani"/>
                        </xs:restriction>
                    </xs:simpleType>
                </xs:attribute>
            </xs:complexType>
            <xs:complexType name="MamdaniRuleBaseType">
                <xs:complexContent>
                <xs:extension base="RuleBaseType">
                    <xs:sequence>
                        <xs:element        name="Rule"
type="MamdaniFuzzyRuleType"   minOccurs="0"   maxOccurs="unbound-
ed"/>
                    </xs:sequence>
                </xs:extension>
                </xs:complexContent>
            </xs:complexType>
            <xs:complexType name="AntecedentType">
                <xs:sequence>
```

```xml
            <xs:element name="Clause" type="ClauseType"
maxOccurs="unbounded"/>
        </xs:sequence>
    </xs:complexType>
    <xs:complexType name="MamdaniConsequentType">
        <xs:sequence>
            <xs:element name="Clause" type="ClauseType"
maxOccurs="unbounded"/>
        </xs:sequence>
    </xs:complexType>
    <xs:complexType name="ClauseType">
        <xs:sequence>
            <xs:element name="Variable">
                <xs:simpleType>
                    <xs:restriction        base="x-
s:string">
                        <xs:whiteSpace    val-
ue="collapse"/>
                        <xs:pattern       val-
ue="((([A-Z])|([a-z]))+([A-Z]|[a-z]|[0-9])*"/>
                    </xs:restriction>
                </xs:simpleType>
            </xs:element>
            <xs:element name="Term" type="xs:string">
            </xs:element>
        </xs:sequence>
        <xs:attribute name="modifier" use="optional">
            <xs:simpleType>
                <xs:restriction base="xs:string">
                    <xs:pattern
```

```
                                            value="above|-
below|extremely|intensify|more_or_less|norm|not|plus|slight-
ly|somewhat|very"/>

                        </xs:restriction>

                  </xs:simpleType>

             </xs:attribute>

        </xs:complexType>

        . . . . . . . . . .

</xs:schema>
```

Fml Synthesis

Since an FML program realizes only a static view of a fuzzy system, the so-called eXtensible Stylesheet Language Translator (XSLT) is provided to change this static view to a computable version. In particular, the XSLT technology is used convert a fuzzy controller description into a general-purpose computer language to be computed on several hardware platforms. Currently, a XSLT converting FML program in runnable Java code has been implemented. In this way, thanks to the transparency capabilities provided by Java virtual machines, it is possible to obtain a fuzzy controller modeled in high level way by means of FML and runnable on a plethora of hardware architectures through Java technologies. However, XSLT can be also used for converting FML programs in legacy languages related to a particular hardware or in other general purpose languages.

Intelligent Control

Intelligent control is a class of control techniques that use various artificial intelligence computing approaches like neural networks, Bayesian probability, fuzzy logic, machine learning, evolutionary computation and genetic algorithms.

Overview

Intelligent control can be divided into the following major sub-domains:

- Neural network control
- Bayesian control
- Fuzzy control

- Neuro-fuzzy control

- Expert Systems

- Genetic control

- Intelligent agents (Cognitive/Conscious control)

New control techniques are created continuously as new models of intelligent behavior are created and computational methods developed to support them.

Neural Network Controllers

Neural networks have been used to solve problems in almost all spheres of science and technology. Neural network control basically involves two steps:

- System identification

- Control

It has been shown that a feedforward network with nonlinear, continuous and differentiable activation functions have universal approximation capability. Recurrent networks have also been used for system identification. Given, a set of input-output data pairs, system identification aims to form a mapping among these data pairs. Such a network is supposed to capture the dynamics of a system.

Bayesian Controllers

Bayesian probability has produced a number of algorithms that are in common use in many advanced control systems, serving as state spaceestimators of some variables that are used in the controller.

The Kalman filter and the Particle filter are two examples of popular Bayesian control components. The Bayesian approach to controller design often requires an important effort in deriving the so-called system model and measurement model, which are the mathematical relationships linking the state variables to the sensor measurements available in the controlled system. In this respect, it is very closely linked to the system-theoretic approach to control design.

Computational Intelligence

The expression computational intelligence (CI) usually refers to the ability of a computer to learn a specific task from data or experimental observation. Even though it is commonly considered a synonym of soft computing, there is still no commonly accepted definition of computational intelligence.

Generally, computational intelligence is a set of nature-inspired computational method-ologies and approaches to address complex real-world problems to which mathematical or traditional modelling can be useless for a few reasons: the processes might be too com-plex for mathematical reasoning, it might contain some uncertainties during the process, or the process might simply be stochastic in nature. Indeed, many real-life problems can-not be translated into binary language (unique values of 0 and 1) for computers to process it. Computational Intelligence therefore provides solutions for such problems.

The methods used are close to the human's way of reasoning, i.e. it uses inexact and incomplete knowledge, and it is able to produce control actions in an adaptive way. CI therefore uses a combination of five main complementary techniques. The fuzzy logic which enables the computer to understand natural language, artificial neural networks which permits the system to learn experiential data by operating like the biological one, evolutionary computing, which is based on the process of natural selec-tion, learning theory, and probabilistic methods which helps dealing with uncertainty imprecision.

Except those main principles, currently popular approaches include biologically in-spired algorithms such as swarm intelligence and artificial immune systems, which can be seen as a part of evolutionary computation, image processing, data mining, natural language processing, and artificial intelligence, which tends to be confused with Com-putational Intelligence. But although both Computational Intelligence (CI) and Artifi-cial Intelligence (AI) seek similar goals, there's a clear distinction between them.

Computational Intelligence is thus a way of performing like human beings. Indeed, the characteristic of "intelligence" is usually attributed to humans. More recently, many products and items also claim to be "intelligent", an attribute which is directly linked to the reasoning and decision making.

History

The notion of Computational Intelligence was first used by the IEEE Neural Net-works Council in 1990. This Council was originally founded in the 1980s by a group of researchers interested in the development of biological and artificial neural net-works. On November 21, 2001, the IEEE Neural Networks Council became the IEEE Neural Networks Society, to become the IEEE Computational Intelligence Society two years later by including new areas of interest such as fuzzy systems and evolu-tionary computation, which they related to Computational Intelligence in 2011 (Dote and Ovaska).

But the first clear definition of Computational Intelligence was introduced by Bezdek in 1994: a system is called computationally intelligent if it deals with low-level data such as numerical data, has a pattern-recognition component and does not use knowledge in the AI sense, and additionally when it begins to exhibit computational adaptively, fault tolerance, speed approaching human-like turnaround and error rates that approximate human performance.

Bezdek and Marks (1993) clearly differentiated CI from AI, by arguing that the first one is based on soft computing methods, whereas AI is based on hard computing ones.

Difference between Computational and Artificial Intelligence

Although Artificial Intelligence and Computational Intelligence seek a similar long-term goal: reach general intelligence, which is the intelligence of a machine that could perform any intellectual task that a human being can; there's a clear difference between them. According to Bezdek (1994), Computational Intelligence is a subset of Artificial Intelligence.

There are two types of machine intelligence: the artificial one based on hard computing techniques and the computational one based on soft computing methods, which enable adaptation to many situations.

Hard computing techniques work following binary logic based on only two values (the Booleans true or false, 0 or 1) on which modern computers are based. One problem with this logic is that our natural language cannot always be translated easily into absolute terms of 0 and 1. Soft computing techniques, based on fuzzy logic can be useful here. Much closer to the way the human brain works by aggregating data to partial truths (Crisp/fuzzy systems), this logic is one of the main exclusive aspects of CI.

Within the same principles of fuzzy and binary *logics* follow crispy and fuzzy *systems*. Crisp logic is a part of artificial intelligence principles and consists of either including an element in a set, or not, whereas fuzzy systems (CI) enable elements to be partially in a set. Following this logic, each element can be given a degree of membership (from 0 to 1) and not exclusively one of these 2 values.

The Five Main Principles of CI and its Applications

The main applications of Computational Intelligence include computer science, engineering, data analysis and bio-medicine.

Fuzzy Logic

As explained before, fuzzy logic, one of CI's main principles, consists in measurements and process modelling made for real life's complex processes. It can face incompleteness, and most importantly ignorance of data in a process model, contrarily to Artificial Intelligence, which requires exact knowledge.

This technique tends to apply to a wide range of domains such as control, image processing and decision making. But it is also well introduced in the field of household appliances with washing machines, microwave ovens, etc. We can face it too when using a video camera, where it helps stabilizing the image while holding the camera unsteadily. Other areas such as medical diagnostics, foreign exchange trading and business strategy selection are apart from this principle's numbers of applications.

Fuzzy logic is mainly useful for approximate reasoning, and doesn't have learning abilities, a qualification much needed that human beings have. It enables them to improve themselves by learning from their previous mistakes.

Neural Networks

This is why CI experts work on the development of artificial neural networks based on the biological ones, which can be defined by 3 main components: the cell-body which processes the information, the axon, which is a device enabling the signal conducting, and the synapse, which controls signals. Therefore, artificial neural networks are doted of distributed information processing systems, enabling the process and the learning from experiential data. Working like human beings, fault tolerance is also one of the main assets of this principle.

Concerning its applications, neural networks can be classified into five groups: data analysis and classification, associative memory, clustering generation of patterns and control. Generally, this method aims to analyze and classify medical data, proceed to face and fraud detection, and most importantly deal with nonlinearities of a system in order to control it. Furthermore, neural networks techniques share with the fuzzy logic ones the advantage of enabling data clustering.

Evolutionary Computation

Based on the process of natural selection firstly introduced by Charles Robert Darwin, the evolutionary computation consists in capitalizing on the strength of natural evolution to bring up new artificial evolutionary methodologies. It also includes other areas such as evolution strategy, and evolutionary algorithms which are seen as problem solvers... This principle's main applications cover areas such as optimization and multi-objective optimization, to which traditional mathematical one techniques aren't enough anymore to apply to a wide range of problems such as DNA Analysis, scheduling problems...

Learning Theory

Still looking for a way of "reasoning" close to the humans' one, learning theory is one of the main approaches of CI. In psychology, learning is the process of bringing together cognitive, emotional and environmental effects and experiences to acquire, enhance or change knowledge, skills, values and world views (Ormrod, 1995; Illeris, 2004). Learning theories then helps understanding how these effects and experiences are processed, and then helps making predictions based on previous experience.

Probabilistic Methods

Being one of the main elements of fuzzy logic, probabilistic methods firstly introduced by Paul Erdos and Joel Spencer (1974), aim to evaluate the outcomes of a Computation Intelligent system, mostly defined by randomness. Therefore, probabilistic methods bring out the possible solutions to a reasoning problem, based on prior knowledge.

Lateral Computing

Lateral computing is a lateral thinking approach to solving computing problems. Lateral thinking has been made popular by Edward de Bono. This thinking technique is applied to generate creative ideas and solve problems. Similarly, by applying lateral-computing techniques to a problem, it can become much easier to arrive at a computationally inexpensive, easy to implement, efficient, innovative or unconventional solution.

The traditional or conventional approach to solving computing problems is to either build mathematical models or have an IF- THEN -ELSE structure. For example, a brute-force search is used in many chess engines, but this approach is computationally expensive and sometimes may arrive at poor solutions. It is for problems like this that lateral computing can be useful to form a better solution.

A simple problem of truck backup can be used for illustrating lateral-computing. This is one of the difficult tasks for traditional computing techniques, and has been efficiently solved by the use of fuzzy logic (which is a lateral computing technique). Lateral-computing sometimes arrives at a novel solution for particular computing problem by using the model of how living beings, such as how humans, ants, and honeybees, solve a problem; how pure crystals are formed by annealing, or evolution of living beings or quantum mechanics etc.

Logical Thinking and Artificial Intelligence

Chess position analysis can be used to illustrate the logical thinking. The following board position describes a chess problem which has to be solved with two moves.

The white has several options to make a move and checkmate the black. The move Rd5 × Rd7 or Rf7 × Rd7 will immediately provide material advantage to white. There are similar moves which capture pieces and provide immediate material advantages to the white. But a knight move Nc6 which does not provide any material advantage, provides a solution for checkmate for black in two moves.

	..	Nc6		
1	...	Kxf7	2	g8Q++
1	...	Kxd5	2	Qa2++
1	...	Rdxd5	2	Re7++
1	...	Rfxd5	2	Rf6++
1	...	Rdxf7	2	Rd6++
1	...	Rfxf7	2	Re5++

This is an example which illustrates the use of logical thinking. The logical thinking in chess progresses by evaluating the immediate material gain in each move. This will result in a solution which will require more number of moves or failure to checkmate. However, the not so obvious move of knight results in a very powerful checkmate. Even though this move does not look logical, it is the solution to two-move checkmate problem. A computer programmed to play chess might miss out some good opportunities if it does a material-based search to find moves. Several attempts have been made to build the powerful chess computers in history. But these chess computers have been defeated by Grandmaster human chess players.

Logic Programming

The attempts to use logic programming such as prolog to represent knowledge and build artificial intelligent systems has not provided the anticipated thrust to solving interesting problems. The lack of generalization and learning capability of these systems and exponential growth of the IF-THEN ELSE rules has made this approach unpopular. An example to illustrate the failure of the rule-based system is the following flawed proof:

Start with $81/4 = 81/4$
Adding -20 to LHS and RHS gives: $-20 + 81/4 = -20 + 81/4$
Splitting -20 as ($-36 + 16$) on the LHS and ($-45 + 25$) on the RHS: $16 + 81/4 - 36 = -45 + 81/4 + 25$
Now expressing the terms 16, 25 and $81/4$ as squares of 4, 5 and $9/2$ respectively: $4^2 + (9/2)^2 - 2 * (9/2) * 4 = 5^2 + (9/2)^2 - 2 * (9/2) * 5$
Expressing this as $a^2 + b^2 - 2*a*b = (a-b)^2$ gives: $(4 - 9/2)^2 = (5 - 9/2)^2$
Taking the square roots, $4 - 9/2 = 5 - 9/2$

This would imply that $4 = 5$, which a wrong result. While taking the square roots, the step of considering the signs has been missed. This has resulted in an absurd outcome. A rule-based system, even if it missed a simple rule in its database may yield such an unacceptable output.

Another interesting mathematical proof gone wrong is as follows:

Let a = b

Multiply both sides by b

ab = b²

Subtract a² from both sides

ab - a² = b² - a²

Factor each side

a(b - a) = (b+a)(b - a)

cancelling (b-a) from both sides

a = b+a

If a = 1, then we get an absurd result of 1 = 2

From Lateral-thinking to Lateral-computing

Lateral thinking is technique for creative thinking for solving problems. The brain as center of thinking has a self-organizing information system. It tends to create patterns and traditional thinking process uses them to solve problems. The lateral thinking technique proposes to escape from this patterning to arrive at better solutions through new ideas. Provocative use of information processing is the basic underlying principle of lateral thinking,

The provocative operator (PO) is something which characterizes lateral thinking. Its function is to generate new ideas by provocation and providing escape route from old ideas. It creates a provisional arrangement of information.

Water logic is contrast to traditional or rock logic. Water logic has boundaries which depends on circumstances and conditions while rock logic has hard boundaries. Water logic, in someways, resembles fuzzy logic.

Transition to Lateral-computing

Lateral computing does a provocative use of information processing similar to lateral-thinking. This is explained with the use of evolutionary computing which is a very useful lateral-computing technique. The evolution proceeds by change and selection. While random mutation provides change, the selection is through survival of the fittest. The random mutation works as a provocative information processing and provides a new avenue for generating better solutions for the computing problem.

Lateral computing takes the analogies from real-world examples such as:

- How slow cooling of the hot gaseous state results in pure crystals (Annealing)

- How the neural networks in the brain solve such problems as face and speech recognition

- How simple insects such as ants and honeybees solve some sophisticated problems

- How evolution of human beings from molecular life forms are mimicked by evolutionary computing

- How living organisms defend themselves against diseases and heal their wounds

- How electricity is distributed by grids

Differentiating factors of "lateral computing":

- Does not directly approach the problem through mathematical means.

- Uses indirect models or looks for analogies to solve the problem.

- Radically different from what is in vogue, such as using "photons" for computing in optical computing. This is rare as most conventional computers use electrons to carry signals.

- Sometimes the Lateral Computing techniques are surprisingly simple and deliver high performance solutions to very complex problems.

- Some of the techniques in lateral computing use "unexplained jumps". These jumps may not look logical. The example is the use of "Mutation" operator in genetic algorithms.

Convention – lateral

It is very hard to draw a clear boundary between conventional and lateral computing. Over a period of time, some unconventional computing techniques become integral part of mainstream computing. So there will always be an overlap between conventional and lateral computing. It will be tough task classifying a computing technique as a conventional or lateral computing technique as shown in the figure. The boundaries are fuzzy and one may approach with fuzzy sets.

Formal Definition

Lateral computing is a fuzzy set of all computing techniques which use unconventional computing approach. Hence Lateral computing includes those techniques which use semi-conventional or hybrid computing. The degree of membership for lateral computing techniques is greater than 0 in the fuzzy set of unconventional computing techniques.

The following brings out some important differentiators for lateral computing.

Conventional computing

- The problem and technique are directly correlated.

- Treats the problem with rigorous mathematical analysis.

- Creates mathematical models.

- The computing technique can be analyzed mathematically.

Lateral computing

- The problem may hardly have any relation to the computing technique used

- Approaches problems by analogies such as human information processing model, annealing, etc.

- Sometimes the computing technique cannot be mathematically analyzed.

Lateral Computing and Parallel Computing

Parallel computing focuses on improving the performance of the computers/algorithms through the use of several computing elements (such as processing elements). The computing speed is improved by using several computing elements. Parallel computing is an extension of conventional sequential computing. However, in lateral computing, the problem is solved using unconventional information processing whether using a sequential or parallel computing.

A Review of Lateral-computing Techniques

There are several computing techniques which fit the Lateral computing paradigm. Here is a brief description of some of the Lateral Computing techniques:

Swarm Intelligence

Swarm intelligence (SI) is the property of a system whereby the collective behaviors of (unsophisticated) agents, interacting locally with their environment, cause coherent functional global patterns to emerge. SI provides a basis with which it is possible to explore collective (or distributed) problem solving without centralized control or the provision of a global model.

One interesting swarm intelligent technique is the Ant Colony algorithm:

- Ants are behaviorally unsophisticated; collectively they perform complex tasks. Ants have highly developed sophisticated sign-based communication.

- Ants communicate using pheromones; trails are laid that can be followed by other ants.

- Routing Problem Ants drop different pheromones used to compute the "shortest" path from source to destination(s).

Agent-based Systems

Agents are encapsulated computer systems that are situated in some environment and are capable of flexible, autonomous action in that environment in order to meet their design objectives. Agents are considered to be autonomous (independent, not-controllable), reactive (responding to events), pro-active (initiating actions of their own volition), and social (communicative). Agents vary in their abilities: they can be static or mobile, or may or may not be intelligent. Each agent may have its own task and/or role. Agents, and multi-agent systems, are used as a metaphor to model complex distributed processes. Such agents invariably need to interact with one another in order to manage their inter-dependencies. These interactions involve agents cooperating, negotiating and coordinating with one another.

Agent-based systems are computer programs that try to simulate various complex phenomena via virtual "agents" that represent the components of a business system. The behaviors of these agents are programmed with rules that realistically depict how business is conducted. As widely varied individual agents interact in the model, the simulation shows how their collective behaviors govern the performance of the entire system - for instance, the emergence of a successful product or an optimal schedule. These simulations are powerful strategic tools for "what-if" scenario analysis: as managers change agent characteristics or "rules," the impact of the change can be easily seen in the model output

Grid Computing

By analogy, a computational grid is a hardware and software infrastructure that provides dependable, consistent, pervasive, and inexpensive access to high-end computational capabilities. The applications of grid computing are in:

- Chip design, cryptographic problems, medical instrumentation, and supercomputing.

- Distributed supercomputing applications use grids to aggregate substantial computational resources in order to tackle problems that cannot be solved on a single system.

Autonomic Computing

The autonomic nervous system governs our heart rate and body temperature, thus freeing our conscious brain from the burden of dealing with these and many other low-level, yet vital, functions. The essence of autonomic computing is self-management, the intent of which is to free system administrators from the details of system operation and maintenance.

Four aspects of autonomic computing are:

- Self-configuration
- Self-optimization
- Self-healing
- Self-protection

This is a grand challenge promoted by IBM.

Optical Computing

Optical computing is to use photons rather than conventional electrons for computing. There are quite a few instances of optical computers and successful use of them. The conventional logic gates use semiconductors, which use electrons for transporting the signals. In case of optical computers, the photons in a light beam are used to do computation.

There are numerous advantages of using optical devices for computing such as immunity to electromagnetic interference, large bandwidth, etc.

DNA Computing

DNA computing uses strands of DNA to encode the instance of the problem and to manipulate them using techniques commonly available in any molecular biology laboratory in order to simulate operations that select the solution of the problem if it exists.

Since the DNA molecule is also a code, but is instead made up of a sequence of four bases that pair up in a predictable manner, many scientists have thought about the possibility of creating a molecular computer. These computers rely on the much faster reactions of DNA nucleotides binding with their complements, a brute force method that holds enormous potential for creating a new generation of computers that would be 100 billion times faster than today's fastest PC. DNA computing has been heralded as the "first example of true nanotechnology", and even the "start of a new era", which forges an unprecedented link between computer science and life science.

Example applications of DNA computing are in solution for the Hamiltonian path problem which is a known NP complete one. The number of required lab operations using DNA grows linearly with the number of vertices of the graph. Molecular algorithms have been reported that solves the cryptograhic problem in a polynomial number of steps. As known, factoring large numbers is a relevant problem in many cryptographic applications.

Quantum Computing

In a quantum computer, the fundamental unit of information (called a quantum bit or qubit), is not binary but rather more quaternary in nature. This qubit property arises as a direct consequence of its adherence to the laws of quantum mechanics, which differ radically from the laws of classical physics. A qubit can exist not only in a state corresponding to the logical state 0 or 1 as in a classical bit, but also in states corresponding to a blend or quantum superposition of these classical states. In other words, a qubit can exist as a zero, a one, or simultaneously as both 0 and 1, with a numerical coefficient representing the probability for each state. A quantum computer manipulates qubits by executing a series of quantum gates, each a unitary transformation acting on a single qubit or pair of qubits. In applying these gates in succession, a quantum computer can perform a complicated unitary transformation to a set of qubits in some initial state.

Reconfigurable Computing

Field-programmable gate arrays (FPGA) are making it possible to build truly reconfigurable computers. The computer architecture is transformed by on the fly reconfiguration of the FPGA circuitry. The optimal matching between architecture and algorithm improves the performance of the reconfigurable computer. The key feature is hardware performance and software flexibility.

For several applications such as fingerprint matching, DNA sequence comparison, etc., reconfigurable computers have been shown to perform several orders of magnitude better than conventional computers.

Simulated Annealing

The Simulated annealing algorithm is designed by looking at how the pure crystals form from a heated gaseous state while the system is cooled slowly. The computing problem is redesigned as a simulated annealing exercise and the solutions are arrived at. The working principle of simulated annealing is borrowed from metallurgy: a piece of metal is heated (the atoms are given thermal agitation), and then the metal is left to cool slowly. The slow and regular cooling of the metal allows the atoms to slide progressively their most stable ("minimal energy") positions. (Rapid cooling would have "frozen" them in whatever position they happened to be at that time.) The resulting structure of the metal is stronger and more stable. By simulating the process of annealing inside a computer program, it is possible to find answers to difficult and very complex problems. Instead of minimizing the energy of a block of metal or maximizing its strength, the program minimizes or maximizes some objective relevant to the problem at hand.

Soft Computing

One of the main components of "Lateral-computing" is soft computing which approaches problems with human information processing model. The Soft Computing technique comprises Fuzzy logic, neuro-computing, evolutionary-computing, machine learning and probabilistic-chaotic computing.

Neuro Computing

Instead of solving a problem by creating a non-linear equation model of it, the biological neural network analogy is used for solving the problem. The neural network is trained like a human brain to solve a given problem. This approach has become highly successful in solving some of the pattern recognition problems.

Evolutionary Computing

The genetic algorithm (GA) resembles the natural evolution to provide a universal optimization. Genetic algorithms start with a population of chromosomes which represent the various solutions. The solutions are evaluated using a fitness function and a selection process determines which solutions are to be used for competition process. These algorithms are highly successful in solving search and optimization problems. The new solutions are created using evolutionary principles such as mutation and crossover.

Fuzzy Logic

Fuzzy logic is based on the fuzzy sets concepts proposed by Lotfi Zadeh. The degree of membership concept is central to fuzzy sets. The fuzzy sets differ from crisp sets since they allow an element to belong to a set to a degree (degree of membership). This approach finds good applications for control problems. The Fuzzy logic has found enormous applications and has already found a big market presence in consumer electronics such as washing machines, microwaves, mobile phones, Televisions, Camcoders etc.

Probabilistic/Chaotic Computing

Probabilistic computing engines, e.g. use of probabilistic graphical model such as Bayesian network. Such computational techniques are referred to as randomization, yielding probabilistic algorithms. When interpreted as a physical phenomenon through classical statistical thermodynamics, such techniques lead to energy savings that are proportional to the probability p with which each primitive computational step is guaranteed to be correct (or equivalently to the probability of error, (1–p). Chaotic Computing is based on the chaos theory.

Fractals

Fractal Computing are objects displaying self-similarity at different scales. Fractals generation involves small iterative algorithms. The fractals have dimensions greater than their topological dimensions. The length of the fractal is infinite and size of it cannot be measured. It is described by an iterative algorithm unlike a Euclidean shape which is given by a simple formula. There are several types of fractals and Mandelbrot sets are very popular.

Fractals have found applications in image processing, image compression music generation, computer games etc. Mandelbrot set is a fractal named after its creator. Unlike the other fractals, even though the Mandelbrot set is self-similar at magnified scales, the small scale details are not identical to the whole. I.e., the Mandelbrot set is infinitely complex. But the process of generating it is based on an extremely simple equation. The Mandelbrot set M is a collection of complex numbers. The numbers Z which belong to M are computed by iteratively testing the Mandelbrot equation. C is a constant. If the equation converges for chosen Z, then Z belongs to M. Mandelbrot equation:

Randomized Algorithm

A Randomized algorithmmakes arbitrary choices during its execution. This allows a savings in execution time at the beginning of a program. The disadvantage of this method is the possibility that an incorrect solution will occur. A well-designed randomized algorithm will have a very high probability of returning a correct answer. The two categories of randomized algorithms are:

- Monte Carlo algorithm

- Las Vegas algorithm

Consider an algorithm to find the kth element of an array. A deterministic approach would be to choose a pivot element near the median of the list and partition the list around that element. The randomized approach to this problem would be to choose a pivot at random, thus saving time at the beginning of the process. Like approximation algorithms, they can be used to more quickly solve tough NP-complete problems. An advantage over the approximation algorithms, however, is that a randomized algorithm will eventually yield an exact answer if executed enough times

Machine Learning

Human beings/animals learn new skills, languages/concepts. Similarly, machine learning algorithms provide capability to generalize from training data. There are two classes of Machine Learning (ML):

- Supervised ML

- Unsupervised ML

One of the well known machine learning technique is Back Propagation Algorithm. This mimics how humans learn from examples. The training patters are repeatedly presented to the network. The error is back propagated and the network weights are adjusted using gradient descent. The network converges through several hundreds of iterative computations.

Support Vector Machines

This is another class of highly successful machine learning techniques successfully applied to tasks such as text classification, speaker recognition, image recognition etc.

Example Applications

There are several successful applications of lateral-computing techniques. Here is a small set of applications that illustrates lateral computing:

- Bubble sorting: Here the computing problem of sorting is approached with an analogy of bubbles rising in water. This is by treating the numbers as bubbles and floating them to their natural position.

- Truck backup problem: This is an interesting problem of reversing a truck and parking it at a particular location. The traditional computing techniques have found it difficult to solve this problem. This has been successfully solved by Fuzzy system.

- Balancing an inverted pendulum: This problem involves balancing and inverted pendulum. This problem has been efficiently solved by neural networks and fuzzy systems.

- Smart volume control for mobile phones: The volume control in mobile phones depend on the background noise levels, noise classes, hearing profile of the user and other parameters. The measurement on noise level and loudness level involve imprecision and subjective measures. The authors have demonstrated the successful use of fuzzy logic system for volume control in mobile handsets.

- Optimization using genetic algorithms and simulated annealing: The problems such as traveling salesman problem have been shown to be NP complete problems. Such problems are solved using algorithms which benefit by heuristics. Some of the applications are in VLSI routing, partitioning etc. Genetic algorithms and Simulated annealing have been successful in solving such optimization problems.

- Programming The Unprogrammable (PTU) involving the automatic creation of computer programs for unconventional computing devices such as cellular automata, multi-agent systems, parallel systems, field-programmable gate arrays, field-programmable analog arrays, ant colonies, swarm intelligence, distributed systems, and the like.

Summary

Above is a review of lateral-computing techniques. Lateral-computing is based on the lateral-thinking approach and applies unconventional techniques to solve computing problems. While, most of the problems are solved in conventional techniques, there are problems which require lateral-computing. Lateral-computing provides advantage of computational efficiency, low cost of implementation, better solutions when compared to conventional computing for several problems. The lateral-computing successfully tackles a class of problems by exploiting tolerance for imprecision, uncertainty and partial truth to achieve tractability, robustness and low solution cost. Lateral-computing techniques which use the human like information processing models have been classified as "Soft Computing" in literature.

Lateral-computing is valuable while solving numerous computing problems whose mathematical models are unavailable. They provide a way of developing innovative solutions resulting in smart systems with Very High Machine IQ (VHMIQ). This article has traced the transition from lateral-thinking to lateral-computing. Then several lateral-computing techniques have been described followed by their applications. Lateral-computing is for building new generation artificial intelligence based on unconventional processing.

Distribution Management System

In the recent years, utilization of electrical energy increased exponentially and customer requirement and quality definitions of power were changed enormously. As the electric energy became an essential part of the daily life, its optimal usage and reliability became important. Real-time network view and dynamic decisions have become instrumental for optimizing resources and managing demands, thus making a distribution management system which could handle proper work flows, very critical.

Overview

A Distribution Management System (DMS) is a collection of applications designed to monitor & control the entire distribution network efficiently and reliably. It acts as a decision support system to assist the control room and field operating personnel with the monitoring and control of the electric distribution system. Improving the reliability and quality of service in terms of reducing outages, minimizing outage time, maintaining acceptable frequency and voltage levels are the key deliverables of a DMS.

Most distribution utilities have been comprehensively using IT solutions through their Outage Management System (OMS) that makes use of other systems like Customer Information System (CIS), Geographical Information System (GIS) and Interactive Voice Response System (IVRS). An outage management system has a network component/connectivity model of the distribution system. By combining the locations of outage calls from customers with knowledge of the locations of the protection devices (such as circuit breakers) on the network, a rule engine is used to predict the locations of outages. Based on this, restoration activities are charted out and the crew is dispatched for the same.

In parallel with this, distribution utilities began to roll out Supervisory Control and Data Acquisition (SCADA) systems, initially only at their higher voltage substations. Over time, use of SCADA has progressively extended downwards to sites at lower voltage levels.

DMSs access real-time data and provide all information on a single console at the control centre in an integrated manner. Their development varied across different geographic territories. In the USA, for example, DMSs typically grew by taking Outage Management Systems to the next level, automating the complete sequences and providing an end to end, integrated view of the entire distribution spectrum. In the UK, by contrast, the much denser and more meshed network topologies, combined with stronger Health & Safety regulation, had led to early centralisation of high-voltage switching operations, initially using paper records and schematic diagrams printed onto large wallboards which were 'dressed' with magnetic symbols to show the current running states. There, DMSs grew initially from SCADA systems as these were expanded to allow these centralised control and safety management procedures to be managed electronically. These DMSs required even more detailed component/connectivity models and schematics than those needed by early OMSs as every possible isolation and earthing point on the networks had to be included. In territories such as the UK, therefore, the network component/connectivity models were usually developed in the DMS first, whereas in the USA these were generally built in the GIS.

The typical data flow in a DMS has the SCADA system, the Information Storage & Retrieval (ISR) system, Communication (COM) Servers, Front-End Processors (FEPs) & Field Remote Terminal Units (FRTUs).

Why DMS?

- Reduce the duration of outages

- Improve the speed and accuracy of outage predictions.

- Reduce crew patrol and drive times through improved outage locating.

- Improve the operational efficiency

- Determine the crew resources necessary to achieve restoration objectives.

- Effectively utilize resources between operating regions.

- Determine when best to schedule mutual aid crews.

- Increased customer satisfaction

- A DMS incorporates IVR and other mobile technologies, through which there is an improved outage communications for customer calls.

- Provide customers with more accurate estimated restoration times.

- Improve service reliability by tracking all customers affected by an outage, determining electrical configurations of every device on every feeder, and compiling details about each restoration process.

DMS Functions

In order to support proper decision making and O&M activities, DMS solutions should support the following functions:

- Network visualization & support tools

- Applications for Analytical & Remedial Action

- Utility Planning Tools

- System Protection Schemes

The various sub functions of the same, carried out by the DMS are listed below:-

Network Connectivity Analysis (NCA)

Distribution network usually covers over a large area and catering power to different customers at different voltage levels. So locating required sources and loads on a larger GIS/Operator interface is often very difficult. Panning & zooming provided with normal SCADA system GUI does not cover the exact operational requirement. Network connectivity analysis is an operator specific functionality which helps the operator to identify or locate the preferred network or component very easily. NCA does the required analyses and provides display of the feed point of various network loads. Based on the status of all the switching devices such as circuit breaker (CB), Ring Main Unit (RMU) and/or isolators that affect the topology of the network modeled, the prevailing network topology is determined. The NCA further assists the operator to know operating state of the distribution network indicating radial mode, loops and parallels in the network.

Switching Schedule & Safety Management

In territories such as the UK a core function of a DMS has always been to support safe switching and work on the networks. Control engineers prepare switching schedules to isolate and make safe a section of network before work is carried out, and the DMS validates these schedules using its network model. Switching schedules can combine telecontrolled and manual (on-site) switching operations. When the required section has been made safe, the DMS allows a Pemit To Work (PTW) document to be issued. After its cancellation when the work has been finished, the switching schedule then facilitates restoration of the normal running arrangements. Switching components can also be tagged to reflect any Operational Restrictions that are in force.

The network component/connectivity model, and associated diagrams, must always be kept absolutely up to date. The switching schedule facility therefore also allows 'patches' to the network model to be applied to the live version at the appropriate stage(s) of the jobs. The term 'patch' is derived from the method previously used to maintain the wallboard diagrams.

State Estimation (SE)

The state estimator is an integral part of the overall monitoring and control systems for transmission networks. It is mainly aimed at providing a reliable estimate of the system voltages. This information from the state estimator flows to control centers and database servers across the network. The variables of interest are indicative of parameters like margins to operating limits, health of equipment and required operator action. State estimators allow the calculation of these variables of interest with high confidence despite the facts that the measurements may be corrupted by noise, or could be missing or inaccurate.

Even though we may not be able to directly observe the state, it can be inferred from a scan of measurements which are assumed to be synchronized. The algorithms need to allow for the fact that presence of noise might skew the measurements. In a typical power system, the State is quasi-static. The time constants are sufficiently fast so that system dynamics decay away quickly (with respect to measurement frequency). The system appears to be progressing through a sequence of static states that are driven by various parameters like changes in load profile. The inputs of the state estimator can be given to various applications like Load Flow Analysis, Contingency Analysis, and other applications.

Load Flow Applications (LFA)

Load flow study is an important tool involving numerical analysis applied to a power system. The load flow study usually uses simplified notations like a single-line diagram and focuses on various forms of AC power rather than voltage and current. It analyzes the power systems in normal steady-state operation. The goal of a power flow study is to obtain complete voltage angle and magnitude information for each bus in a power system for specified load and generator real power and voltage conditions. Once this information is known, real and reactive power flow on each branch as well as generator reactive power output can be analytically determined.

Due to the nonlinear nature of this problem, numerical methods are employed to obtain a solution that is within an acceptable tolerance. The load model needs to automatically calculate loads to match telemeter or forecasted feeder currents. It utilises customer type, load profiles and other information to properly distribute the load to each individual distribution transformer. Load-flow or Power flow studies are important for planning future expansion of power systems as well as in determining the best operation of existing systems.

Volt-VAR Control (VVC)

Volt-VAR Control or VVC refers to the process of managing voltage levels and reactive power (VAR) throughout the power distribution systems. These two quantities are related, because as reactive power flows over an inductive line (and all lines have some inductance) that line sees a voltage drop. VVC encompasses devices that purposely inject reactive power into the grid to alter the size of that voltage drop, in addition to equipment that more directly controls voltage.

In the legacy grid, there are three primary tools for carrying out voltage management: Load Tap Changers (LTCs), voltage regulators, and capacitor banks. LTCs and voltage regulators refer to transformers with variable turns ratios that are placed at strategic points in a network and adjusted to raise or lower voltage as is necessary. Capacitor banks manage voltage by "generating" reactive power, and have thus far been the primary tools through which true Volt/VAR control is carried out. These large capacitors are connected to the grid in shunt configuration through switches which, when closed, allow the capacitors to generate VARs and boost voltage at the point of connection. In the future, further VVC might be carried out by smart inverters and other distributed generation resources, which can also inject reactive power into a distribution network. A VVC application helps the operator mitigate dangerously low or high voltage conditions by suggesting required action plans for all VVC equipment. The plan will give a required tap position and capacitor switching state to ensure the voltage stays close to its nominal value and thus optimize Volt-VAR control function for the utility.

Beyond maintaining a stable voltage profile, VVC has potential benefits for the ampacity (current-carrying capacity) of power lines. There could be loads that contain reactive components like capacitors and inductors (such as electric motors) that strain the grid. This is because the reactive portion of these loads causes them to draw more current than an otherwise comparable, purely resistive load would draw. The extra current can result in heating up of equipment like transformers, conductors, etc. which might then need resizing to carry the total current. An ideal power system needs to control current flow by carefully planning the production, absorption and flow of reactive power at all levels in the system.

Load Shedding Application (LSA)

Electric Distribution Systems have long stretches of transmission line, multiple injection points and fluctuating consumer demand. These features are inherently vulnerable to instabilities or unpredicted system conditions that may lead to critical failure. Instability usually arises from power system oscillations due to faults, peak deficit or protection failures. Distribution load shedding and restoration schemes play a vital role in emergency operation and control in any utility.

An automated Load Shedding Application detects predetermined trigger conditions in the distribution network and performs predefined sets of control actions, such as opening or closing non-critical feeders, reconfiguring downstream distribution or sources of injections, or performing a tap control at a transformer. When a distribution network is complex and covers a larger area, emergency actions taken downstream may reduce burden on upstream portions of the network. In a non-automated system, awareness and manual operator intervention play a key role in trouble mitigation. If the troubles are not addressed quickly enough, they can cascade exponentially and cause major catastrophic failure.

DMS needs to provide a modular automated load shedding & restoration application which automates emergency operation & control requirements for any utility. The application should cover various activities like Under Frequency Load Shedding (UFLS), limit violation and time of day based load shedding schemes which are usually performed by the operator.

Fault Management & System Restoration (FMSR)

Reliability and quality of power supply are key parameters which need to be ensured by any utility. Reduced outage time duration to customer, shall improve over all utility reliability indices hence FMSR or automated switching applications plays an important role. The two main features required by a FMSR are: Switching management & Suggested switching plan

The DMS application receives faults information from the SCADA system and processes the same for identification of faults and on running switching management application; the results are converted to action plans by the applications. The action plan includes switching ON/OFF the automatic load break switches / RMUs/Sectionalizer .The action plan can be verified in study mode provided by the functionality .The switching management can be manual/automatic based on the configuration.

Load Balancing via Feeder Reconfiguration (LBFR)

Load balancing via feeder reconfiguration is an essential application for utilities where they have multiple feeders feeding a load congested area. To balance the loads on a network, the operator re-routes the loads to other parts of the network. A Feeder Load Management (FLM) is necessary to allow you to manage energy delivery in the electric distribution system and identify problem areas. A Feeder Load Management monitors the vital signs of the distribution system and identifies areas of concern so that the distribution operator is forewarned and can efficiently focus attention where it is most needed. It allows for more rapid correction of existing problems and enables possibilities for problem avoidance, leading to both improved reliability and energy delivery performance.

On a similar note, Feeder Reconfiguration is also used for loss minimization. Due to several network and operational constraints utility network may be operated to its maximum capability without knowing its consequences of losses occurring. The overall energy losses and revenue losses due to these operations shall be minimized for effective operation. The DMS application utilizes switching management application for this, the losses minimization problem is solved by the optimal power flow algorithm and switching plans are created similar to above function

Distribution Load Forecasting (DLF)

Distribution Load Forecasting (DLF) provides a structured interface for creating, managing and analyzing load forecasts. Accurate models for electric power load forecasting are essential to the operation and planning of a utility company. DLF helps an electric utility to make important decisions including decisions on purchasing electric power, load switching, as well as infrastructure development.

Load forecasting is classified in terms of different planning durations: short-term load forecasting or STLF (up to 1 day, medium-term load forecasting or MTLF (1 day to 1 year), and long-term load forecasting or LTLF (1-10years). To forecast load precisely throughout a year, various external factors including weathers, solar radiation, population, per capita gross domestic product seasons and holidays need to be considered. For example, in the winter season, average wind chill factor could be added as an explanatory variable in addition to those used in the summer model. In transitional seasons such as spring and fall, the transformation technique can be used. For holidays, a holiday effect load can be deducted from the normal load to estimate the actual holiday load better.

Various predictive models have been developed for load forecasting based on various techniques like multiple regression, exponential smoothing, iterative reweighted least-squares, adaptive load forecasting, stochastic time series, fuzzy logic, neural networks and knowledge based expert systems. Amongst these, the most popular STLF were stochastic time series models like Autoregressive (AR) model, Autoregressive moving average model (ARMA), Autoregressive integrated moving average (ARIMA) model and other models using fuzzy logic and Neural Networks.

DLF provides data aggregation and forecasting capabilities that is configured to address today's requirements and adapt to address future requirements and should have the capability to produce repeatable and accurate forecasts.

Standards Based Integration

In any integrated energy delivery utility operation model, there are different functional modules like GIS, Billing & metering solution, ERP, Asset management system that are operating in parallel and supports routine operations. Quite often, each of these functional modules need to exchange periodic or real time data with each other for assessing present operation condition of the network, workflows and resources (like crew, assets, etc.). Unlike other power system segments, distribution system changes or grows every day, and this could be due to the addition of a new consumer, a new transmission line or replacement of equipment. If the different functional modules are operating in a non-standard environment and uses custom APIs and database interfaces, the engineering effort for managing shall become too large. Soon it will become difficult to manage the growing changes and additions which would result in making system integrations non- functional. Hence utilities cannot make use of the complete benefit of functional modules and in some cases; the systems may even need to be migrated to suitable environments with very high costs.

As these problems came to light, various standardization processes for inter application data exchanges were initiated. It was understood that a standard based integration shall ease the integration with other functional modules and that it also improves the operational performance. It ensures that the utility can be in a vendor neutral environment for future expansions, which in turn means that the utility can easily add new functional modules on top of existing functionality and easily push or pull the data effectively without having new interface adapters.

IEC 61968 Standards Based Integration

IEC 61968 is a standard being developed by the Working Group 14 of Technical Committee 57 of the IEC and defines standards for information exchanges between electrical distribution system applications. It is intended to support the inter-application integration of a utility enterprise that needs to collect data from different applications which could be new or legacy.

As per IEC 61968, a DMS encapsulates various capabilities like monitoring and control of equipment for power delivery, management processes to ensure system reliability, voltage management, demand-side management, outage management, work management, automated mapping and facilities management. The crux of IEC 61968 standards is the Interface Reference Model (IRM) that defines various standard interfaces for each class of applications. Abstract (Logical) components are listed to represent concrete (physical) applications. For example, a business function like Network Operation (NO) could be represented by various business sub-functions like Network Operation Monitoring (NMON), which in turn will be represented by abstract components like Substation state supervision, Network state supervision, and Alarm supervision.

IEC 61968 recommends that system interfaces of a compliant utility inter-application infrastructure be defined using Unified Modelling Language (UML). UML includes a set of graphic notation techniques that can be used to create visual models of object-oriented software-intensive systems. The IEC 61968 series of standards extend the Common Information Model (CIM), which is currently maintained as a UML model, to meet the needs of electrical distribution. For structured document interchange particularly on the Internet, the data format used can be the Extensible Markup Language (XML). One of its primary uses is information exchange between different and potentially incompatible computer systems. XML is thus well-suited to the domain of system interfaces for distribution management. It formats the message payloads so as to load the same to various messaging transports like SOAP (Simple Object Access Protocol), etc.

References

- Kosko, Bart (1992). Neural Networks and Fuzzy Systems: A Dynamical Systems Approach to Machine Intelligence. Englewood Cliffs, NJ: Prentice Hall. ISBN 0-13-611435-0.

- William R. Taylor: Lethal American Confusion (How Bush and the Pacifists Each Failed in the War on Terrorism), 2006, ISBN 0-595-40655-6 (FCM application in chapter 14) Archived September 30, 2007, at the Wayback Machine.

- Benjoe A. Juliano, Wylis Bandler: Tracing Chains-of-Thought (Fuzzy Methods in Cognitive Diagnosis), Physica-Verlag Heidelberg 1996, ISBN 3-7908-0922-5

- W. B. Vasantha Kandasamy, Florentin Smarandache: Fuzzy Cognitive Maps and Neutrosophic Cognitive Maps, 2003, ISBN 1-931233-76-4

- Leondes, Cornelius T. (2002). Expert systems: the technology of knowledge management and decision making for the 21st century. pp. 1–22. ISBN 978-0-12-443880-4.

- Russell, Stuart; Norvig, Peter (1995). Artificial Intelligence: A Modern Approach (PDF). Simon & Schuster. pp. 22–23. ISBN 0-13-103805-2. Retrieved 14 June 2014.

- Hayes-Roth, Frederick; Waterman, Donald; Lenat, Douglas (1983). Building Expert Systems. Addison-Wesley. pp. 6–7. ISBN 0-201-10686-8.

- Orfali, Robert (1996). The Essential Client/Server Survival Guide. New York: Wiley Computer Publishing. pp. 1–10. ISBN 0-471-15325-7.

- Hurwitz, Judith (2011). Smart or Lucky: How Technology Leaders Turn Chance into Success. John Wiley & Son. p. 164. ISBN 1118033787. Retrieved 29 November 2013.

- Hayes-Roth, Frederick; Waterman, Donald; Lenat, Douglas (1983). Building Expert Systems.

Addison-Wesley. ISBN 0-201-10686-8.

- Hayes-Roth, Frederick; Waterman, Donald; Lenat, Douglas (1983). Building Expert Systems. Addison-Wesley. p. 6. ISBN 0-201-10686-8.

- Kendal, S.L.; Creen, M. (2007), An introduction to knowledge engineering, London: Springer, ISBN 978-1-84628-475-5, OCLC 70987401

- Feigenbaum, Edward A.; McCorduck, Pamela (1983), The fifth generation (1st ed.), Reading, MA: Addison-Wesley, ISBN 978-0-201-11519-2, OCLC 9324691

- Siddique, Nazmul; Adeli, Hojjat (2013). Computational Intelligence: Synergies of Fuzzy Logic, Neural Networks and Evolutionary Computing. John Wiley & Sons. ISBN 978-1-118-53481-6.

- Rutkowski, Leszek (2008). Computational Intelligence: Methods and Techniques. Springer. ISBN 978-3-540-76288-1.

- Palit, Ajoy K.; Popovic, Dobrivoje (2006). Computational Intelligence in Time Series Forecasting : Theory and Engineering Applications. Springer Science & Business Media. p. 4. ISBN 9781846281846.

- Computational Intelligence: A Logical Approach by David Poole, Alan Mackworth, Randy Goebel. Oxford University Press. ISBN 0-19-510270-3

- Computational Intelligence: A Methodological Introduction by Kruse, Borgelt, Klawonn, Moewes, Steinbrecher, Held, 2013, Springer, ISBN 9781447150121

- Hsu, F. H. (2002). Behind Deep Blue: Building the Computer That Defeated the World Chess Champion. Princeton University Press. ISBN 0-691-09065-3.

- Hwang, K. (1993). Advanced Computer Architecture: Parallelism, Scalability, Programmability. McGraw-Hill Book Co., New York. ISBN 0-07-031622-8.

- Bonabeau, E.; Dorigo, M.; THERAULUZ, G. (1999). Swarm Intelligence: From Natural to Artificial Systems. Oxford University Press. ISBN 0-19-513158-4.

- Goldberg, D. E. (2000). Genetic Algorithms in search, optimization and Machine Learning. Addison Wesley Publishers. ISBN 0-201-15767-5.

- Kosko, B. (1997). Neural Networks and Fuzzy Systems: A Dynamical Systems Approach to Machine Intelligence. Prentice Hall Publishers. ISBN 0-13-611435-0.

Permissions

We would like to thank the editorial team for lending their expertise to make the book truly unique. They have played a crucial role in the development of this book. Without their invaluable contributions this book wouldn't have been possible. They have made vital efforts to compile up to date information on the varied aspects of this subject to make this book a valuable addition to the collection of many professionals and students.

This book was conceptualized with the vision of imparting up-to-date and integrated information in this field. To ensure the same, a matchless editorial board was set up. Every individual on the board went through rigorous rounds of assessment to prove their worth. After which they invested a large part of their time researching and compiling the most relevant data for our readers.

The editorial board has been involved in producing this book since its inception. They have spent rigorous hours researching and exploring the diverse topics which have resulted in the successful publishing of this book. They have passed on their knowledge of decades through this book. To expedite this challenging task, the publisher supported the team at every step. A small team of assistant editors was also appointed to further simplify the editing procedure and attain best results for the readers.

Apart from the editorial board, the designing team has also invested a significant amount of their time in understanding the subject and creating the most relevant covers. They scrutinized every image to scout for the most suitable representation of the subject and create an appropriate cover for the book.

The publishing team has been an ardent support to the editorial, designing and production team. Their endless efforts to recruit the best for this project, has resulted in the accomplishment of this book. They are a veteran in the field of academics and their pool of knowledge is as vast as their experience in printing. Their expertise and guidance has proved useful at every step. Their uncompromising quality standards have made this book an exceptional effort. Their encouragement from time to time has been an inspiration for everyone.

The publisher and the editorial board hope that this book will prove to be a valuable piece of knowledge for students, practitioners and scholars across the globe.

Index

www.ingramcontent.com/pod-product-compliance
Lightning Source LLC
Chambersburg PA
CBHW061930190326
41458CB00009B/2704